The Double Helix and
the Law of Evidence

The Double Helix and the Law of Evidence

DAVID H. KAYE

HARVARD UNIVERSITY PRESS
Cambridge, Massachusetts
London, England
2010

Library of Congress Cataloging-in-Publication Data

Kaye, D. H. (David H.), 1947–
    The double helix and the law of evidence / David H. Kaye.
        p. cm.
    Includes bibliographical references and index.
    ISBN 978-0-674-03588-1
    1. DNA fingerprinting—Law and legislation—United States.   2. Forensic
genetics.   I. Title.
    KF9666.5.K39 2009
    347.73'64—dc22   2009023848

Contents

Figures

Preface

THIS BOOK is part history, part legal analysis, part popular science, and part applied statistics. I hope that the whole is greater than the sum of the parts because I have not tried to write an encyclopedic history, a legal treatise, a genetics textbook, or a statistical primer. Instead, I have tried to illuminate a series of fascinating and important developments in law and science regarding the admissibility of DNA evidence in courts across the United States. This is a tale of scientific egos, journalistic hype, lawyerly maneuvering, and judicial doctrine and disposition.

I do not hesitate to take sides on these matters and to criticize where I think that criticism is warranted. As an academic lawyer—a professor rather than a litigator—I focus on the appropriate application of the rules of evidence to the findings of scientists and technicians. Naturally, there is room for differences of opinions on these matters. Inevitably, my conclusions are colored by my own biases and beliefs, including the conviction that it is possible to grade, at least roughly, the truth of competing claims about what is scientifically known and unknown. In addition, I would not wish to give the impression that I am writing entirely as an academic observer with no involvement in the events themselves. I have filed affidavits or prepared written arguments for defense lawyers in a handful of cases involving the admissibility of DNA evidence. I also played a minor role in the history of DNA evidence as a member of some of the governmental or scientific commissions or committees that are mentioned in this book.

Because the law has had to apply and make judgments based on subtle (and sometimes not-so-subtle) principles of genetics and statistics, I have presented these principles in sufficient detail to evaluate the performance of the actors in the legal dramas—the judges, prosecutors, defense counsel, expert witnesses, and researchers who have debated whether DNA identification evidence should be admissible in criminal and other cases. In law, as in life, vague generalities are of limited value. The devil is in the details, and oversimplifications of the details of population genetics and statistics created a devil of a time for the legal system. I have tried to be as simple and entertaining as possible, but no more than that. I have shied away from Greek symbols and algebraic expressions except where they are especially helpful in presenting and clarifying statistical ideas.

Finally, I have tried to be accurate in my recapitulations of the facts of cases and the surrounding events, but mistakes surely remain. Discrepancies abound in the many published accounts of some of the cases and occurrences discussed here. These usually are just the details that give a story color and vividness—things like people's ages, the timing and locations of meetings, the order in which witnesses testified, or even their names. A determined historian would go back to the original records or contact individuals with firsthand knowledge of these matters, but I am not a historian. I have left some details unspecified, noted the presence of discrepancies in the accounts from other authors, or chosen the most plausible version in light of the quality and date of the source of the information.

My hope is that this history will illuminate not just the past but also the current controversies over DNA evidence in the courtroom, and that the explication of evidence law and legal procedure will be useful in settling emerging and future issues. These include the interpretation of mixtures, the analysis of low-copy-number DNA, the effect on the probative value of a match of finding a suspect by trawling a database of the DNA profiles of offenders, and the implications of the (superficially) puzzling numbers of partial matches that occur when one compares all possible pairs of DNA profiles in the offender databases. This book examines the first two issues to some small degree and leaves the remaining ones—which have been the subject of considerable misunderstanding in the case law and the popular and academic press (Kaye 2009a, 2009b)—for another volume. Also beyond the reach of this book are a host of issues of constitutional law and criminal procedure arising from the more aggressive methods of acquiring DNA samples and storing the DNA profiles of suspects, previous offenders, victims, and other individuals. These matters also deserve more balanced and probing scrutiny than they have received to date, and they are the topic of a second forthcoming book. The present work focuses on the developments in genetics and the rules and procedures of American law through which tests of genetic markers and, most recently, DNA itself became admissible evidence.

Acknowledgments

MANY INDIVIDUALS and institutions contributed to this book. I am indebted to Marc Allard, Jay Aronson, Bruce Budowle, John Butler, James Crow, Bernard Devlin, Ian Evett, Brandon Garrett, Paul Giannelli, Eliot Goldstein, H. Robert Horvitz, Edward Imwinkelried, Randolph Jonakait, Jonathan Jay Koehler, Jennifer Mnookin, George Sensabaugh, William Thompson, and Bruce Weir for personal communications on some of the topics addressed here. Lisa Forman and Valerie Hans were kind enough to read drafts of one or more chapters. George "Woody" Clarke, Rockne Harmon, and Laurence Mueller responded to inquiries about particular cases. James Crow, Roger Park, and an anonymous reviewer read and made both kind and insightful comments on a draft of most of the manuscript. Sir Alec Jeffreys and John Butler provided graphics that were adapted for use in Figures 3.6 and 9.2, respectively, and Nature Publishing Group and Elsevier Academic Press granted permission to reproduce them here. Jenny Bishop and Abby Chicoine helped with the production of other figures. The Sandra Day O'Connor College of Law at Arizona State University supported this work with summer research grants and a stipend for Alice Jones, who worked diligently as a research assistant in 2008–2009. Also providing wonderful last-minute assistance in perfecting the list of references were Serene Rock at ASU and the following Pennsylvania State University Dickinson School of Law librarians: Laura Ax-Fultz, Kevin Gray, Steven Hinckley, Kimberli Morris, Gail Partin, Mark Podvia, and Judy Swarthout.

When I started to write this book, I thought that I could stitch together the best of my previous essays and articles on DNA evidence to form a seamless tapestry. Although I was sorely mistaken, bits and pieces of the following articles appear in various chapters: "DNA Evidence: Probability, Population Genetics, and the Courts," *Harvard Journal of Law and Technology* 7 (1993): 101–172; "The DNA Chronicles: Is Simpson Really Collins?" 1994 WL 592117; "The DNA Chronicles: The Meaning of General Acceptance," 1994 WL 595559; "DNA Identification in Criminal Cases: Lingering and Emerging Evidentiary Issues," in *Proceedings of the Seventh International Symposium on Human Identification* (Madison, Wis.: Promega Corp., 1997); and "*Bible* Reading: DNA Evidence in Arizona," *Arizona State Law Journal* 28 (1997): 1035–1078.

The Double Helix and
the Law of Evidence

Introduction

FOR MUCH of the twentieth century, the scientific icon was the atom. This was the "atomic age"—a period of atomic bombs, atomic submarines, atomic clocks, and nuclear medicine. Atomic power plants were expected to usher in an age of abundant energy and to shrink in size to the point where automobiles like Ford's 1958 concept car, the Nucleon, would travel for thousands of miles propelled by a pint-size fission reactor in the trunk.

As the twenty-first century unrolls, a new image has emerged as the dominant scientific icon. It is the double helix—the backbone of the DNA molecule. James Watson, the codiscoverer of this structure, once remarked, "We used to think our fate was in our stars. Now we know, in large measure, our fate is in our genes" (Jaroff 1989). Likewise, some bioethicists portray these molecules as "secret future diaries" of human beings (Annas 1993), while others rightly bemoan the tendency to regard the collection of these molecules in human cells as "the true essence of human nature" (Mauron 2001). This imagery and this essentialism surely are exaggerated (as explained in, e.g., Lewontin 2000 and Juengst 2004), but the power of DNA technology to identify murderers, rapists, and robbers and to confirm or refute claims of parentage or other kinship is beyond dispute. Sequences of DNA base pairs that are of little or no importance in medicine have become exceedingly important in law. Testimony about these genetic features has been prominent and persuasive not just in cases of violent crimes but also in cases of auto theft, breaking and entering, child support, domestic relations, immigration and naturalization, smuggling, slander, stealing of babies for international adoptions, importation of the products of protected species, animal-control laws, and even judicial discipline for sexual harassment. It has contributed to the enforcement of laws as mundane as those requiring dog owners to scoop their pets' poop and to international prosecutions of the most heinous war criminals. Outside legal proceedings, various techniques of DNA identification have been applied to detect fraudulently bottled wine, to aid wildlife-conservation efforts, to reunite Holocaust survivors, and to investigate such historical disputes as whether President Thomas Jefferson's offspring include the descendants of his slave, Sally Hemings; whether the bodies in a mass grave unearthed in Siberia in 1991 were those of Tsar Nicholas II and his family; whether a body exhumed in Embu, Brazil, was that of Josef Mengele, the Nazi doctor known as Auschwitz's "Angel of Death"; whether the outlaw Jesse James really was resting in the grave that bore his name (Lee and Tirnady 2003); and whether a skeleton of an elderly man found under an altar in Poland's Frombork Cathedral was that of Nicolaus Copernicus.

But it is the criminal justice system that has benefited the most from forensic DNA identification, in terms both of exonerating the innocent and of convicting the guilty. When judges and lawyers first heard that the "almost magical technique" (Erzinçlioglu 2002, 100) of DNA profiling could identify the source of biological trace evidence such as blood or saliva, they were stunned. In 1988, a trial judge in New York wrote in *People v. Wesley* that "if DNA Fingerprinting works and receives evidentiary acceptance, it can constitute the single greatest advance in the 'search for truth' and the goal of convicting the guilty and acquitting the innocent since the advent of cross-examination" (533 N.Y.S.2d at 644). Considering that the legal profession's leading commentator on the law of evidence regarded cross-examination as "the greatest legal engine ever invented for the discovery of truth" (Wigmore 1974, 5:32), this is high praise indeed.

The path to stable judicial acceptance of DNA identification evidence, however, was far from smooth. A timeline of the key scientific and legal developments can be found in the Appendix. For half a century, statutes and cases slowly expanded the use of genetic markers—proteins manufactured according to instructions in the genes. When it became possible to analyze DNA—the genetic material itself—directly, initial judicial acceptance came quickly, largely through one-sided pretrial hearings. In a few years, the defense bar was able to mobilize a group of scientists who directed a barrage of criticisms at the work of the forensic laboratories. Some of the leading scientists of our time were drawn into the controversy and found themselves debating arcane issues of population genetics in courtrooms, as well as scientific journals. Despite two extensive reports from the National Research Council—a degree of scrutiny and assistance that is unique in the history of scientific evidence—the courts still had considerable difficulty recognizing the points of disagreement and their implications. Today, as new methods and refinements are introduced and old disputes flare up, the process of gaining legal acceptance continues.

This book examines the legal and scientific controversies that swirled around DNA and earlier forms of genetic evidence. It shows how the adversarial process affected arguments not only in courtrooms but also in the scientific community. It delves into evidence law, genetics, and statistics to ascertain which of these arguments were correct and applies the resulting lessons to propose procedures to facilitate the transition of new scientific methods from the laboratory to the courtroom.

The story starts even before the 1943 discovery by Oswald Avery and his colleagues that the DNA molecule carries hereditary information in bacteria (Avery, MacLeod, and McCarty 1944). Chapter 1 examines the use of "genetic markers" such as blood groups in criminal or civil cases dating back to the 1930s. In several ways, the legal battles over the introduction of genetic-marker evidence presaged those that were to come over modern DNA evidence. This chapter describes some of those battles and the standards that courts apply to decide whether a purportedly scientific finding is admissible in a court of law.

Chapter 2 examines the statistical aspects of the genetic-marker evidence. It describes how estimates of the frequencies of the genetic markers in the general population were transformed into probabilities and presented in court, the abuses and concerns that motivated courts to limit "probability evidence," and the resulting legal doctrine in criminal identification cases. To clarify the ubiquitous issues of the relevance and prejudice of evidence, it introduces the likelihood-ratio interpretation of probative value.

Chapter 3 traces the transition in forensic science from genetic markers to genetic material. It discusses the inheritance of nuclear DNA and the discovery of the variations in these molecules that permitted the exquisitely discriminating forms of DNA typing that captivated the courts in the late 1980s and that prompted bitter disputes among world-class experts—disputes that began in courtrooms, spilled over into academic journals, and returned to the courts. These disputes of the early 1990s are the subject matter of Chapters 4 through 7. They became central to admissibility of DNA evidence because quantitatively oriented scientists led courts to believe that proof of a match between a defendant and a trace of DNA at a crime scene should not be accepted without a numerical estimate of the frequency of the matching type in the general population or some sub-population. Chapter 8 examines some of the pretrial sparring over these estimates in the so-called trial of the century, *People v. Simpson,* which was the high-water mark for arguments against admitting DNA evidence.

As the population-genetics debate over such numbers drew to a close in the mid-1990s, objections to admitting DNA matches and frequency or probability estimates centered on the fact that frequencies and "random-match" probabilities do not incorporate the probability of a false match due to laboratory error or the possibility that the explanation for the matching DNA is that a close relative of the defendant is the true source of the crime-scene DNA. Chapter 8 analyzes these arguments as they were deployed in *Simpson.* Chapter 9 sketches the relatively calm shift to new genetic systems, culminating in the current technique of short tandem repeat profiling. Chapter 10 discusses the continued use of racial categories in estimating the frequencies of different DNA types and the role of likelihood ratios in making sense of samples that are a mixture of DNA from several individuals.

Chapters 11 and 12 take us outside the cell nucleus and into the mitochondria—the descendants of ancient bacteria that now live within human cells. These chapters describe how and why the puny mitochondrial genome has proved to be a powerful tool in certain situations. It also notes the current gaps in our knowledge that should affect how this evidence is presented in court.

Chapter 13 closes with reflections on what the legal and scientific history of DNA evidence tells us about the integration of science into law and how the process might be improved. It discusses the production and disadvantages of adversarial science, procedures for obtaining representative as opposed to extreme scientific opinions, and the value of instituting better systems for producing scientific findings for the investigation and prosecution of crimes.

CHAPTER ONE

Before DNA:
Genetic Markers

GENETICS ENTERED the courtroom long before modern science deciphered the structure and biology of deoxyribonucleic acid (DNA). In criminal cases, "genetic markers" such as blood groups frequently were introduced to link suspects to crimes. In several ways, the legal battles over the introduction of genetic-marker evidence presaged those that were to come over modern DNA evidence. Defendants in these antediluvian days denied that the technology was scientifically accepted; judges had to deal with numbers rather than words; and courtrooms become a stage for rancorous performances by the same cast of expert witnesses. The resulting judicial opinions guided the courts when DNA-typing results first were offered into evidence. They influenced the developers of forensic DNA technology, who were keen to ensure that this new and powerful source of evidence would be embraced in the courtroom. They served as both obstacles and signposts for prosecutors and defense counsel, who jockeyed to maximize or minimize the impact of DNA matches or exclusions.

These are reasons enough to study the pre-DNA years and cases, but there is another. Because the genetics and statistics behind the immunogenetic evidence are relatively simple, they provide a foundation for understanding similar issues with DNA testing that proved far more controversial. Thus the saga of what might be called "the protein wars" is well worth rehearsing. I begin with a short description of the ABO red blood cell groups and their initial rejection in the courts. Then I move to the broader range of genetic markers that were the subject of conflicting judicial opinions. Finally, I consider the legal rules for introducing probabilities or statistics to express the significance of genetic markers that match those of the defendant.

Immunogenetics of the ABO Group

For hundreds of years, physicians carried out experiments with blood transfusions, often with fatal consequences. As early as 1667, French and English physicians were transferring blood from lambs to humans. The results were so disastrous that within ten years, laws were adopted to prohibit animal-to-human transfusions. The first human-to-human blood transfusion occurred more than a century later, in Philadelphia in 1795 (American Association of Blood Banks 2006). Human transfusions sometimes worked, but often the red blood cells would clump together and crack open, spewing their contents and causing toxic reactions.

At the start of the twentieth century, Karl Landsteiner discovered why. Born in Austria and educated in medicine and organic chemistry, he introduced the idea of antibodies and antigens (*antibody generators*). Antigens mark the cells of the donor's blood. Clumping occurs when the recipient of this blood has antibodies that attach themselves to these antigens. This sounds simple enough, but what are these antibodies and antigens that become locked in this deadly embrace?

They are what can be called sugarized proteins. A protein is a chain of smaller molecules—building blocks known as amino acids. A sugar is a carbohydrate—an assembly of carbon, hydrogen, and oxygen atoms. Landsteiner's antigens are combinations of proteins and sugars that stick out of the surface of cells, somewhat like the spikes on a studded snow tire. Landsteiner's antibodies are larger sugarized proteins floating in the blood plasma. They grab on to the studs and rupture the tire. The crucial point is that each type of antibody is shaped so that it binds to a particular antigen—one antigen, one opposing antibody.

It takes but a few of these antigen-antibody pairs to produce the common A, B, and O blood types. Individuals with type A blood, for example, have the A antigen on their red blood cells and the anti-B antibody floating in their blood plasma.[1] By adding the different antibodies one at a time to see which one would cause clumping, Landsteiner could determine the type of antigen on the individual's red blood cells. This discovery was a major medical breakthrough because it permitted blood transfusions to be conducted more safely by matching the ABO blood types.

But why would this be of interest to the criminal courts? The answer is that one's blood type is an observable, inherited trait—a *phenotype*—that is not altered by the environment. It is governed entirely by the versions of the relevant genes—the *genotype*—that the individual carries. Consequently, the ABO type is a stable genetic marker. This phenotype supplies information about the underlying genotype.

Although it took some twenty-five years to establish what turned out to be a simple pattern of inheritance for the ABO blood group, by the 1930s, it was clear that the antigens were the product of a single gene. The gene has three major forms, or *alleles,* also designated A, B, and O (Crow 1993).[2]

Everyone has two copies of this ABO gene—one inherited from each parent. For instance, a mother and a father who have blood group O (indicating the OO genotype) each will pass one of their two O alleles to their child, who will also be type OO. But the agglutination test on blood does not detect the alleles directly. It reveals only the phenotype, that is, the blood group. Suppose, then, that a mother and a child both have blood group O. This tells us that the father could not have blood group AB. The child's second O allele must have come from his father, and type AB men do not carry this allele. Hence they are all excluded as possible fathers. (Readers who enjoy logic puzzles can work out the exclusionary implications of all possible combinations of blood groups among mothers, children, and excluded men.)

Is the Theory Generally Accepted?

In 1931, a South Dakota man accused of rape wanted to use these principles of immunogenetics to demonstrate his innocence. The time seemed ripe. One year earlier, Landsteiner had been awarded the Nobel Prize in Medicine and Physiology for his 1900–1901 discovery of immunological reactions. Some continental courts were accepting blood-test evidence in civil paternity cases, and articles in the *American Bar Association Journal* called attention to the practice (Lee 1926). Letters and reports in leading medical journals endorsed such testing (e.g., Schiff 1929; Weiner 1930).

The issue came before the South Dakota Supreme Court in 1933, in *State v. Damm,* "a matter . . . of first impression in the courts of this country" (252 N.W. at 8). Clement Damm and his wife were married in 1913. After twelve childless years, they adopted Ruby and Ruth Wilson, twin seven-year-old girls. In 1931, as Ruby entered her teens, she became pregnant. She told the authorities that her adoptive father was responsible, and

Damm was charged with second-degree rape. After Ruby gave birth to a baby girl, Damm requested that the trial court order blood testing, which, his expert testified, might exclude Damm as a possible father. The court denied the request. Damm was tried, convicted, and sentenced to sixteen years in prison. He appealed to the Supreme Court of South Dakota, alleging seventy-two errors at trial. The court rejected these claims but paused at "several somewhat novel and interesting questions" (id. at 10). It summarized the testimony of the defendant's "medical expert" in support of his motion:

> It was not the contention of this witness that paternity could in any case be affirmatively proved by such blood test, but he did contend that in quite a percentage of cases the impossibility of claimed paternity could be demonstrated by blood test. It was, in substance, the testimony of this witness that human blood is divided into four recognized types or groups [A, B, O, and AB], and that, if . . . the blood groups of a mother and child are known, it can be said what must have been the blood group of the father, and consequently the impossibility of certain paternity may be effectively demonstrated. Appellant in this case offered to submit himself to such blood test, and asked the court to require prosecutrix and her infant child to submit thereto. The blood test itself is a laboratory procedure, and requires but a drop or two of blood, which can be taken from the subject without pain or danger. (Id. at 8)

Apparently, the state produced no opposing medical testimony, but the supreme court upheld the trial judge's refusal to order the blood testing. It pointed to the fact that not all jurisdictions admitted blood tests and to accounts in the *Irish Law Times* of family-law cases in which blood tests were conducted. In one case, a "Dr. Stephens, for the plaintiff, said he knew that a great amount of research and experiments had been made, and were still being made, by eminent pathologists the world over in connection with [the] theory [of inherited blood types], but the medical profession had not as yet universally accepted the results as infallible" (id. at 11). The *Irish Law Times* contained no other expert or judicial criticism of the procedure, but the author of one of its reports concluded that "[l]awyers, however, may be excused for feeling some scepticism as to the positive value of such tests, in the present stage of this branch of medical science" (id. at 12). The only ground given for the lawyerly skepticism was the fact that "English courts have never attached much importance to evidence of resemblance or alleged hereditary characteristics in pedigree and legitimacy suits and there is no reported case of a blood test being acted upon" (id.). On this basis, the South Dakota Supreme Court concluded that the learned trial judge did not abuse his discretion [because] it does not sufficiently appear from the record in this case that modern medical science is agreed upon the transmissibility of blood characteristics to such an extent that it can be accepted as an unquestioned scientific fact that, if the blood groupings of the parents are known, the blood group of the offspring can be necessarily determined, or that, if the blood groupings of the mother and child are known, it can be accepted as a positively established scientific fact that the blood group of the father could not have been a certain specific characteristic group. In other words, we think it insufficiently appears that the validity of the proposed test meets with such generally accepted recognition as a scientific fact among medical men as to say that it constituted an abuse of discretion for a court of justice to refuse to take cognizance thereof, as would undoubtedly be the case if a court to-day should refuse to take cognizance of the accepted scientific fact that the finger prints of no two individuals are in all respects identical. (Id.)

Apparently, the supreme court's rule was that a trial court may exclude all scientific evidence that has not yet met with general acceptance in the scientific community. The only time the court must hear the evidence is when the "scientific fact" is so clearly established as to be subject to judicial notice—a doctrine that allows a court to act on facts that are not part of the record but that could not reasonably be disputed.

This seems to be a weak version of the requirement of "general acceptance" that had been injected into American law just ten years earlier, in the now-famous case *Frye v. United States*. After several days of police grilling in the summer of 1920, a young black man named James Frye confessed to shooting Dr. R. W. Brown, but he repudiated his confession just before the trial. To buttress his protestations of innocence, he proposed to have a psychologist, William Moulton Marston, testify. Frye's appointed counsel had invited Marston to examine Frye with the aid of Marston's discovery, the "systolic blood pressure test" for deception. Marston concluded that Frye had not committed the murder.

Marston's accomplishments were protean—he held degrees in psychology and in law from Harvard, and he created the comic-strip character Wonder Woman, whose golden lasso compelled anyone within its thrall to speak the truth. The courts were less enthralled. The trial court sustained the prosecutor's objection, and the Court of Appeals for the District of Columbia Circuit affirmed. The court of appeals observed that "[j]ust when a scientific principle or discovery crosses the line between the experimental and the demonstrable stages is difficult to define" (293 F. at 1014). In drawing this line, the court was not content to rely solely on the assertions of the test's inventor, Marston, that his experiments had demonstrated that systolic blood pressure could indicate truthfulness. Neither did it inquire directly into whether Marston's work was sufficient to establish the validity of the technique. Rather, in a novel move, it affirmed the exclusion of the evidence on the ground that other psychologists had yet to accept Marston's claim that he could verify honesty by measuring the speaker's blood pressure. Although no previous cases explicitly had held general acceptance to be indispensable, the court wrote, "Somewhere in this twilight zone [between the experimental and the demonstrable] the evidential force of the principle must be recognized, and while courts will go a long way in admitting expert testimony deduced from a well-recognized scientific principle or discovery, the thing from which the deduction is made must be sufficiently established to have gained general acceptance in the particular field in which it belongs" (id.). Concluding, with no further discussion or documentation, that the deception test lacked the requisite "standing and scientific recognition among physiological and psychological authorities," the court of appeals upheld the exclusion of the psychologist's testimony (id.).

For a time, the general-acceptance standard achieved a dominant status in U.S. courts. It suffered a major setback in 1993, when the U.S. Supreme Court held in *Daubert v. Merrell Dow Pharmaceuticals, Inc.,* that the adoption of the Federal Rules of Evidence in 1975 "superseded" *Frye* and "displaced" general acceptance as "the exclusive test for admitting expert scientific testimony" (509 U.S. at 589). *Daubert* was not a criminal case and did not involve criminalistics. It was a product-liability case brought on behalf of two young children born with missing or malformed limbs. The children and their parents sued for damages from the maker of Bendectin, a drug approved by the Food and Drug Administration as safe and effective for the relief of nausea and vomiting during pregnancy. Their case foundered when they were unable to point to any published epidemiological studies concluding that Bendectin causes limb-reduction defects. The federal district court granted summary judgment for the drug's manufacturer, Merrell Dow. The Ninth Circuit Court of Appeals affirmed on the theory that under *Frye,* there could be no admissible expert testimony of causation without some peer-reviewed, published studies showing a statistically significant association between exposure to Bendectin and limb-reduction defects. Having determined that *Frye* no longer governed, however, the Supreme Court reversed and remanded the case to the court of appeals for further proceedings.[3]

The Supreme Court did not simply hold, as had the courts in a significant minority of federal and state jurisdictions, that with *Frye*'s demise, the relevancy-plus standard governs scientific expert testimony. Instead, it read into the phrase "scientific . . . knowledge" in Rule 702 a requirement of a "body of known facts or . . . ideas inferred from such facts or ac-cepted as truths on good grounds" in accordance with "the methods and procedures of science" (509 U.S. at 590). The Court then offered an ab-stract discussion of how the requirement of scientifically "good grounds" might be satisfied. It suggested inquiring into such matters as the degree to which a theory has been tested empirically, the extent to which it has been "subjected to peer review and publication," the rate of errors associated with a particular technique, and the extent of acceptance in the scientific community (id. at 594).

Although *Daubert* is a Supreme Court case, it applies only to the federal rules of evidence. States are free to use other rules. Even in those states whose rules for expert testimony tracked the wording of the federal ones, state courts often declined to follow the Supreme Court's lead. Conse-quently, *Frye* continues to hold sway as the special test for scientific evi-dence in many state jurisdictions.

Whatever one thinks of the *Frye* test—and it has been strongly criticized over the years (Kaye, Bernstein, and Mnookin 2004, 191–192)—the early application of its variant in *Damm* is disappointing. Technically, the re-cord before the trial court consisted of the unopposed testimony of a qualified medical expert that the ABO test had a fair chance to exclude the defendant as a possible father if, in fact, he was not the father. If the trial judge had doubts about the validity or reliability of this test, he could have asked the defendant to address the point, or he could have gone outside this record to consult the scientific literature. Had the literature been ex-amined, the flimsy misgivings in the *Irish Law Times,* which the South Dakota Supreme Court cited as indicative of the lack of general accep-tance, would not have emerged. The Irish reports were not published until the following year. Furthermore, the reports do little to demonstrate a lack of general scientific acceptance. That a scientific test is not yet in use in the courts of every country hardly shows that it lacks acceptance in the scien-tific community. Similarly, the view of one physician, hired by one party, that "the medical profession had not as yet universally accepted the results [of blood tests] as infallible" is a far cry from testimony that, in general, the profession is not persuaded that ABO testing can exclude some men as possible fathers. More fundamentally, a proper review of the scientific literature as of 1931 would have shown no dispute over the validity of serological tests for ABO types and no remaining controversy over the single-gene, three-allele theory used to rule out certain blood types on the part of the father. By 1925, there was an "enormous literature on blood group frequencies throughout the world," and in that year, the German mathematician Felix Bernstein showed, in one population after another, that the frequencies matched those expected for a three-allele locus in a randomly mating population (Crow 1993, 4). Bernstein's article (F. Bernstein 1925), together with an earlier publication (F. Bernstein 1924), drove a stake through the heart of an alternative "two-locus" theory propounded by Von Dungern and Hirzfeld (1910) that the blood groups reflected two genes with two alleles. Thus by the time *Damm* was tried, there was ample general acceptance of the blood test that the defendant sought.[4]

Indeed, in a sequel to its 1933 opinion, the South Dakota Supreme Court conceded as much. The court granted Damm a rehearing and issued a new opinion in 1936. This time, the court expressed its "considered opinion that the reliability of the blood test is definitely, and indeed unanimously, established as a matter of expert scientific opinion entertained by authorities in the field" (266 N.W. at 668). The court now recognized that "the time has undoubtedly arrived when the results of such tests, made by competent persons and properly offered in evidence, should be deemed admissible in a court of justice whenever paternity is in issue" (id.). Nevertheless, the court adhered to its original decision that the trial judge properly declined to order blood tests largely because Damm's expert witness did not explicitly discuss general acceptance. It was a triumph of form over substance.

More Immunogenetic Markers

The discovery of the ABO blood group was just the beginning. There are hundreds of other red blood cell antigens, and many of them can be, and were, used in parentage testing. Indeed, the opinion on rehearing in *Damm* suggested that the defendant was seeking testing for several additional antigens. Furthermore, red blood cells are not the only ones that possess antigens. Antigens of another category, human leukocyte antigens (HLA types), are found on the surface of most human cells. The full set of antigens that a cell possesses thus distinguishes it from the cells of other organisms, and understanding the biochemical mechanisms by which a multicellular organism distinguishes between self and nonself—between its own cells and foreign substances—is fundamental to understanding how the body responds to infections from microorganisms, to grafts of foreign tissues or materials, and to blood transfusions. Moreover, it is central to the study of allergies, tumors, and autoimmune diseases.

In forensic work involving traces of dried blood or other bodily fluids left at crime scenes, however, the quantities of the biological material are normally too small to permit extended typing of antigens, some antigens are less stable than the ABO markers, and many antigens are hardly worth typing for criminal investigations because they exhibit very little variation in a given population. Indeed, the ABO system has limited powers of discrimination. The O blood type, for instance, is very common around the world. About 63% of humans share it. Among the indigenous populations of Central and South America, the frequency approaches 100%. To enhance the power of the genetic markers to discriminate among individuals, forensic scientists turned to enzymes and proteins found in blood serum. Serum proteins and enzymes tend to be more variable within a population, and the particular enzymes that an individual has are determined by genes and thus can serve as genetic markers. By the late 1970s, a large number of polymorphic proteins (a set of proteins having distinct molecular forms but the same biochemical function) were identified in dried blood samples. For instance, phosphoglucomutase (PGM) is an enzyme that catalyzes a reaction in the metabolism of sugars. One person may have the variant designated PGM 1, while another has PGM 2-1. Even though the two molecules are slightly different, they are both good catalysts, so neither person has any medical concerns over the PGM type he or she inherited, but each can be differentiated from the other on the basis of this genetic marker in the blood.

Which version of each such protein is in the blood can be ascertained by a technique called "gel electrophoresis." The sample material is placed on a rectangular slab of gelatinous or starchy material (a *gel*), and an electric field is applied to pull the molecules down or up the length of the gel. The direction and rate of this movement depend on the size and charge of the protein. If the field is turned off at the right time, proteins of different masses or shapes will have migrated different distances on the gel. The positions are ascertained by adding dyes that combine with specific proteins.

The Thin-Gel-Multisystem Controversy

Although electrophoresis applied to one type of protein at a time works well with ample amounts of fresh blood, law-enforcement authorities wanted a method that could quickly handle small samples of aged or dried bloodstains. A "thin-gel multisystem" that simultaneously analyzed several protein polymorphisms emerged from experiments in London and in Berkeley and Anaheim, California, and was promptly deployed in forensic laboratories across the United States (Aronson 2006). But this technique was quite unlike traditional blood-group evidence, which used the same technology that clinical laboratories relied on every day to supply physicians with information for life-and-death decisions. Thin-gel-multisystem testing was employed only in crime laboratories, there had been little outside investigation of the effects of aging and environmental contamination, and the crime laboratories did not submit to routine proficiency testing. The courtroom battles were about to begin.[5]

By and large, appellate courts applied the *Frye* standard and concluded, without much effort, that the multisystem (and the more traditional) electrophoretic procedures were scientifically accepted. *State v. Dirk,* decided by the South Dakota Supreme Court in 1985, is representative. The defendant was accused of breaking into a store by breaking a window (and cutting himself in the process). A forensic serologist from the state crime laboratory testified that the stains at the store and a sample of the defendant's blood matched in two enzyme systems, as well as in the ABO system, and that 18% to 19% of the total population had blood with the matching characteristics. The tests, he added, "were recognized as being scientifically reliable and accurate" (364 N.W.2d at 120). The defense did not call an expert, but it cited a law review article (Jonakait 1982) that reviewed the scientific literature on the accuracy of electrophoretic typing of enzymes in dried blood and argued that the research was inadequate to establish general scientific acceptance. The court responded with a single sentence stating that the serologist's "testimony regarding the acceptance by forensic serologists of the tests that he employed in this case constituted an adequate basis for the trial court's determination that the results of those tests should be presented to the jury" (id. at 121). For good measure, the court also suggested jettisoning the pesky *Frye* test.

A few courts were more thorough. The first appellate opinion on multisystem electrophoresis had come from nearby Kansas in 1981, in *State v. Washington.* A woman's body was found, naked below the waist, in a pool of blood from eighteen stab wounds, in her Topeka apartment. The rear window was open, and the screen was missing. Adrian Washington lived down the street and came to the attention of the police when he appeared at a hospital with stab wounds some hours after the murder. Before long, he was charged with murder, rape, and burglary.

The state's case was strong. Washington's fingerprints were on the outside of the rear screen door of the apartment and on a window screen found in bushes down the street from the apartment. Two "foreign pubic hairs found on the victim" were microscopically similar to Washington's (622 P.2d at 988). Also, blood samples from Washington matched drops of blood in the apartment leading to the back door.

The defendant attacked some of the serological evidence with much more than a law review article, and his objection to this evidence became the first point raised on the appeal of his conviction in *State v. Washington*. The evidence in question was the testimony of Eileen Burnau, a criminalist employed by the Kansas Bureau of Investigation. She analyzed blood samples from the apartment, the victim, and the defendant, using ABO typing and multisystem typing of six enzyme or protein systems. Burnau found that Washington's types matched those of various samples in the apartment and that these samples did not match the types of the victim. Then she maintained "that only 6/10ths of 1% of the population would have defendant's combination of blood factors" (id. at 989).

Washington attacked only the "reliability" of the results of multisystem typing of aged bloodstains. The central figure in this assault was Benjamin Grunbaum, a research biochemist at the University of California at Berkeley whose "impressive string of credentials" (id.) included developing an automated electrophoresis device for rapidly analyzing blood proteins for the National Aeronautics and Space Administration (Woodfill 2000), and who was a leader in the federally funded effort to develop a similar system for law enforcement. Grunbaum not only disputed the criminalist's interpretation of the multisystem results but asserted that the system was "inherently unreliable" and that "apart from use in crime laboratories, [it] was not accepted within the scientific community" (622 P.2d at 990).

The state responded with the testimony of Mark Stolorow. Stolorow's qualifications consisted of "a bachelor's degree in chemistry and a master's in forensic chemistry" followed by law-enforcement work in forensic serology and participation in the research with Grunbaum that culminated in the gel multisystem used in *Washington* (id.). The full story of Stolorow's and Grunbaum's involvement in the research project is convoluted and is told well by Aronson (2006). In a few words, Grunbaum, as the project's leader, favored a different material for the gel than the two technicians, namely, Stolorow, who was recruited from a crime laboratory in Illinois, and Brian Wraxall, who came to the project from the Metropolitan Police Laboratory in London. Eventually, Grunbaum dropped out of the project, Stolorow returned to Illinois, and Wraxall moved to a commercial laboratory in Anaheim, California, to finish developing the system. Wraxall never published a validation study of his thin-gel multisystem in the scientific literature, and other scientists did not fill the gap.[6]

Although the Kansas Supreme Court recognized the nature of the scientific disagreement in *Washington*—a distinct improvement over cases like *Dirk*—its resolution of the charges and countercharges ultimately was shallow. Perhaps taking a leaf from the dictum about *Frye* in *Dirk*, the government "suggested that the *Frye* test should be abolished" (622 P.2d at 992). Reasonably enough, the court declined this invitation, but its analysis of general acceptance was not fully convincing. The court was very "impressed by the testimony that the Multi-System analysis is [in] use in over 100 criminal laboratories in this country and that the FBI research laboratory . . . approves it" (id.). The opinion did address some of Grunbaum's stated reasons for believing that the system was totally unreliable, but it ignored the fact that there was no published research to validate multisystem testing.

This situation illustrates a recurring problem with the *Frye* standard. Assessing "general acceptance" works fairly well when there is a large scientific community with a tradition of critical peer review and replication of research results. It is less of a barrier to premature and overenthusiastic deployment of a technology in disciplines or professions that have no such tradition. "Voiceprints" to identify speakers, bite-mark analysis to associate teeth marks with a given set of teeth, polygraphs to detect liars, neutron-activation analysis to individualize human hair, and the compositional analysis of bullet lead to link a bullet fragment to a box of ammunition are just a few of the dubious methods that, at one time or another, were widely used or accepted in the field of law enforcement (Faigman et al. 2006). In these circumstances, it is vital to ask not merely whether the law-enforcement community generally accepts a new scientific test, but whether this community can present a satisfactory scientific basis for believing that the test works as advertised.

This additional inquiry is clearly mandated under the alternative "scientific-validity standard" for admitting scientific evidence adopted in *Daubert v. Merrell Dow Pharmaceuticals* and other cases. *Daubert* and its progeny require courts to inquire into such matters as the existence of peer-reviewed literature and the efforts that have been made to validate a theory and characterize the risk of error of a test in order to determine whether the forensic application of science possesses sufficient intellectual rigor.

Still, some courts in *Frye* jurisdictions have achieved a roughly similar result by defining the relevant scientific community to consist of scientists outside the law-enforcement establishment. This move led to the rejection of the thin-gel multisystem in Michigan in 1986. In *People v. Young*, the Michigan Supreme Court carefully examined the testimony of Grunbaum, several geneticists, and a forensic scientist. Noting that "the few reported cases involving electrophoresis of evidentiary bloodstains . . . might be described as reflecting and reporting a debate between Stolorow and Grunbaum" (391 N.W.2d at 275), the court focused on the subject of that debate, "Wraxall's thin-gel multisystem" (id. at 283). The court was unwilling to predicate a finding of general acceptance on the behavior of crime laboratories and the studies and assertions of the developers of this test. "[S]elf-verification," the court insisted, "is not a sufficiently reliable procedure" (id. at 280), and police serologists could not constitute the kind of "disinterested and impartial experts in the scientific community . . . necessary to assure that the technique is trustworthy" (id. at 274). In the court's view, experts like Stolorow, whose "livelihood was . . . intimately connected with the new technique," were not within this community (id. at 276). Furthermore, while the *Washington* court described Stolorow as "a forensic chemist" (id. at 274) and Grunbaum "as a biochemist, both specializ[ing] in microanalysis and serology" (id.), the *Young* court demoted Stolorow to the rank of "police detective who did the electrophoresis in the instant case" (id.). As such, he was excluded from "the relevant scientific community [of] scientists not technicians" (id. at 274–275). However, this maneuver left hardly any "disinterested" scientists who had studied and worked with the forensic technology. The court therefore concluded that the group that needed to accept the thin-gel multisystem comprised both these few scientists and "a larger number of nonforensic scientists using [other forms of] electrophoresis who are capable of evaluating the reliability of electrophoresis of evidentiary bloodstains if presented with the information they need to fill the gaps in their own knowledge and experience" (id. at 271–272).

With the relevant scientific community defined in this restrictive way and with Wraxall's "self-verification" study placed out of bounds, the Michigan Supreme Court was able to find a lack of general acceptance in view of the absence of other validation studies. The court noted that when the prosecution asked the nonforensic geneticists why they believed the thin-gel multisystem was reliable, "[t]heir collective response could be summarized in the following comment . . . 'I have no reason to suppose it wouldn't work.' They testified that they had seen no study demonstrating that the multisystem was unreliable" (id. at 281). The court was not impressed. "This line of reasoning," it wrote, "would be adequate if the burden of establishing general acceptance of unreliability were placed on the defense. The burden of establishing general acceptance of reliability is, however, on the prosecution" (id.). Having perceived "substantial unanswered questions respecting the reliability of Wraxall's thin-gel multisystem" raised by the conflicting expert testimony, the court concluded that "until independent verification tests have been conducted regarding the thin-gel multisystem, general agreement in the scientific community on the reliability of that multisystem is unlikely" (id.).

Young was the high-water mark of the challenges to electrophoresis. Courts outside Michigan refused to exclude police serologists from the definition of the relevant scientific community.[7] Grunbaum continued to testify not only in cases involving Wraxall's multisystem but also in cases involving more standard forms of electrophoresis, questioning the ability of analysts to interpret patterns from aged or contaminated samples. Joining in these criticisms was a young geneticist then at Emory University, Diane Juricek.[8] Prosecutors responded with testimony from a panoply of forensic serologists and geneticists to the effect that skilled analysts could recognize potential threats to accurate typing due to contamination or aging (e.g., People v. Morris (Cal. Ct. App. 1988); People v. Reilly (Cal. Ct. App. 1987)). Therefore, courts consistently rejected as idiosyncratic Grunbaum's insistence that in the absence of stricter protocols and quality-assurance measures, the interpretations of electropherograms from police serologists could not qualify as generally accepted science. The conclusion of an intermediate appellate court in California that "Dr. Grunbaum stands virtually alone in his opposition to electrophoretic typing of dried bloodstain evidence" reverbrated across the country (*Reilly,* 242 Cal. Rptr. at 509). Some courts were further reassured after Federal Bureau of Investigaion (FBI) researchers (Budowle and Allen 1987) published a study of the effect of contaminants.[9]

The successful marginalization of Grunbaum's position on contamination and aging not only ensured the admissibility of genetic-marker evidence obtained with conventional electrophoretic procedures but also may well have dampened the impact of his criticism of the thin-gel multisystem, which the court in *Young* had found compelling. Judges naturally would be less inclined to apply such a demanding version of the general-acceptance test if they thought that the bearer of bad news was imprudent in other criticisms. In any event, only the Michigan courts adhered to the view that the absence of published validation studies of this system precluded a finding of general acceptance in the relevant scientific community (Giannelli and Imwinkelried 1993, 556).

With the emergence of a clear legal consensus, challenges to protein polymorphisms waned. Indeed, they ceased entirely with the emergence of techniques that analyze DNA itself rather than its protein products. As the 1980s drew to a close, this new technology burst onto the legal scene and soon swept away the dogged efforts of criminalists to type the genetic markers in bloodstains. The "protein wars" were but a prelude to the looming "DNA wars."

Before I approach those more consuming battles, however, one further facet of the genetic-marker cases deserves attention. This is the judicial treatment of the quantitative aspects of the testimony. For a time, testimony about the probability and statistics of DNA matches proved to be the Achilles' heel of that evidence. What, then, was the state of the law on "probability evidence" at the dawn of the DNA era?

CHAPTER TWO

Trial by Mathematics

GENETIC-MARKER TESTS became commonplace in two medicolegal areas: identifying the perpetrators of violent crimes or sexual offenses from traces of blood or semen and ascertaining parentage in child-support, criminal, and immigration cases (*McCormick on Evidence* 2006, 1:854–855). In using blood and semen stains for identification, it was recognized that if any of the suspect's antigens or serum proteins departed from those found at the crime scene, then the trace evidence could not have come from the suspect. No statistics were needed to appreciate the meaning of this exclusion. But what of an inclusion? Should the prosecution be permitted to prove that the defendant possessed the same genetic markers as those that were deposited at a crime scene when, say, 60% of the population carry the markers, or 10%, or 1%? What do such numbers imply about the chance that the crime-scene material came from the defendant rather than someone else? Will jurors use this information correctly, or will they give too much weight to genetics and statistics, prejudicing the defendant? This chapter describes how courts groped for the answers to these questions.

Inclusions versus Exclusions

Because of the limited discriminatory power of the ABO system, for a considerable time, a match in that system was not regarded as admissible evidence of the suspect's presence. As late as 1970, the New York Court of Appeals, that state's highest court, thought it so obvious that inclusionary results were inadmissible that it devoted but a single sentence to the issue. In its two-sentence opinion in *People v. Robinson,* the Court of Appeals faulted the trial court for admitting such evidence, writing that "[p]roof that defendant had type 'A' blood and that the semen found in and on the body of decedent was derived from a man with type 'A' blood was of no probative value in the case against defendant in view of the large proportion of the general population having blood of this type and, therefore, should not have been admitted" (265 N.E.2d at 543).

On the face of it, this statement is ridiculous. "Probative value" refers to the tendency of an item of evidence to prove a proposition. Here, the proposition in question is whether the defendant is the source of the semen. We can abbreviate this hypothesis as S, for "source." The evidence is the testimony that Robinson has the A antigen, which matches that of the semen. If this evidence were just as likely to arise for Robinson as for anyone else, then the court would be correct—the matching antigen would have no value in proving whether S or not-S is true. This would be the case if everyone who might have committed the crime had type A blood. The blood type then would have "no probative value." But it would be incredible if this were the situation in *Robinson.* More than 60% of blood donors in the United States do not have type A blood (Garraty, Glynn, and McEntire 2004). That Robinson had this type rather than O, or B, or AB blood surely is meaningful evidence against him.

The meaning of probative value can be made more precise by the statistical concept of "likelihood." The premise of the likelihood theory of relevance is that evidence supports the hypothesis S over not-S when the probability of the observed data is larger under S than not-S (Keynes 1921; Edwards 1972; Royall 1997). Here, if a suspect is the source of the semen, then (putting aside the chance of an erroneous typing result, perjury, or the like), the probability of the report that he has type A blood is 1. If the suspect is not the source (and not did become a suspect on the basis of his blood type or any factor correlated with his blood type), then the probability that he has type A blood is simply the proportion of the population of plausible suspects who have this blood type—roughly 40%. The ratio of these two "likelihoods" is $1/(.4) = 2.5$. In other words, it is two-and-one-half times more likely to find a match when a suspect is the source than when he is not. This "likelihood ratio" is a measure of the probative value of the evidence. When the likelihood ratio is unity, the evidence is just as likely to be observed when the suspect is not the source as when he is. Such evidence is of no help in deciding between these two hypotheses. In legal language, it has no probative value and is irrelevant. The more the ratio departs from unity, the more probative the evidence.

That the ABO match in *Robinson* had some nonzero probative value, however, does not mean that it was admissible. Relevance is a necessary but not a sufficient condition for admissibility. If jurors would be unduly impressed with the scientific nature of the evidence and would give it more weight than the likelihood ratio indicates it deserves, then it might be excluded as unfairly prejudicial under the long-standing rule that evidence whose prejudicial impact is likely to substantially outweigh its probative value is inadmissible (*McCormick on Evidence* 2006, 1:736–738). In this way, *Robinson* might be rationalized as reflecting the view that blood-group evidence with such a modest likelihood ratio as 2.5 is too prejudicial to inject into the trial. The risk is too high that the jury will overvalue the evidence.

Thus in later cases the New York Court of Appeals refused to read *Robinson* as establishing a simplistic distinction between exclusionary results and inclusionary ones. In 1980, a man designated by the court only as Abe A. was found "bludgeoned to death in his Manhattan apartment. There was dramatic evidence of a violent struggle. His head contained multiple lacerations, his face was severely contused, his larynx crushed. Blood was spattered throughout the apartment and five teeth, missing from the decedent's mouth, were on the floor. Yet, there were no signs of a forced entry" (*In* re Abe A., 437 N.E.2d at 267). Suspicion fell on Abe's business partner, Jon L., and investigators wanted a sample of his blood. The blood in the apartment "was of two types, one matching that of the deceased and the other a relatively rare one which was to be found in less than 1% of the population" (id.). When Jon refused to provide a sample, the district attorney secured a court order. Jon persisted in his refusals and was held in criminal contempt.

On appeal, he argued that the order to submit to blood sampling violated his Fourth Amendment right to be free from unreasonable searches and seizures because he had not been charged with a crime. In upholding the forced extraction of blood, the Court of Appeals noted that "there can be [no] serious question but that Jon L.'s blood type would constitute material probative evidence. . . . That the incidence of the presumed killer's blood type in the general population is less than one to a hundred is well documented by medical statistics" (id. at 271). As for *Robinson*, the court contrasted "the relative rarity of the assailant's type of blood" in the case at bar with "type A blood, which is found in 40% of the population" (id. at 271, 271 n.4). Evidently, a likelihood ratio of $1/(.01) = 100$ was too good to resist. In this situation, at least, the possible prejudice did not substantially outweigh the obvious probative value of the evidence.

The New York Court of Appeals returned to the statistics of type A blood one more time, in *People v. Mountain*. *Mountain* flatly repudiated *Robinson*. On a Saturday morning in 1981, Schenectady police arrested "a young woman on a minor charge" (486 N.E.2d at 803). They put her in a detention cell for arraignment on Monday morning. A police officer assigned to the police station as a telephone operator took an unusual interest in the prisoner. On Saturday, he suggested that he could help her raise bail and left her his telephone number. A little later, he gave her a handwritten note telling her to call him when she was released. On Sunday afternoon, he entered the cell, raped her, and demanded that she perform oral sodomy on him.

As soon as he left, the woman spit his sperm into a paper cup from her lunch tray. She told a matron what had happened and said she had a "specimen." The matron, it seems, did nothing. When the woman left her cell the following morning, she crushed the cup and put it in her pocket. At her arraignment, she told the judge that she had been raped and had a specimen. The judge, who had already dismissed the charge against her because the complaining witness did not appear, recommended that she inform the police. She reported the rape to an officer in the building and then went to a hospital and produced the cup. The hospital found type A sperm in the cup and in a vaginal smear.

Mountain was indicted for rape, sodomy, and official misconduct. At trial, the prosecution proved that the sperm in the cup was type A and that a blood test had been performed on the defendant. However, the trial judge did not allow the prosecutor to refer to the blood type of the defendant. During deliberations, the jury asked the court whether the defendant's blood type was in evidence and was informed that it was not. The jury found the defendant guilty.

Relying on *Robinson*, the defendant urged on appeal that the trial court erred in permitting the prosecutor to introduce evidence that the assailant's sperm contained type A blood and in allowing "references" to the defendant's blood type. The Court of Appeals took this as an opportunity to announce that "the *Robinson* rule ... is not well founded and should no longer be followed" (id. at 805). The court explained:

When identity is in issue, proof that the defendant and the perpetrator share similar physical characteristics is not rendered inadmissible simply because those characteristics are also shared by large segments of the population. For instance, evidence that the person who committed the crime was white would not be excluded although that may include 80% of the population. Similarly, evidence of a person's sex, which would include roughly 50% of the general population, is routinely accepted as having some probative value with respect to identification. Proof of such common characteristics, of little value individually, may acquire great probative value when considered cumulatively. (Id.)

One might argue that the examples in *Mountain* of relevant, inclusionary evidence are not quite on point. After all, a jury is unlikely to give undue weight to race or sex as identifying factors. These characteristics are within the common experience of jurors. Few jurors know much about blood-group frequencies. The court, however, suggested that the possibility "that the jury may accord [blood-group evidence] undue weight, beyond its probative value, because of its scientific basis . . . can generally be avoided by instructions, where requested, emphasizing the fact that it is only circumstantial evidence and noting, perhaps, the percentage of the population involved."[1] With the opinion in *Mountain,* the rule that exclusionary blood tests were admissible, while inclusionary ones were not, was dead and buried in nearly every jurisdiction (*McCormick on Evidence* 2006, 1:885).[2]

The Admissibility of Percentages and Probabilities

What *Mountain* said "perhaps" could be done—disclosing "the percentage of the population involved" as indicated by matching genetic markers— was, in fact, standard operating procedure in most courts. Indeed, it was genetic-marker evidence that convinced the courts to drop some of their guard against "probability evidence." Such evidence had received a tremendous setback in the California case *People v. Collins* (1968). *Collins* did not involve any real statistics, but it would come to be cited in almost any criminal case in which a party wanted to suppress probability calculations, and generations of law students have been introduced to the rudiments of probability theory by dissecting the California Supreme Court's opinion in that case.

People v. Collins and the Transposition Fallacy

In 1964, Juanita Brooks, a seventy-one-year-old woman, was walking, cane in hand and groceries in tow, down an alley in the San Pedro area of Los Angeles late one morning. Her purse sat atop the packages in her wicker-basket carryall. As George Fisher (2006, 9), recounts the facts, suddenly

she was knocked to the ground and left with a dislocated shoulder and twice-fractured arm. Screaming in pain and fright, she had the presence of mind to look at her attacker and remember the fleeing woman's blond hair. Meanwhile, sixty-six-year-old John Sheridan Bass stood watering his lawn nearby. Mrs. Brooks' screams turned his attention to the bearded [and mustached] African-American man in the yellow car who passed within six feet of Bass as he snatched up the blond woman with the yellow ponytail and sped off.

A few days later, officers investigating the robbery arrested a couple—Malcolm Ricardo Collins and his wife, Janet Louise Collins—on the strength of these descriptions.

At the trial, the prosecution had no scientific evidence to offer. There were no hairs from the robbers, no bloodstains, no fingerprints, no bullets—nothing but an attempt by Malcolm Collins to evade arrest, some incriminating statements made to the police, and, most important, eyewitness accounts. Although this might have sufficed for a conviction, the eyewitness testimony was somewhat equivocal, so the prosecutor (at the suggestion of his brother-in-law, Edward O. Thorpe, who wrote the book *Beat the Dealer,* on card-counting strategies for blackjack) turned to an unlikely quarter for help. He rang up the mathematics department at California State College at Long Beach. According to the California Supreme Court, "The witness [a newly hired assistant professor at the college] testified, in substance to the 'product rule,' which states that the probability of the joint occurrence of a number of mutually independent events is equal to the product of the individual probabilities that each of the events will occur" (438 P.2d at 36–37 (footnotes and italics omitted)). The prosecutor later proposed figures for the frequencies of an interracial couple in a car ($1/1,000$), a girl with a ponytail ($1/10$), a partly yellow automobile ($1/10$), a man with a mustache ($1/4$), and a black man with a beard ($1/10$). The prosecutor applied the product rule to his "conservative estimates" to conclude "that there was but one chance in 12 million that any couple possessed the distinctive characteristics of the defendants," and he argued that "the chances of anyone else besides these defendants being there, . . . having every similarity, . . . is something like one in a billion" (id. at 37).

The California Supreme Court condemned the introduction of this testimony about the "product rule" as "fundamental error" (id. at 38). Cautioning that "[m]athematics, a veritable sorcerer in our computerized society, while assisting the trier of fact in the search for truth, must not cast a spell over him," the court reversed the conviction (id. at 33). In particular, the court saw two "glaring defects" in the calculation—"an inadequate evidentiary foundation" for the numbers that were multiplied and "an inadequate proof of statistical independence" (id. at 38). Indeed, the court quite plausibly believed that the traits were far from independent—"Negroes with beards and men with mustaches obviously represent overlapping categories" (id. at 39 (footnote omitted)). As if this were not enough, the court expressed concern that by insisting that "under [the] theory" that there "was but one chance in 12 million that any couple possessed the distinctive characteristic of the defendants . . . , it was to be inferred that there could be but one chance in 12 million that defendants were innocent and that another equally distinctive couple actually committed the robbery" (id. at 37).

In addition to the concern expressed about mathematics as sorcery, the feature of the opinion that may have given the court's opinion such celebrity status in the law of evidence was an unnecessarily complex and intimidating mathematical appendix. This appendix, which involves the mathematics applicable to a Poisson process, merely demonstrates the obvious fact that with enough couples in a population, the chance of having additional couples with the incriminating traits can be appreciable. If there were, say, three such couples, then, in the absence of other evidence to single out the Collinses, all we can say is that they have a one-third chance of being the guilty couple.[3] In this way, the *Collins* court demonstrated that the chance of finding the incriminating characteristics given an innocent couple (arguably 1 in 12 million) does not necessarily equal the chance that a couple is innocent given that the couple possesses the incriminating characteristics.

The reasoning that equates the two conditional probabilities is known to statisticians as the transposition fallacy. (It also is called the inversion fallacy and, although it is hardly restricted to prosecutors, the prosecutor's fallacy.) It recurs throughout the cases and writing on genetic proof of identity. To see why it called the transposition or inversion fallacy, a few symbols are helpful. The 1 in 12 million figure in *Collins* is supposed to be the frequency of couples with traits that match those mentioned by the witnesses. Let us call the event of observing such a matching couple M. Also, let I be the hypothesis that a given couple is innocent (and just happens to have the matching features). The court cautioned against equating the probability of a match with an innocent couple, $P(M$ given $I)$, with $P(I$ given $M)$. Notice that the evidence M and the hypothesis I have been transposed.

In general, transposing the terms leads to a different probability. For example, the probability that someone is a native English speaker (E) given that he or she is reading this book (R) is not equal to the probability that someone is reading this book given that he or she is a native English speaker. Because this book is written in English, the former probability, $P(E$ given $R)$, is fairly large. However, the number of native English speakers in the world is large, and I regret to say that only a very small fraction of them ever will read this work. Consequently, $P(E$ given $R)$ is much greater than $P(R$ given $E)$.

The *Collins* opinion brought the problem of naive transposition to the attention of the legal community, but it offered no clear solution. The mathematically correct way to obtain the probabilities of the transposed terms appears in a posthumously published essay by the Reverend Thomas Bayes (1702–1761).[4] Recognizing this, an article in the *Harvard Law Review* from a lawyer with a physics background and a young professor of statistics at New York University criticized the *Collins* appendix and proposed "a new approach, based on Bayesian probability analysis" (Finkelstein and Fairley 1970, 490). Laurence Tribe, then a new assistant professor of law at Harvard, had drafted much of the *Collins* opinion while serving as a law clerk on the California Supreme Court (Fisher 2006, 17). Without disclosing his involvement in the case, Tribe expanded his work as a clerk into a brilliant article exposing the dangers of quantification in the legal process (Tribe 1971). The debate on what Tribe provocatively called "trial by mathematics" stimulated much of what came to be denominated "the new evidence scholarship," which sought to understand the logic and process of proof and legal decision making rather than merely to parse the cases and legal rules of evidence (Lempert 1986).

The Impact of *Collins*

The combination of an opinion from one of the nation's most respected courts expressing strong misgivings about the use of probability theory to indicate the significance of evidence and a stream of complicated and sometimes esoteric and polemical studies in law reviews had an impact on the presentation of genetic-marker evidence. Nowhere was this more apparent than in the state of Minnesota. The first case to limit testimony about probabilities was *State v. Carlson* (1978). A twelve-year-old girl was brutally murdered. Her battered body was found in a wooded area. Two foreign pubic hairs were stuck to the skin in the groin area, and head hairs were found clutched in the victim's hand. She had last been seen with David Carlson and another man. Carlson was evasive when the police interviewed him at his home, and he said that a dark stain on his nylon jacket was ketchup. The officers arrested him and took him to the sheriff's office, where the jacket was tested. When Carlson was advised that the stain was blood, he changed his story, telling the officers that his dog had been killed and he had carried it, spilling blood on himself. An officer pointed out that the dog at Carlson's home was very much alive. Carlson rolled with the punch. He replied that the dog had not actually been killed but only injured.

At trial, the state called a laboratory analyst, Mary Ann Strauss, who testified that the bloodstain on the jacket possessed ABO, PGM, and erythrocyte acid phosphatase (EAP) types identical to the victim's blood. She added that 0.85% of the population would have such blood. In addition, Strauss testified that in her microscopic comparison of the hairs on the girl's body, she found a match to Carlson's hairs. An expert on comparative microscopy, Barry Gaudette of the Royal Canadian Mounted Police, confirmed the hair matches and testified that on the basis of his own studies, there was a 1-in-800 chance that the pubic hairs were not Carlson's and a 1-in-4,500 chance that the head hairs clutched in the girl's hands were not Carlson's. I shall return to testimony about hair in Chapter 12. In this case, the jury found Carlson guilty of first-degree murder.

Citing *People v. Collins* and related cases, Carlson argued to the Minnesota Supreme Court that the jury should not have heard the three numbers because "evidence in a criminal trial may not be expressed in the form of statistical probabilities" (267 N.W.2d at 172). The justices observed that the numbers in *Carlson* differed from the made-up numbers in *Collins:* "[T]he foundation for the experts' testimony in the present case was properly laid, based upon empirical scientific data of unquestioned validity."[5] Nevertheless, the court was not satisfied. Relying on Tribe's article, the court explained: "Our concern over this evidence is not with the adequacy of its foundation, but rather with its potentially exaggerated impact on the trier of fact. Testimony expressing opinions or conclusions in terms of statistical probabilities can make the uncertain seem all but proven, and suggest, by quantification, satisfaction of the requirement that guilt be established 'beyond a reasonable doubt' " (267 N.W.2d at 172).

Given this rather general concern with quantification, one might think that the *Carlson* court would have deemed the efforts to explain the import of both the genetic-marker and hair matches inadmissible. But the opinion addresses only Gaudette's figures for matching hair fibers. It is strangely silent about the admissibility of Strauss's percentage for the blood types. Was this because the court believed that percentages are less likely than probabilities to have an exaggerated impact and did not intend to prohibit them? This seems too fine a line to draw. Another possibility is that the court condemned testimony of a "1-in-4,500 chance that the head hairs did not belong to the accused" because it instantiated the transposition fallacy. At most, Gaudette could testify to the probability that hairs known to have come from different heads would match. That is the quantity that his experiments were supposed to estimate. We can abbreviate this probability as $P(M$ given *different source*), where M again stands for "match." As noted in connection with *Collins,* this conditional probability cannot generally be transposed to P(*different source* given M), which is what the opinion suggests Gaudette testified to. In contrast, Strauss's 0.85% figure is much farther removed from any such transposition. Yet this analysis is subtler than anything contained in the opinion, which merely alludes to "the psychological impact of the suggestion of mathematical precision" (id.). *Carlson* is a difficult opinion to decipher.

Later cases dissolved *Carlson*'s ambiguity by imposing a total ban. Half a century earlier, in *State v. Damm,* the South Dakota courts had prevented a defendant from having blood tests conducted that might have exonerated him. In 1983, the state of Minnesota used genetic-marker tests to convict the defendant of statutory rape resulting in a pregnancy. In *State v. Boyd,* Dr. Herbert Polesky, the director of a blood bank and an expert in parentage testing, gave figures for the percentage of the male population that the test of fifteen markers would exclude (namely, 94% to 97%), the mean number of unrelated men who would have to be randomly selected "before another man would be found with all the appropriate genes to have fathered the child" (1,121), a likelihood ratio known as a paternity index (1,121.39), and a so-called probability of paternity (99.911%).[6] The Minnesota Supreme Court read its prior opinion in *Carlson* as rejecting any numbers that a jury conceivably might translate into "a measure of the probability of the defendant's guilt or innocence . . . that . . . will thereby undermine the presumption of innocence, erode the values served by the reasonable doubt standard, and dehumanize our system of justice" (331 N.W.2d at 483). In this case, it held that "the trial court may appropriately limit Dr. Polesky's testimony . . . to the basic theory underlying blood testing and [the finding] that not one of the 15 tests excluded defendant" (id.).

A few years later, the state attempted an end run around the *Carlson-Boyd* exclusionary rule. Together with her husband, a woman described only as "the complainant" worked as the manager of an apartment complex in St. Paul. On a cold December evening in 1984, the two of them quarreled, and the husband stomped out. He first went to the owner of the apartments, Joon Kyu Kim, told Kim of his marital discord, and left the apartment complex. At about 10 P.M., Kim appeared at the couple's apartment. He announced that the complainant was not having enough sex with her husband and that he would show her how. He forced her into the bedroom and raped her. As he left, he gave her a twenty-dollar bill and told her that the next time it would be thirty dollars. He added that she would not call the police because she "needed the job too much."

Kim was wrong. The complainant provided the police with the sheet from the bed and other items. She went to a hospital, where swab samples were taken from her body. The Bureau of Criminal Apprehension Laboratory found semen present on the bed sheet and on the vaginal swabs.

Kim denied having had sexual intercourse, consensual or nonconsensual. He conceded that he had gone to the apartment all right, but only to fire her as a caretaker. He claimed that she was accusing him to retaliate. After filing criminal sexual conduct charges against him, the state obtained a court order for samples of Kim's blood, saliva, and hair. The ABO and PGM types matched the semen found in the complainant's body and on the bed sheet. A laboratory analyst was prepared to testify that 96.4% of males in the Twin Cities metropolitan population, but not Kim, could be excluded by the tests. Kim objected to the scientific evidence at a pretrial hearing. The trial court excluded the percentage figure under *Carlson* and *Boyd*.

The Minnesota Supreme Court agreed to review this ruling in advance of Kim's trial. The state argued, unsuccessfully, that the court should overrule *Carlson* and *Boyd*. Short of that, it tried to distinguish *Boyd*. A statistic disallowed in *Boyd* was the paternity probability of 99.911%. In *State v. Kim,* the lawyers and the court thought that this was the proportion of the population that could be excluded as donors. The exclusion probability that Dr. Polesky testified to was lower, falling between 94% and 97%. In any event, the state suggested that the reason the exclusion probability was objectionable was that jurors were likely to commit the transposition fallacy and misconstrue it as the probability of guilt. Here, the state would merely inform the jury that 3.6% of the population, including Kim, would be identified as possible sources. Framed in this inclusionary way, the statistic would be less likely to be transformed into a probability of innocence, or so the state claimed.

The court did not buy this distinction. It wrote that "faced with an exclusion percentage, a jury will naturally convert it into an inclusion percentage" (398 N.W.2d at 548). Presumably, the court meant the converse— that an allegedly nonprejudicial inclusion probability would be converted into an exclusion probability and then transposed into a probability of guilt. A juror who learns that only 3.6% of the population would be included as a possible source of the semen stain might reason that the chance that Kim was coincidentally included was only 3.6%, leaving a 97.4 percent chance that his inclusion was not coincidental. That this transposition is fallacious can be seen, as in *Collins,* by noting that if the apartment complex housed, say, 200 men, about 7 of them would be included as a possible source of the stain. Kim is just one of these 7 or so. Kim's genotype, standing alone, does not make it 97.4% probable that he was the source.

Evidently doubting that jurors can be trusted to sort out such things, the court stressed the "danger of population frequency statistics" (id.) and explicitly held that "the expert called by the state . . . should not be permitted to [express an] opinion in terms of the percentage of men in the general population with the same frequency of combinations of blood types" (id. at 549). All numbers, it would seem, were banished from the criminal courts in Minnesota. Yet, in a rococo twist, the *Kim* court added an afterthought—the expert could give quantitative testimony confined to "the percentage of people in the general population with each of *the individual* blood types" (id.). Apparently, the court regarded the risk that the jury would multiply to arrive at the probability of inclusion as less than the danger that it would subtract to find the probability of exclusion.

Minnesota's aversion to numbers rests on the premise that they will have a "potentially exaggerated impact on the trier of fact [making] the uncertain seem all but proven" (*Carlson*, 267 N.W.2d at 172). This premise is hardly self-evident. Often, people are less impressed by dry statistics than they are by intuitive stereotypes or juicy anecdotes. A considerable amount of research has been conducted on the ways in which people process and employ probabilistic information. A subset of this work examines the processing of explicitly probabilistic or statistical evidence by mock jurors. These studies do not support the court's assumption. To the contrary, in these experiments, the jurors typically underestimate the impact of the evidence as compared with the effect prescribed by probability theory, and they are not overwhelmed by the statistics for the trace evidence. In a recent study (Kaye et al. 2007), for instance, my colleagues recruited jurors who appeared for jury duty in Delaware but were not selected to sit on a jury to watch a condensed but realistic videotape of a trial and then to deliberate as eight-member juries.[7] The defendant was accused of bank robbery, but the nonscientific evidence against him was weak. The strongest item of evidence was a special kind of DNA (mitochondrial DNA [mtDNA], discussed in Chapters 11 and 12) extracted from hairs in a sweatshirt that the robber discarded while outrunning a pursuing police officer. This DNA matched that of the defendant, who was arrested months later. An FBI analyst testified on direct examination that only 1 in 5,072 Caucasian men have mitochondrial DNA types that match that of hairs from the sweatshirt, meaning that 99.98% of that population would be excluded as a possible source of the hairs. On cross-examination, he agreed that for the metropolitan area in which the robbery occurred, this would leave "six white males as the possible source of those hairs." The defense expert, a genetics professor, questioned the FBI calculation and maintained that "some 10 out of every 5,072 white men in Middletown would have a DNA sequence that could not be excluded from that of the hairs in this case." She concluded that "about 57 Caucasian males in the . . . metro area could not be excluded as the source of the hairs found on the sweatshirt." In a closing statement, the defense attorney claimed that the mtDNA evidence was weak because a substantial number of other people could also be the source of the hairs.

Despite the exclusion probability of 99.98% (according the prosecution) or 99.8% (according to the defense), the jurors were not overwhelmed. Nearly as many voted to acquit (216) as to convict (242). Although the match surely was relevant, when the jurors were asked whether the mtDNA evidence was "completely irrelevant," only half rejected the claim, and 40% agreed with it "because a substantial number of other people also could be the source of the hairs." When the jurors were asked to describe, on a scale of 0% to 100%, their confidence in the proposition that the defendant was the robber, the mean estimate was only 75%. Only about one-third of the jurors placed this source probability in the range of 90% or higher, and only about one-quarter ventured above 95%. These moderate percentages occurred notwithstanding the fact that nearly half the jurors also accepted as true the statement that the mtDNA evidence showed only about a 1% chance that someone else was the robber. These are the kinds of responses that would be expected of jurors who thought that the DNA evidence was powerful standing alone but that the other evidence in the case raised substantial doubt. For these jurors, at least, the probabilities did not "make the uncertain seem all but proven and suggest, by quantification, . . . guilt . . . 'beyond a reasonable doubt,'" as the *Carlson* court feared they would.

Despite Minnesota's state motto, "L'Etoile du Nord," other jurisdictions have not looked to the *Carlson* line of cases as any kind of North Star in navigating their way through the statistics associated with genetic-marker evidence. As a dissenting justice pointed out in *Kim,* "State v. Carlson is one of the few cases that can be found excluding computations that the court considered well-founded" (398 N.W.2d at 552). With genetic markers, almost all courts followed the lead of *State v. Washington,* the first state supreme court case to admit electrophoretic tests for serum enzymes. In that case, the Supreme Court of Kansas described the quantitative portion of the criminalist's testimony as follows:

> Burnau testified to the percentage of the Caucasian population having each of the types present in Cummings's [the deceased's] blood. She also testified to the percentage of the Negroid population which would have each of the types present in defendant's blood. Multiplying the highest of each percentage together, she determined that 3.1% of the Negroid population would have the

same combination of enzymes and proteins as the defendant. Burnau testified that the seminal fluid obtained from the vaginal swab did not show the blood type, indicating that the person who had had intercourse with the victim was a nonsecretor. Saliva samples were tested for both the victim and the defendant, and both were determined to be nonsecretors. Studies show that only 20% of the population are nonsecretors. Adding this factor to her other percentage, Ms. Burnau determined that only 6/10ths of 1% of the population would have defendant's combination of blood factors. (622 P.2d at 989)

People v. Collins was no obstacle to this testimony because the probability in that case was "based on estimations rather than on established facts" (id. at 994). In contrast, "population percentages on the possession of certain combinations of blood characteristics, based upon established facts, are admissible as relevant to identification" (id.).

Although Burnau's numbers are not comparable with the rank speculation in *Collins,* several aspects of the quantitative analysis are noteworthy. First, the multiplication of "the highest of each percentage" from the two racial databases is atypical. Normally, experts would compute a joint frequency within each relevant population rather than picking and choosing from two or more. If the serologist understood that the relevant population consisted of all individuals who might have attacked Cummings, then it is unclear why she limited the statistics to blacks and whites. Why she used the higher frequency among blacks or whites for each genetic marker also was not explained. As we shall see, some fifteen years later, a committee of the National Research Council advanced a similar approach, dubbed the "interim ceiling principle," to calculate "conservative" DNA frequencies. This proposal evoked intense antagonism from forensic scientists and population geneticists, but in its first incarnation, the mix-and-match method of computation passed unnoticed.

Second, even within a single population group, multiplication is valid only if the alleles of each genetic system are statistically independent. Although there are good biological reasons to accept the assumption of stochastic independence for the blood-group and protein and enzyme alleles (the technical buzz words, as explained in Chapter 5, are "Hardy-Weinberg equilibrium" and "linkage equilibrium"), this assumption would become the subject of vehement controversy when it was carried over to the analysis of DNA matches in the 1990s.

Yet the validity of the standard assumption went largely unquestioned in *Washington.* Defense counsel apparently did not raise these points, and the *Washington* court was satisfied that the 0.6% figure was admissible because the "percentage statistics were based on population studies published by the American Association of Blood Banking [*sic*]" (id.). This way of describing the statistics is a bit misleading. Extremely large samples would be needed to ascertain directly the small frequencies of all the possible combinations of phenotypes. The American Association of Blood Banks (AABB) tables simply gave the frequencies for each individual phenotype. They did not cover pairs, let alone septuplets, of blood types. Hence it is misleading to state, as the court did, that "[b]*ased on these studies,* the percentage of blacks in the population having certain characteristics was multiplied by the total number of characteristics found" (id., emphasis added). The studies supplied the numbers that were multiplied, not the justification for multiplying them.

In the few cases in which defense experts did question the independence assumption, the courts blandly asserted that objections to statistics merely went to the weight of the evidence, not to its admissibility. In *People v. Lopez,* Kathy Moorman of the Chicago Police Department gave rather confused testimony about the statistics. She took percentages from a "population study" that she "had never read" and multiplied them, although she conceded that the study itself did not support the multiplication (593 N.E.2d at 654). In response, the defense called "Emmett Harmon, an analytical biochemistry expert [who] testified that there are different enzyme tables for different nationalities since enzyme systems are affected by race and national origin" (id.). He objected to Moorman's use of allele frequencies for Caucasians as a group "[b]ecause the white population is composed of many different nationalities" (id.). A decade later, a professor from Harvard would testify similarly with respect to DNA evidence, and courts would wring their hands over general acceptance. With immunogenetic evidence, however, it was enough that "[t]he American Blood Bank's [*sic*] study on blood characteristic frequency statistics for the general United States population is generally recognized by medical and forensic science professions" (id. at 656). Whether the study established multilocus independence was not discussed.[8]

Likewise, in *Commonwealth v. Gomes,* a case resulting from the slaughter of a mother and two children in Boston, the Massachusetts Supreme Judicial Court brushed aside an objection to the method of computing a multilocus frequency of 1.2%. Diane Juricek "testified that gene frequencies may vary among locations and ethnic or racial groups [and] that . . . simply multiplying the gene frequencies failed to take into account certain variable factors, such as the possibility that some traits may not be independently inherited, possible differences in gene frequency due to differing socioeconomic status, and the lack of genetic 'purity' in American racial groups" (526 N.E.2d at 1280 n.10). Begging the question, the court wrote that "once the witness had been qualified as an expert and it has been shown that the statistics are based on established facts rather than estimates or speculation, such criticisms go only to the weight to be accorded the evidence, not to its admissibility" (id. at 1280).

At this point, the power of the genetic markers to narrow greatly the set of possible sources had become apparent, and courts countenanced the use of immunogenetic markers to include or inculpate defendants as possible perpetrators. Not only were scientists or technicians allowed to testify that a defendant was not excluded, but they also were allowed to estimate the relative frequency of an incriminating set of markers in the relevant population. Occasional objections to the method of computing multilocus genotype frequencies were dismissed without any careful inquiry into the scientific literature on this issue. Then came DNA evidence. The legal world would never be the same.

CHAPTER THREE

The Dawn of DNA Typing

THE PROTEINS and enzymes used by serologists to link individuals to crimes are a window on the genes that contain the instructions for synthesizing these chemicals in cells. Modern DNA typing does not merely peer through this window. It enters the building and examines the DNA—the molecules that contain the genetic information—directly and more efficiently. Moreover, DNA typing is not limited to the tiny fraction of DNA that codes for blood types and protein products. Because DNA typing can examine noncoding DNA as well, it has greater power to discriminate among individuals. For these reasons, DNA typing quickly eclipsed genetic markers in the investigation and prosecution of crimes.

The breakthrough discovery came in the laboratory of a British geneticist who earned a knighthood. The "DNA fingerprinting" technique, as Sir Alec Jeffreys called it (Jeffreys, Wilson, and Thein 1985b), came from a chance discovery with some gray seal meat. To reveal the nomenclature, power, and drawbacks of this early form of DNA typing, we need to know a bit more about the nature of DNA and classical genetics. This chapter surveys these topics. It then describes Jeffreys's discovery of an individualizing multilocus DNA probe and how this discovery soon exonerated a suspect and prompted a confession in a multiple-murder case.

Genes, Chromosomes, DNA, and Sex

When Landsteiner was explaining blood types, the "laws of heredity" propounded by Gregor Mendel between 1856 and 1863 on the basis of experiments with some 28,000 pea plants were being rediscovered (Bowler 1989). Mendel had concluded, among other things, that alternative versions, or *alleles,* of then-unknown particles of heredity account for variations in inherited characters, such as the color and wrinkliness of peas. He knew that gametes (sexual material from each parent) fused to produce progeny. He proposed that when gametes are formed, one member of the allelic pair separates from the other member, and a particular gamete gets one of the separated alleles. This is Mendel's first law of *segregation of traits.*

The alleles for traits that Mendel studied came in only two varieties—dominant and recessive. It takes only one copy of a dominant allele to produce a trait. For example, the presence of long eyelashes is a dominant trait. One gamete from one parent might have the long-eyelash allele, while another gamete in the other parent could have the short-eyelash allele. If these were combined, the offspring would have long eyelashes. To have short eyelashes, a child would have to inherit a short-eyelash allele from both parents.

Mendel's second law states that the emergence of one trait (eyelash length, for example) does not affect the emergence of another (such as the presence of dimples). The first law says that the two members of any one pair of alleles segregate. The second law says that two pairs of alleles (the pair of alleles for eyelash length and the pair for dimples, for example) *assort independently.* Eyelash length and dimpling are not related to one another. Eyelashes and dimples are not a major concern, of course, but some human diseases, such as sickle-cell anemia and Tay-Sachs disease, which are controlled by a single gene, also follow this simple, Mendelian pattern of inheritance.[1]

In the classical genetics of inherited, observable traits, the meaning of a gene was clear. It is some "particulate factor" that is transmitted unchanged from parents to progeny and that largely determines the visible traits—the phenotype—of the offspring (Ridley 2003, 231–236). The Danish botanist and geneticist Wilhelm Johanssen (1857–1927) coined the words "gene," "genotype," and "phenotype" 100 years ago (Johanssen 1909), remarking that "[t]he word 'gene' is completely free from any hypothesis," making it "a very applicable little word" (Zimmer 2008).

Today, we explain Mendelian traits (and other inherited ones as well) in terms of many hypotheses about DNA, ribonucleic acid (RNA), and protein molecules. The once-mysterious genes are known to be part of an organic compound, deoxyribonucleic acid. Found in the cells of all organisms, from the humblest amoeba to the most arrogant human being, the DNA molecule is made of subunits that include four chemical structures known as nucleotide bases, whose names (adenine, thymine, guanine, and cytosine) are abbreviated to A, T, G, and C. The physical structure of DNA is often described as a double helix because the molecule has two spiraling strands connected to one another by weak bonds between the nucleotide bases. As shown in Figure 3.1, A pairs only with T and G only with C. Thus the order of the single bases on either strand reveals the order of the pairs from one end of the molecule to the other, and it suffices to say that the DNA molecule is like a long sequence of As, Ts, Gs, and Cs.

Most human DNA is tightly packed into structures known as chromosomes, which are located in the nuclei of most cells. The chromosomes are numbered (in descending order of size) 1 through 22, and the remaining chromosome is an X or a much smaller Y. If the bases are like letters, then each chromosome is like a book written in this four-letter alphabet, and the nucleus is like a bookshelf in the interior of the cell. All the cells in one individual contain identical copies of the same collection of books. This personal library is the individual's nuclear genome. It contains two versions of chromosome number 1, two of number 2, and so on.

Other organisms carry wildly different numbers of chromosomes. The fruit flies that geneticists studying patterns of inheritance have exploited since the early 1900s have only 8 chromosomes. A house fly has 12. A dog has 78. A king crab has 208. Some species of newt have 226. For almost forty years, scientists were convinced that men had 47 and women 48 (Sykes 2004, 36, 48–50).

In human beings, the process that produces billions of cells with the same genome starts with the production of gametes (sperm or egg cells). When the germ-line cell that splits to form the mother's egg cell divides, the members of each pair of her chromosomes briefly come into contact with one another. Promiscuously, they swap some sections at random in a process called *crossing over* or *recombination*. Let us focus on just one pair of chromosomes, say, chromosome number 1. After recombination, each chromosome number 1 is a mosaic of the original two versions. The female germ-line cells divide so that egg cells contain only one set of 23 chromosomes—one randomly selected chromosome number 1, one randomly selected number 2, and so on. This is consistent with Mendel's first law. Likewise, sperm cells have 23 single, recombined chromosomes. When a sperm cell and an egg cell combine, the fertilized cell therefore contains the full complement of 46 chromosomes.

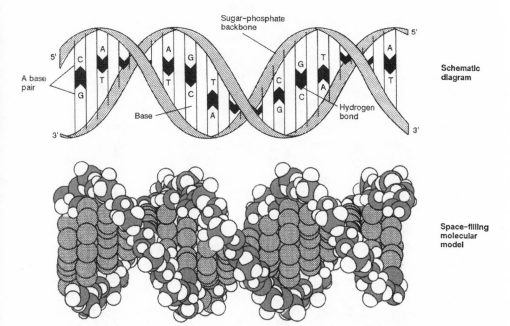

Figure 3.1. Two ways of representing the DNA double helix. Diagrams are of a very short section of the DNA molecule in each chromosome. The human genome contains about 200 million times the amount of DNA shown. The two strands of the DNA double helix run in opposite directions (as indicated by the 3' and the 5' ends) and are paired to each other by the complementary nucleotide pairs. Reprinted with permission from Alberts et al. (1983, 101).

The effect of this rigmarole is to create new genetic combinations in the offspring. This kind of sexual system apparently has an evolutionary advantage. By producing offspring that are genetically novel, the parents produce some individuals that may be better adapted to their environment and more likely to pass their genes on to future generations. But sex is not inevitable. Some species, including some plants, insects, worms, and lizards, do not bother with it. They reproduce asexually in multifarious and sometimes astonishing ways (Sykes 2004).

For better or worse, however, humans still rely on sex to produce a fertilized ovum. Then comes gestation. During pregnancy, this single cell divides to form two cells, each of which has an identical copy of the new combination of the 46 mixed-up chromosomes drawn from the parents. The two cells then divide to form four, the four form eight, and so on. Before long, various cells specialize to form different tissues and organs. In this way, each human being has some 100 trillion copies of the original

23 pairs of chromosomes from the fertilized egg, one member of each pair having come from the mother and one from the father. If we could uncoil the chromosomes in all these cells from just one person and lay them end to end, they would reach to the sun and back over 600 times.

Genes lie at various positions (loci) along the chromosomes. The genes usually are from 1,000 to 10,000 base pairs long. The sequence of these base pairs encodes most of the information used by cells in manufacturing proteins, such as the blood-group antigens described in Chapter 1, that underlie all kinds of observable traits.[2] A protein is a series of chemical units, known as amino acids, that are hooked together into a chain. Any one of twenty different amino acids can be attached at each point in the protein chain. The "coding" DNA in a gene is "read," three base pairs at a time, to determine which amino acid goes next in a protein.[3] In addition to the coding DNA that specifies the physical structure of the particular proteins that are synthesized in cells, genes contain noncoding sequences that regulate their operation, affecting how much protein a gene produces. In brief, at the mechanistic, molecular level, Mendel's "factors" and Johanssen's "genes" are sequences of the nucleotide base pairs in the DNA in the chromosomes.

From Protein Polymorphisms to Restriction Fragment Length Polymorphisms

The paradigm of genetic identification before DNA analysis was to study variations in the proteins expressed by genes. As we saw in Chapters 1 and 2, this testing of genetic markers is what serologists and parentage testers did. This form of genetic testing was the highest state of the art in the genetic proof of identity as of the mid-1980s. As molecular biology came of age, however, the possibility of using DNA for individualization entered the scientific and legal consciousness (Ellman and Kaye 1979, 1138 n.37). The shift in perspective is at once obvious but radical. Traits such as blood types are inherited because DNA is inherited and small parts of it (some 1% of the entire genome) are "expressed." The details of gene expression are complicated, and even the definition of a "gene" grows murky at this molecular level (Snyder and Gerstein 2003; Zimmer 2008). We can get by with a few highlights.

As we have seen, the inherited "particulate factor"—the classical conception of a gene—corresponds to a sequence of DNA base pairs that provides the information used by the cellular machinery to start, stop, and specify the rate of protein synthesis (the regulatory regions) and to sequences that specify the structure of the protein (the coding regions).[4] But it is the long DNA molecules that are copied and transmitted from parent to child. The sequences we call genes go along for the ride. So do any other base-pair sequences in the DNA. Therefore, we can treat the DNA sequences as if they were phenotypes inherited from one generation to the next according to Mendel's laws.

We can call this perspective the "inherited DNA approach" because the focus is on the inheritance of specific DNA sequences rather than on the expressed traits. It is, in a sense, a DNA-eye view of the biological world, reminiscent of the quip that "a scholar is just a library's way of making another library" (Dennett 1995, 328). From the standpoint of DNA, a human is just a way of making more DNA. The DNA sequences are swimming in a vast sea of humanity, and as new generations of humans come and go, mutations arise, and the relative proportions of different sequences ebb and flow.

From the standpoint of forensic science, the change in focus from external trait to internal sequences of base pairs is significant for three reasons. First, it opens up the possibility of technologies to characterize DNA variations that are faster and cheaper and require less trace material than the techniques for analyzing the protein products of genes.

Second, at the DNA level, genes can be more variable than their protein products. Some mutations within genes will not interfere with or alter the structure of the proteins. In principle, all these neutral mutations can be detected by direct DNA analysis even though they have no external effects.

Finally, most human DNA is not part of the coding and control regions of genes. The "extra" DNA is found both inside and between genes. Intervening DNA sequences (*introns* that may or may not have any physiological function)—split up genes.[5] (In Chapter 4, we shall see how one of the two scientists who discovered introns in 1977 was drawn into the early courtroom controversies over DNA evidence.) For example, the human beta globin gene extends for 1,600 base pairs, but the coding and control regions come in three chunks (called *exons*) that collectively extend for only 626 base pairs. The other 974 base pairs are in the introns between the beginning and the end of the gene. In general, noncoding, nonregulatory sequences that no longer participate in protein synthesis constitute about 23% of the base pairs within genes (Goodwin, Linacre, and Hade 2007, 10). In addition, there are huge stretches of DNA (about 75% of the genome) between genes (ibid.).

But introns and intergenic DNA still are inherited in a Mendelian fashion. If they are not functional, then mutations can accumulate in these regions without harm. As a result, these mutations will be propagated in new generations more readily than deleterious mutations in functional regions. (Mutations that can harm an organism before it reproduces tend to disappear in populations because the organisms bearing the mutation tend to have fewer progeny.) Hence introns and intergenic DNA are likely to be more variable than the coding and regulatory sequences. If differences in the DNA molecules themselves can be measured, then we can use these *DNA genotypes* to differentiate and individualize people. (The term *genotype* for DNA features that have nothing to do with genes is too well established to dislodge, but it is unfortunate. Not only does it confuse students of genetics, but it can lead policy makers to think that the DNA loci that are of forensic interest carry much more socially, medically, or personally meaningful information than they do.)

Implementing the shift in perspective from expressed traits to DNA sequences is not as easy as it sounds. Despite intensive research into physical or chemical techniques that might rapidly and cheaply "read out" the sequence of base pairs in a DNA molecule, sequencing remains a cumbersome and complicated procedure. A much-anticipated international project to sequence the entire genomes of 1,000 people was launched in January 2008. Even with the most advanced sequencing technology, the 1,000 Genomes Project is expected to cost "just US$30 million to $50 million" (Hayden 2008, 378). The DNA scanner available to *Star Trek*'s Dr. McCoy is still science fiction because of the sheer size of the genome. All told, the DNA in the 23 chromosomes consists of more than 3 billion base pairs of genetic "text." If the letters for the bases on one strand of the double helix were printed as a string of letters, the full listing would stretch for about 7,000 miles. If the letters were organized on pages of paper, and the papers were placed in a single stack, the pile would be as high as the Washington Monument.

One might think that almost any single such page would uniquely identify an individual. After all, the first letter could be an A, T, C, or G. Adding the second letter produces $4 \times 4 = 16$ possible pairs; a string of three letters yields $16 \times 4 = 64$ possibilities, and so on. Three billion letters can accommodate $4^{3,000,000,000}$, or about $10^{1,806,180}$, possibilities. However, the letters in most of the positions on most pages are the same. At least 99.5% of the base pairings are the same in everyone (Goodwin, Linacre, and Hade 2007, 11; Wade 2007). This similarity is not really surprising. It accounts for the common features that make humans a viable and identifiable species—and one that has not been around long enough to develop a great deal of genetic variability. The remaining 0.5% or less of the order of the base pairs is particular to an individual (identical twins excepted). This 0.5% variation makes each person genetically unique. The trick is to find some short, specific regions of DNA that are highly polymorphic—that tend to vary from one person to another.

Restriction Fragment Length Polymorphisms

This feat was performed in Sir Alec Jeffreys's laboratory in 1984 (Jeffreys, Wilson, and Thein 1985a, 1985b). The key to Jeffreys's discovery was an established technique for detecting DNA differences that give rise to restriction fragment length polymorphisms (RFLPs). The method of RFLP analysis in use at the time works as follows. After DNA is extracted from cells, a bacterial enzyme (called a restriction enzyme) is added. The enzyme binds to DNA when it encounters a certain short sequence of base pairs and cleaves the DNA at a specific site within that sequence. Digesting a sample of DNA with such an enzyme usually gives rise to fragments ranging from several hundred to several thousand base pairs in length.

These restriction fragments can be separated according to size by gel electrophoresis—the same method that was used to characterize the serum proteins and enzymes we encountered in Chapter 1. In this procedure, the broken pieces of DNA are loaded into small holes cut into one end of a slab of gel. Because DNA fragments have a negative charge, applying an electric field to the gel pulls all the fragments toward the positive pole. Larger fragments have more difficultly moving through the gel, so after a while, the smaller fragments migrate farther. When the electric current is turned off, long pieces of DNA will lie in bands near their starting point, and short pieces will be in bands at the other end of the gel. The length of any particular fragment can be measured by comparing the distance it has traveled with the distances traversed by standard fragments of known size placed in a parallel slot in the gel.

The resulting array of fragments is transferred for manageability to a sheet of nylon by a process known as Southern blotting. Either before or during this transfer, the double-stranded DNA is "unzipped" into two single strands by alkali treatment, which breaks the weak bonds that connect the two members of a base pair but leaves intact the stronger bonds that hold a base to the backbone.

A quantity of a "probe" is applied to the DNA strands on the membrane. Probes are short, single strands of DNA with a radioactive component attached. When the sequence of bases in the probe line up with their complementary sequence (according to the A-T, C-G rule) of a DNA strand in the membrane, they bind together, reconstituting the duplex DNA.

Finally, the position of the specifically bound probe is made visible by autoradiography. That is, the washed nylon membrane is placed between two sheets of photographic film. Over time, the radioactive probe material exposes the film where the biological probe has hybridized with the DNA fragments. The result is an autoradiograph, or autorad—a visual pattern of bands representing DNA fragments that contain the specific base-pair sequence targeted by the probe. The huge number of other DNA fragments remain invisible. Only those fragments containing the tagged sequences contribute to the image. An autorad that shows two bands in a single lane indicates that the individual who is the source of the DNA is a heterozygote at that locus. If the autorad shows only one band, the person may be homozygous for that allele (that is, each parent contributed the same allele), or the second band may be present but invisible for technical reasons. The band pattern thus defines the person's "DNA genotype" at the locus associated with the probe.

Most enzyme-probe systems in the early 1980s detected single nucleotide polymorphisms (SNPs, pronounced "snips"). That is, the probe responded to a difference in a single base pair (a substitution, a deletion, or an insertion) within the short sequence that the particular enzyme recognized and cut. For example, the restriction enzyme *Alu I* cleaves the sequence AGCT in the DNA between the G and the C on both strands. Suppose that initially everyone's genome is such that there is a stretch of DNA in which the distance between two of these cleavage sites is the same, say, 3,200 base pairs, and that a probe binds to a part of this 3.2-kilobase (kb) region.[6] At this time, there is no polymorphism to observe. As shown in Figure 3.2, everyone tested would have a 3.2 kb fragment. Because the restriction sites are the same on both chromosomes, RFLP analysis should yield a single band at the 3.2 kb position on all autoradiographs. The order of some bases within this fragment might differ from person to person, but the length of the restriction fragments, which is the only thing measured, would be the same in everyone. The RFLP locus is monomorphic.

Suddenly a point mutation occurs in one individual. Within the 3.2 kb sequence, an A is substituted for a G, changing what was an AGCT into an AACT and thereby creating a new restriction site. Some of the individual's children inherit the mutated DNA. As generations go by, the population comes to include a variation in the genome. Now there is a detectable DNA polymorphism. The people who inherit the mutation have a cleavage site within the 3.2 kb region. Their DNA differs from everyone else's at this location. Instead of a single 3.2 kb restriction fragment, they will have two short fragments, say, a 1.2 and a 2.0 kb fragment. Two alleles are detected in the population because the restriction enzyme cuts one allele (the

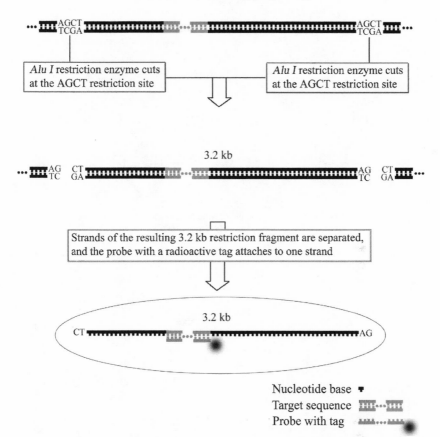

Figure 3.2. Experiment detects a DNA allele with restriction sites 3.2 kb apart by producing a radioactively tagged 3.2 kb restriction fragment. Only one DNA molecule is shown. A sample with a large number of DNA molecules with such restriction sites would yield many such restriction fragments that could be separated by gel electrophoresis and made visible by exposure to x-ray film.

mutant allele with the extra *Alu I* site) but not the other allele. As shown in Figure 3.3, a probe that hybridizes with some of the fragments detects this difference. In general, where a base substitution happens to create or destroy a restriction site, an RFLP indicative of the two alleles results.

Testing individuals in the population now can produce three distinct patterns on an autorad. A person with the new restriction site on both chromosomes will have a single band at the 1.2 kb position. A heterozygous individual with the mutant allele on one chromosome and the original allele on the other will have a band at the 1.2 kb position and another, clearly separated band at the 3.2 kb position. Finally, a person who

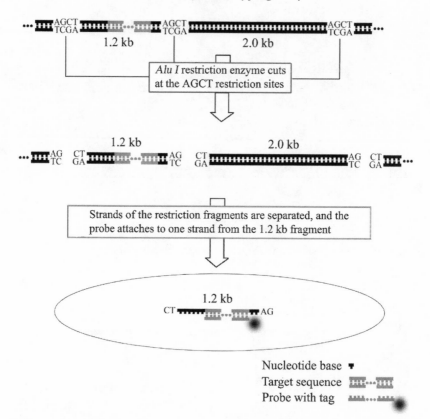

Figure 3.3. Experiment detects a DNA allele with an interior restriction site by producing a tagged 1.2 kb fragment. Only one DNA molecule is shown. A sample with a large number of DNA molecules with the same restriction sites would yield many 1.2 kb restriction fragments that could be separated by gel electrophoresis and made visible by exposure to x-ray film.

is homozygous for the original allele will have a single band at the 3.2 kb position. Figure 3.4 illustrates how this polymorphism would appear on an autorad of all three DNA types.

Although the underlying polymorphism in our example consists of a single substitution in the sequence of nucleotide bases, it is the length of the fragments that is ascertained via the particular restriction enzyme and probe. It is common to speak of the DNA "alleles" for an RFLP as the different lengths of the restriction fragments rather than the underlying variations within single base pairs in the original restriction site.

Simple biallelic RFLP loci such as this one could be extremely useful in searching for the locations of disease-related genes (Botstein et al. 1980; Gusella et al. 1983; White et al. 1985), but they were not up to the task of

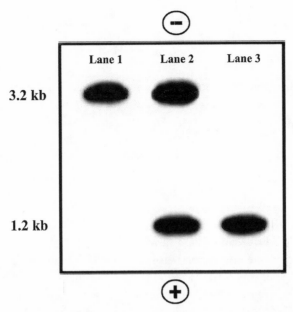

Figure 3.4. Sketch of an RFLP autorad indicative of a SNP. The band in lane 1 is from a person without the point mutation that produced the interior restriction site. Such homozygotes have the sequence leading to a 3.2 kb fragment on both chromosomes. The bands in lanes 2 and 3 are from individuals with the mutation. The bands in lane 2 are from a heterozygote who has the mutation on one chromosome (resulting in the 1.2 kb band) and the original sequence without the mutation (resulting in the 3.2 kb band). The band in lane 3 is from a person who is homozygous for the mutation. No 2.0 kb band appears because the radioactive probe binds only to a shorter stretch of the DNA within the first 1.2 kilobases of the 3.2 kilobases between the original two restriction sites.

human individualization from trace evidence. A locus with two common alleles—for instance, a 60–40 split—would do little to individualize because, at best, the crime-scene sample would match 40% of the population. A locus with a rare allele—say, a 99–1 split—could lead to a more powerful result, but only rarely. In 99 cases out of 100, a crime-scene sample would have the ubiquitous allele rather than the rare variant. Of course, if enough loci were tested, the combined result might be highly discriminating. If twenty independent, biallelic loci were typed, the combination of alleles would occur in not more than $1/2^{20} = .000095\%$ of the population. But each enzyme-probe test consumes some previous crime-scene DNA. It would require copious quantities of this trace DNA to perform twenty RFLP tests. Something different was needed before RFLP testing could be used in criminal cases.

Multilocus Variable Number Tandem
Repeat Probes

Sir Alec Jeffreys was not looking for a way to use RFLP testing in criminal investigations, but by the kind of chance that favors the prepared mind, he found one in the flesh of an Antarctic seal. When Jeffreys joined the Department of Genetics at the University of Leicester in 1977, he wanted "to marry together genomics with the classic discipline of human genetics and try to detect heritable variation directly in human DNA" (Jeffreys 2005, 1035). In other words, Jeffreys wanted to follow the "inherited DNA approach" to better understand the genetics of diseases or other phenotypes. To do this, he and his colleagues started to develop RFLPs to track inherited DNA features in families. "We got our first SNP in 1978," he explained in one interview. "Before that we knew about heritable variation in gene products, such as blood groups, but here we had examples of inherited variation in DNA, the most fundamental level of all" (Newton 2004a).

Still, "RFLPs . . . were difficult to find and to assay, and did not tell you much about variation between people—you either had the change or you didn't." So Jeffreys started looking for pieces of DNA that would be more variable than SNPs. At this point, two geneticists at the University of Utah using RFLP analysis on DNA samples from Mormon families discovered "a highly polymorphic locus in human DNA" (Wyman and White 1980). Unlike the one- or two-banded pattern for RFLPs resulting from an SNP that produced a new restriction site or deleted an old one, the DNA fragments at this one locus had at least eight distinguishable lengths.

Wyman and White had fortuitously found the first variable number tandem repeat (VNTR) locus—a type of "repetitive DNA" that consists of contiguous repetitions of a particular short sequence. Figure 3.5 illustrates this structure.

The repeated sequence of a VNTR is usually 15 to 50 base pairs long and is duplicated over and over in a sort of genetic stutter that results in alleles that typically are thousands of base pairs long (National Research Council Committee on DNA Forensic Science: An Update 1996, 65). The alleles observed by Wyman and White, for example, ranged from roughly 14,000 to 29,000 base pairs in length.

The function, if any, of these length polymorphisms is unclear. One reason for the great variability of VNTRs is their high mutation rate, as much as 1% per generation. The repeated units predispose the chromosomes to mistakes in the process of replication and crossing over. The result is a

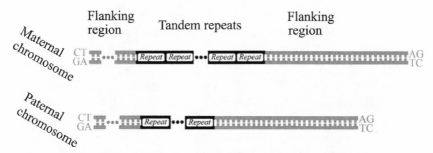

Figure 3.5. Schematic diagram of a VNTR locus. Between two flanking regions of DNA within adjacent restriction sites are many repeats of the same small sequence of base pairs. The number of repeats often varies, as within the pair of chromosomes in an individual (shown here) and as among the chromosomes from different people.

small increase or decrease in fragment length in some children in each succeeding generation (NRC Committee 1996, 65). Therefore, the precise number of the tandemly repeating units at a given location is highly variable in the population, and analysis of these loci can be especially informative. For instance, at a locus with 20 alleles, which is on the low side with modern methods for electrophoresis, there are 20 homozygous genotypes plus $(20)(19)/2 = 190$ heterozygous ones, for a total of 210 possibilities. The corresponding number of possible genotypes at a locus with 50 alleles is 1,275.

Although Jeffreys suspected that "regions of tandemly repeated DNA would be open to mutation processes such as duplication and recombination [that would make them] extremely informative genetic markers," his initial efforts to find these regions failed (Newton 2004a). A few other VNTRs were discovered in other laboratories, but detection was a hit-or-miss proposition. No systematic method for locating VNTRs was apparent to Jeffreys or anybody else.

Elsewhere in Jeffreys's laboratory, a search was on for the human myoglobin gene. The hemoglobin protein in red blood cells conveys oxygen from the lungs to the tissues and facilitates the return of carbon dioxide from the tissues back to the lungs. Myoglobin stores the oxygen released by hemoglobin and transports it to the mitochondria—the energy-producing organelles in cells. Jeffreys's group decided to look for the myoglobin gene in gray seals. Why seals? To dive and swim underwater, seals produce lots of myoglobin, which means that they must have high levels of myoglobin messenger RNA (mRNA). Molecular biologists can work backward from this mRNA to produce the coding DNA sequences that are being transcribed. If Jeffreys could determine the location and structure of seal myoglobin, he figured that it would be a short step to locating the corresponding human gene.

The plan worked (Blanchetot et al. 1983; Jeffreys et al. 1984), and it had an unexpected dividend—"DNA fingerprinting." As Jeffreys explains, "The true story of DNA fingerprinting starts at the headquarters of the British Antarctic Survey in Cambridge. I collected a big lump of seal meat from their lock-up freezer, and, to cut a long story short, we got the seal myoglobin gene, had a look at human myoglobin gene, and there, inside an intron in that gene, was tandem repeat DNA" (Newton 2004a). Having stumbled on this bit of DNA "purely by chance" (ABC Radio International 2002), Jeffreys observed that "[c]uriously, there were sequence similarities between its repeat unit and the repeats of the few other [VNTRs] described at the time" (Jeffreys 2005, 1035). Further experiments revealed that the VNTRs had a repeating sequence of ten to fifteen base pairs. "Even today, the significance of this core sequence remains unknown," but Jeffreys "immediately" saw in it "a general method for isolating hypervariable loci." He selected the core sequence for "a probe that should latch onto lots of these [VNTRs] at the same time" (Newton 2004a).

To test this idea, Jeffreys and his colleagues dumped the new probe on a Southern blot of DNA from a mother, father, and child, as well as "DNAs from various species including baboon, lemur, seal, cow, mouse, rat, frog, and tobacco." When Jeffreys came into the lab on the morning of Monday, September 10, 1984, he pored over an "indistinct and messy" autoradiograph, a "horrible, smudgy, blurry mess" (Zagorski 2006, 8919). Amid the blobs and smears, "the results were amazing—emerging from the gloom were what seemed to be highly variable profiles of DNA that looked as though they were simply inherited in the family" (Jeffreys 2005, 1036). Figure 3.6 is a photograph of the original autorad.

For Jeffreys, "it was very much a 'Eureka' moment; my life literally changed in five minutes flat, in a darkroom when I pulled out that first DNA fingerprint and saw just what we'd stumbled upon" (ABC Radio International 2002). "I took one look, thought 'what a complicated mess,' then suddenly realized we had patterns. There was a level of individual specificity that was light years beyond anything that had been seen before" (Newton 2004a). Jeffreys rushed out of the x-ray developing room and proclaimed to his group, "I think we're onto something really exciting here" (Zagorski 2006, 8919). "Within minutes, we had drawn up a list of possible applications, including criminal investigations, paternity disputes, zygosity testing in twins, monitoring transplants, wildlife forensics, con-

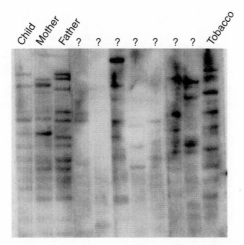

Figure 3.6. Jeffreys's autoradiograph of the first "DNA fingerprints" showing a human family trio plus a range of DNAs from various species including baboon, lemur, seal, cow, mouse, rat, frog, and tobacco (no longer labeled as such). Reprinted with permission from Sir Alec Jeffreys and Macmillan Publishers Ltd., *Nature Medicine,* Jeffreys (2005, fig. 3), copyright 2005.

servation biology and the like. My wife Sue added another that evening—immigration disputes" (Jeffreys 2005, 1036).

There was, however, some way to go before these forensic applications could be realized. The clarity of the autoradiographs had to be improved, the Mendelian inheritance pattern had to be confirmed, the success of the technique with the kinds of samples from crime scenes had to be verified, and the discriminating power of the system had to be measured. Jeffreys's lab attacked some of these problems. In marked contrast to the development of the thin-gel-multisystem method of typing serum proteins and enzymes (Chapter 1), a stream of articles appeared in science journals. A mere six months after the "Eureka moment," Jeffreys, Wilson, and Thein (1985a, 67) announced in *Nature* that a probe based on "the myoglobin 33-bp repeat" "can detect many highly variable loci simultaneously and can provide an individual-specific DNA 'fingerprint' of general use in human genetic analysis." The claim of individual specificity was based on the observation that length mutations are relatively common and on one set of more direct tests, namely, that "in an extensive Asian-Indian pedigree of Gujerati origin," all the autoradiographs could be distinguished, including the children of a marriage of two first cousins. Within this family group of fifty-four people, the bands appeared to be inherited independently. For Jeffreys, this was enough to warrant the assertion of "an individual-specific DNA 'fingerprint.'"

Four months later, the research team followed up with the announcement of more multilocus probes that "can detect additional sets of hypervariable [VNTRs] to produce somatically stable DNA 'fingerprints' which
are completely specific to an individual (or to his or her identical twin) and
can be applied directly to problems in human identification" (Jeffreys, Wilson, and Thein 1985b, 76). Whereas the first article had the obscure title
"Hypervariable 'Minisatellite' Regions in Human DNA" (a "minisatellite"
being another term for a VNTR), this letter to *Nature* was titled "Individual-
Specific 'Fingerprints' of Human DNA," making the point evident to the
most casual reader.

The Probability of Individuality

This second publication in an elite journal also went beyond the first study
of a Gujerati family by examining "the variability of DNA fingerprints [in]
a panel of 20 unrelated British caucasians" (Jeffreys, Wilson, and Thein
1985b, 77). Although the article refers to this group as "a random sample"
(ibid., 76), and Jeffreys later presented it as representative of "North Europeans" (Jeffreys, Brookfield, and Semeonoff 1985, 818; Jeffreys 1987,
314), it actually was a convenience sample of twenty students at his university (Cohen 1990, 361). On the basis of these twenty autorads, Jeffreys
arrived at a probability of a match for two unrelated individuals of less
than 3×10^{-11} (3 in 100 billion) for one probe and 5×10^{-19} (5 in 10 quin-
tillion) for two probes combined.[7] These numbers are not nearly as ex-
treme as the absurdly estimated random-match probability of 10^{-97} that
the FBI produced in litigation involving ordinary (dermal) fingerprints
(Kaye 2003), but one might well wonder how it is possible to generate so
many digits from a study of only twenty DNA samples. Because the figures
continue to be cited without question (Jobling and Gill 2004, 740; Aronson 2005, 127), it may be worthwhile to describe Jeffreys's reasoning so
that readers can decide for themselves how precise these numbers are. I
shall simplify Jeffreys's original presentation, but the essentials of the reasoning are unchanged.

Jeffreys began by counting the number of DNA fragments (as revealed
in the bands showing on the autorads) that clearly matched in two adjacent lanes of an autorad. (If a band from individual A was at the same
position as one from individual B, but one was much lighter than the
other, he excluded the pair of bands from the analysis.) With the probe
designated 33.15, for instance, there were, on average, just under 15 bands
that could be distinguished on the autorads, but most of these did not

match between the pairs of adjacent individuals. The proportion p of matching, or shared bands, was about one in five. If the position of every allele is statistically independent (as Jeffreys's family studies led him to believe), then observing a band from an unrelated person in the same position as a given band in the adjacent lane is akin to rolling a five-sided die that has an equal chance of showing any face. If we roll the dice once, the chance of a particular face (the match) is just $1/5$, but if we roll it twice, the chance is $(1/5) \times (1/5) = 1/25$. By an extension of this reasoning, the probability of observing exactly $x = 15$ matching alleles is simply $p^x = (1/5)^{15} = 3 \times 10^{-11}$. Because the world's population is numbered in the billions (10^{-9}), Jeffreys felt confident in describing the patterns obtained with this one probe as "fingerprints." Using the same procedure for the average amount of band sharing and the mean number of about 14 bands discernible with the other probe (designated 33.6), and multiplying the probabilities for the two probes, Jeffreys arrived at the 5×10^{-19} figure.

As Jeffreys was careful to point out, this simple multiplication assumes that every band is inherited independently and that the individual band frequencies are given by the average over all n bands. It also applies only to the situation of twenty-nine equal-intensity bands. With some forensic samples, fewer bands will be scored. Also, of course, a different grab-bag sample of twenty individuals would have led to different figure for p, the mean proportion of shared bands. Five years after Jeffreys claimed uniqueness in light of the calculation of 5×10^{-19}, Joel Cohen, a renowned mathematical biologist at Rockefeller and Columbia universities, challenged the calculation as resting on inadequate data and reflecting a "straightforward mathematical error" that understated the probability of a random match (Cohen 1990, 365).[8]

How wide of the mark might Jeffreys's estimated probability have been? If we were to assume that as many as ten alleles were perfectly correlated (and thus contribute no more information than a single allele), and that the true mean band-sharing proportion is one-half—more than twice the number in Jeffreys's sample—then the estimated average probability of a matching VNTR pattern would be $(1/2)^{20}$, which is less than 1 in 1 million. Given the billions of people now living, the technique no longer would warrant the sobriquet of "DNA fingerprinting."

Nevertheless, these modified assumptions are extremely unkind. Jeffreys and his colleagues continued to gather and analyze data. Jeffreys (1987, 314) soon reported that "[o]n average, 36 bands > 3 kb long can be scored per individual, using two DNA fingerprint probes," and that "[t]he mean level of band sharing . . . between unrelated individuals is 25%." Jeffreys, Turner, and Debenham (1991) turned to over 1,700 Caucasian paternity cases submitted to Cellmark Diagnostics, a commercial firm originally established in the United Kingdom to commercialize Jeffreys's probes. The mean band sharing between genetically unrelated individuals (mothers and alleged fathers) was about 14%, resulting in even smaller random-match probabilities than had been reported in the first *Nature* article. The mean probability of a match for the two probes combined was said to be 6×10^{-20}, and the "typical" probability was estimated at 1×10^{-30} (ibid., 835, Table 5) Of course, these infinitesimally small numbers still presume complete independence in the inheritance of the alleles, but if the bands are even roughly independent—a matter that we shall take up in later chapters—then Jeffreys's bold initial claim of "fingerprinting" seems plausible.

Forensic Applications

By the end of 1985, England's Forensic Science Service was ecstatic about the prospect of multilocus VNTR testing. In a third *Nature* article written with Jeffreys, its scientists enumerated all the difficulties with traditional protein and blood-group markers—they degrade with age, bacterial activity can cause erroneous results, in rape cases vaginal fluid can mask the semen markers, and even in unambiguous cases the traditional markers have limited power to discriminate among individuals (Gill, Jeffreys, and Werrett 1985, 577). The researchers reported that semen DNA could be isolated and that a few aged DNA samples either were typeable or gave no results. In addition to these "preliminary results," they repeated the figure of 5×10^{-19} to conclude that "DNA fingerprints . . . are completely specific to an individual" and "will revolutionize forensic biology."

However, it is not clear how often Jeffreys's patented multilocus-probe system itself was admitted as evidence in criminal cases.[9] Certainly, the use of the system was short lived, for it soon was superseded by more easily interpretable single-locus probes. Rather than use one or two generic probes that detected many VNTRs at once to generate complex-pattern autoradiographs, as Jeffreys had done, laboratories—including the Leicester lab (Wong et al. 1987)—began to develop and validate more sensitive enzyme and probe RFLP systems to detect one VNTR at a time. In the midst of the efforts to complete the foundational research for forensic applications of the original multilocus probes, however, Jeffreys became involved in two legal proceedings that captured the imagination and attention of the public.

The Immigration Appeal of Andrew Sarbah

A London immigration lawyer, having learned of the new multilocus VNTR test from a newspaper article (Aronson 2005, 8), contacted Jeffreys to ask for his help in an unusual case. A Ghanian boy born in the United Kingdom had emigrated to Ghana to join his father. When he returned to London at age thirteen to rejoin his mother, brother, and sisters, authorities suspected that his British passport had been doctored. They believed that the young man trying to enter the country was actually one of Andrew's cousins. Conventional immunogenetic and serologic testing was inconclusive, but Jeffreys found that the degree of band sharing between the mother and the boy was so high that "X [the boy] and M [the putative mother] must be related" (Jeffreys, Brookfield, and Semeonoff 1985, 819). Having calculated that "the odds that M is a sister of X's true mother but by chance contains all 25 of X's maternal-specific bands are $0.62^{25} = 6 \times 10^{-6}$," Jeffreys and his colleagues concluded that "beyond any reasonable doubt, M must be the true mother of X." The immigration tribunal that was hearing the appeal pointedly refused to rest its decision on "startling but untried scientific evidence." Instead, it used more conventional evidence to rule in the boy's favor (Aronson 2005, 127). Nevertheless, "DNA fingerprinting" had scored a major victory in the press, and it may have influenced the Home Office, which decided not to challenge the decision. "It was," as Jeffreys remarked years later, "a good news story of 'science fighting bureaucracy and helping families' " (Newton 2004a).

The *Pitchfork* Case

Jeffreys turned his scientific sleuthing to a double-murder case in 1986. On a cold November morning in 1983, a hospital porter was walking to work along a shady footpath in the village of Narborough, near Leicester. The man caught sight of the savagely raped and strangled body of a fifteen-year-old girl. One hundred fifty police looked for clues. Analysis of semen indicated that the killer was a type A secretor with the enzyme marker PGM1+, a combination that occurred in 10% of the adult male population. Beyond that, however, the investigation led nowhere.

Three years later, in the adjoining village of Enderby, another fifteen-year-old girl went missing. Two days later, police officers found her strangled and sexually assaulted body hidden in some bushes within a mile of the site where the first body had been discovered. Again, the biological evidence pointed to a type A, PGM1+ secretor.

This time, the police had a suspect—a seventeen-year-old kitchen porter at a local mental hospital located close to where the body was found. Described by one detective as "the flippin village idjit you always hear about," the boy ultimately confessed to the second murder but stubbornly insisted that he was not involved in the first one (Wambaugh 1989, 188). Although the boy was not a PGM1+ secretor, the detectives did not regard that finding as definitive. In an effort to link their suspect to the first murder and rape, Leicestershire police turned to Jeffreys. They gave him a blood sample from the boy and an old semen sample from the unsolved 1983 case.

The latter sample was apparently too small or degraded to yield results with the multilocus "fingerprinting" probes, but Jeffreys produced satisfactory autorads with two of his newer, more sensitive single-locus probes (Gill and Werrett 1987, 146; Newton 2004b; Jeffreys 2005). With these, Jeffreys "could see the signal of the rapist, and it was *not* the person whose blood sample was given to me" (Wambaugh 1989, 194). So the police delivered a semen sample from the first case. Jeffreys telephoned in his finding—the prime suspect was innocent of this assault as well. The detective's response was "not repeatable." But, Jeffreys said, there was good news—"You only have to catch one killer. The same man murdered both girls." The police fed the kitchen porter sandwiches and set him free. It was the world's first DNA exoneration.[10]

A squad of fifty men continued to pursue the investigation. Some of them thought that their superiors were obsessed with the kitchen porter and that "[o]ur job was to get enough evidence to finally make an airtight case against that lad, genetic fingerprinting or no" (ibid., 213). However, a more open-minded detective persuaded his superiors to adopt a radical plan. They would ask the male residents of Narborough and Enderby between the ages of seventeen and thirty-four years to supply blood and saliva samples voluntarily to "eliminate" them as suspects. Samples from nearly 5,000 men went to the Forensic Science Service laboratory. Some 90% were eliminated off the bat by conventional serology; the remaining 500 were DNA tested. When the dust cleared (or the blood settled), no one matched the killer's DNA profile.

Finally, police got the break they needed. The manager of a bakery outlet in Leicester told them that during a pub lunch, a baker named Ian Kelly had revealed that he had provided DNA samples for Colin Pitchfork, a fellow employee. The police arrested both men. Kelly soon confessed and explained how Pitchfork had altered his passport by putting a photograph of Kelly in it. The police described him as extremely gullible, and he received a suspended prison sentence of eighteen months for conspiracy to pervert the course of justice. Pitchfork also confessed and pled guilty. He received concurrent sentences of life imprisonment for each murder and additional time for the rapes and other offenses. The judge observed that "if it wasn't [*sic*] for DNA, you might still be at large today, and other women would be in danger" (ibid., 369).

DNA Evidence circa 1986

RFLP testing involves digesting a DNA sample with bacterial enzymes, running the fragments through a gel with electrophoresis, blotting to a nylon membrane, applying a probe that attaches to fragments with a DNA sequence of interest, and exposing x-ray film to yield an autoradiograph of the fragments. Geneticists were using this procedure with biallelic loci (having only two possible fragment sizes) to find the locations of genes with disease-causing mutations when Wyman and White hit on the first VNTR locus—DNA sequences that were like trains with many boxcars. Jeffreys soon discovered a kind of DNA boxcar that produced DNA trains of varying lengths at many locations in the genomes of human beings and other animals. Although his initial studies of the multilocus probes were limited to a small number of conveniently located volunteers, he was quick to announce that "DNA fingerprinting" had been achieved.

The press and the British police took note. Before long, this new form of DNA analysis had resolved two difficult legal cases. But no court had considered the admissibility of either the "individual-specific" multilocus DNA "fingerprinting" or the less revealing single-locus VNTR testing.[11] The courtroom dramas were yet to come.

CHAPTER FOUR

The Emergence of
VNTR Profiling

D ESPITE their extraordinary discriminating power, Jeffreys's multi-locus probes for RFLP testing never really caught on in criminal cases. They were used for a time in civil parentage disputes, but even there, they were supplanted by multiple single-locus probes. The terminology is confusing, and it may be helpful to restate how these two types of VNTR probes differ. There are many VNTR loci scattered across the chromosomes. RFLP testing with the proper probes is a way to measure the lengths of the restriction fragments containing these variable numbers of tandem repeats. Jeffreys's contribution to forensic science was a multilocus probe that simultaneously attaches at many different VNTR loci across the genome. Because the resulting autoradiographs show so many bands (coming from many different places in the genome), they have been analogized to supermarket bar codes, although they are never as clear and sharp as these printed patterns. An example of an autorad produced with one of Jeffreys's probes is shown in Figure 4.1.

The variability at each locus and the sheer number of loci make the patterns distinctive, but there are costs to this multilocus approach. For one thing, enough high-quality DNA must be present in the sample to give responses from many loci at once. For another, the exact number of loci that are represented in an autorad is unknown, and it is impossible to know which fragments come from which loci.

A single-locus probe overcomes these problems. These bits of a DNA strand are constructed to bind to the alleles at one specific locus (Naka-

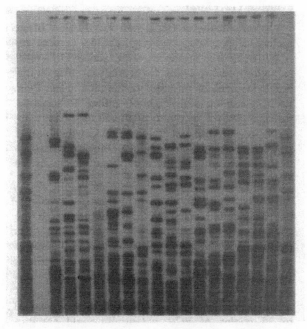

Figure 4.1. Autoradiograph of restriction fragment length alleles from many loci in seventeen individuals obtained with the 33.15 multilocus probe. Source: Cellmark Diagnostics, 1990, as reproduced in Office of Technology Assessment (1990, 47).

mura et al. 1987; Odelberg and White 1987), which requires much less DNA in the sample. Because a known locus is detected, ambiguity is reduced, and data about how often different alleles of this locus occur in a given population can be collected. For example, the D2S44 locus is a region of chromosome number 2 that varies in length because of differences in the numbers of repeats of a nine-base-pair core sequence.[1] A probe for this single locus generally will reveal two fragments. (If the maternal and paternal chromosomes happen to have fragments of the same length, only one band will appear.) An autoradiograph of the alleles of twelve individuals is shown in Figure 4.2. The two single-banded vertical strips show VNTR fragments from homozygotes (or heterozygotes whose repeat numbers are too close to be distinguished by this kind of electrophoresis).

This single-locus probe approach has its own drawbacks. Although the twelve lanes all appear to be distinguishable, there is far less information on each individual in this autorad. A thirteenth person might well have fragments that would migrate about the same distance and hence line up with the bands in one of the twelve lanes in the picture. The chance of this occurring would depend on how many distinguishable alleles are present in the population at the D2S44 locus. Greater individualizing power can

Figure 4.2. Autoradiograph of restriction fragment length alleles at the D2S44 locus for twelve individuals obtained with a single-locus probe. Source: FBI 1988, as reproduced in Office of Technology Assessment (1990, 47).

be achieved by using a series of different single-locus probes. The D1S80 locus, for instance, contains a sixteen-nucleotide core sequence that is repeated between sixteen and forty times on chromosome number 1. The first probe (for D2S44) can be washed off and this second probe (for D1S80) applied to the membrane to provide a second autorad. Some DNA may be lost in the probe-stripping process, but in the early years of VNTR testing, it usually was possible to probe three, four, or even five single loci sequentially. This multiple single-locus probing can be quite revealing. If there are fifty distinguishable alleles per locus, it follows that the number of possible banding patterns for four loci combined exceeds 2 trillion (National Research Council Committee on DNA Forensic Science: An Update 1996, 65). To obtain multiple single-locus genotypes, however, could take weeks. For each probe, the x-ray film had to be in contact with the weak radioactive signal long enough to produce a visibly dark spot.

Into the Courtroom

If Jeffreys's "DNA fingerprinting" with multilocus probes constituted the zeroth generation of forensic DNA technology, "DNA profiling" with multiple single-locus probes was the first-generation technology.[2] This was the technique that came before the courts in the late 1980s. The initial judicial response was uniformly positive. Expert testimony for the prosecution rarely was countered, and courts readily admitted RFLP findings. Indeed, in *People v. Wesley*, which is said to be "the first time that a defense attorney brought in expert witnesses to challenge prosecution claims about the technique" (Aronson 2007, 45), the trial judge, quoting "a mean power of certainty of identification for American Whites of 1 in 840,000,000; [and] for American Blacks, 1 in 1.4 billion," not only rejected the views of those witnesses but heralded DNA evidence as potentially "the single greatest advance in the 'search for truth' . . . since the advent of cross-examination," (533 N.Y.S.2d at 644).

The first case to affirm such a ruling was *Andrews v. State* (1988).[3] In the early morning of a February day in 1987, a woman in Orlando, Florida, awoke to a nightmare. A man was on top of her holding what felt like a straight-edge razor to her neck. Clamping his hand over her mouth, he told her to keep quiet and threatened to kill her if she saw his face. She struggled and "for her efforts was cut on her face, neck, legs and feet" (533 So. 2d at 842).

Semen was recovered from a vaginal smear, but conventional immunogenetic testing was not very helpful. However, the intruder apparently had entered through a broken window and left fingerprints on the window screen. Similar rapes were occurring both before and after this one, and police suspected that a serial rapist was responsible. In March, they received a call about a prowler and gave chase to a blue car that was speeding away. Their quarry turned a sharp corner and crashed into a telephone pole. The driver was Tommy Lee Andrews, and his fingerprints matched the ones on the window screen (Erzinçlioglu 2002, 101; Wilson and Wilson 2003, 307). The prosecution turned to Lifecodes Corporation of Valhalla, New York, one of the commercial outfits that had sprung up to provide forensic DNA testing in the wake of Jeffreys's discovery.[4] When Lifecodes found that Andrews's VNTRs matched those of the semen and estimated that only 0.0000012% of the population would be similarly incriminated, Andrews was charged in the series of rapes and related offenses.

Before trial in this case, the court conducted a hearing on the general acceptance of single-locus VNTR profiling. The prosecution presented testimony from three witnesses: David Housman, a biology professor at the Massachusetts Institute of Technology (MIT); Alan Giusti, "a forensic scientist" at Lifecodes with a bachelor's degree from Yale University, who did the RFLP testing; and Michael Baird, a University of Chicago Ph.D. in genetics, who was Lifecodes' manager of forensic testing (533 So. 2d at 847). Housman, as a research scientist, testified not only about the nature of RFLPs in general, but also that he had "reviewed Dr. Guisti's results [and] testified that in his opinion the test was accurately and properly performed" (id. at 849). No one testified for the defense. Later, the defense attorney explained that he had called many biology departments but was unable to locate anyone who would question Housman's judgment (Mnookin 2006, 214). On this record, the Florida court of appeals had no difficulty upholding Andrews's convictions. It had some trouble deciding whether *Frye* still was good law in Florida, but it held that the DNA evidence satisfied both the general-acceptance and the ordinary relevance standards for admitting scientific evidence.

The Maryland Court of Special Appeals soon relied on *Andrews* to uphold the admission of a DNA match in *Cobey v. State*. In 1985, a man forced a woman off a jogging trail and into the woods. There, as the court put it, he "ravished" her and then drove away in her car. His driving left something to be desired, for an officer pulled the car over and issued a traffic citation to the driver, Kenneth Cobey. At trial, Robin Cotton, director of research and development at Cellmark Diagnostics's laboratory in Germantown, Maryland, testified that the DNA in Cobey's blood sample matched DNA in semen stains on the victim's underclothing.[5] On appeal, Cobey contended that Cellmark's single-locus VNTR tests lacked general acceptance, but "[s]ignificantly, Cobey produced no expert testimony challenging the validity of the locus probe."[6] His attack on the database that Cellmark used to obtain allele frequencies also suffered because "four experts said at trial [that the database] fell within generally acceptable scientific criteria" and "[t]here was no expert testimony contradicting [this] proposition" (559 A.2d at 398). Citing the research literature on VNTR typing and the outcomes in *Andrews* and one other case, the court concluded that the trial judge "did not err in finding that DNA fingerprinting was generally acceptable in the scientific community" (id.). Interestingly, even though the opinion contained pages on the structure of DNA, the steps in gel electrophoresis, Southern blotting, and autoradiography, the *Cobey* court confused multilocus and single-locus testing. It took as proof of general acceptance a journalist's report that "Cellmark Diagnostics . . . claims its 'DNA fingerprint' test can identify a suspect with 'virtual certainty,' and that the chances that any two people having [*sic*] the same DNA fingerprint are one in 30 billion" (id. at 392 n.7, citing "D. Moss, 'DNA—The New Fingerprints,' 74 ABA Journal 66 (1988)"). These statements applied only to the multilocus probes that Cellmark licensed from Jeffreys but never used in reported criminal cases in the United States.[7]

It is instructive to compare the states' cases for DNA in *Andrews* and *Cobey* to the earlier hearings on thin-gel-multisystem protein electrophoresis (Chapter 1). In some of those cases, an academic expert testified that the system was defective, there were no scientific publications attesting to the validity of the forensic procedures, and the only witnesses supporting it were technicians, police officers, or its developers. In these circumstances, a few courts held that general acceptance in the relevant scientific community was lacking. Here, an "outside" academic expert confirmed the quality and validity of the forensic work. In addition, Lifecodes, Cellmark, the FBI, and academic researchers had been publishing one study after another on various aspects of DNA typing (Andrews v. State, 533 So. 2d at 847, noting the existence of publications by Giusti and Baird). The DNA testers had learned a lesson from the protein wars.

In a few early cases, defendants recruited some experts, but the hearings remained one sided. Take the Texas case *Kelly v. State.*[8] Mary Copeland lived with her sixty-three-year-old widowed mother in Fort Worth. She returned home just after midnight in the fall of 1987 to find her mother, her mother's truck, and some other possessions of her mother's gone. Barry Dean Kelly, Mary's former boyfriend, was seen driving the missing truck that evening. The next day, he sold a set of the mother's wedding rings to the owner of a Texas tire store. A week and a half later, the police located the mother's badly decomposed body near a dry creek bed off a dirt road not far from Kelly's last place of employment. The medical examiner could not say if she had been sexually assaulted, but he believed that she had been strangled.

The forensic investigation at the home was none too thorough. It was Mary who, on the day her mother was buried, noticed a yellow stain on her mother's bedspread. (The bedspread had been purchased after Mary had stopped seeing Kelly.) The stain was identified as semen from a type O secretor with PGM type 1+1−. About 10% of white males have these phenotypes, and Kelly was included in this group.

This evidence—the truck, the jewelry, and the serology of the semen stain—probably would have been enough to convict Kelly, but the Fort Worth police also contracted with Lifecodes for DNA testing of the semen. VNTR testing confirmed the serology, and Lifecodes computed the random-match probability for another white male to be 1 in 13.5 million. The jury found Kelly guilty of murder, and he was sentenced to life in prison.

On appeal, Kelly contended that the trial court erred in overruling his "'*Frye* Motion' and admitting expert testimony on DNA because the DNA evidence was not reliable under either the relevancy standard or the '*Frye*' standard" (792 S.W.2d at 582). The court of appeals summarized the pretrial hearing on these questions. The state produced five expert witnesses. Philip Hartman, an associate professor of biology at Texas Christian University, assured the court "that there existed no possibility of a false DNA match between known and unknown samples" and insisted that "all of the scientific procedures in DNA testing are 'straight-forward,' known, and accepted scientific techniques" (id. at 583). Joseph Sambrook, then chairman of the Biochemistry Department at Southwestern Medical School, was not shy in describing his qualifications. In the court's words, "Dr. Sambrook claimed to have written over 100 research papers, some of which have been published in various journals . . . as well as a book called *Molecular Cloning* which he describes as the technical 'Bible' of the field" (id.). Sambrook, the court noted, "is not associated with Lifecodes but has read theses and articles written by the PhD's affiliated with Lifecodes" (id.). Likewise, there was "Robert C. Benjamin, assistant professor of biological science at the University of North Texas and holder of a PhD in biology from Harvard," who had "seen published protocols from Lifecodes and found them to be accepted, state-of-the-art procedures" (id.). Again, the analyst who performed the testing was Alan Giusti, who had moved to a new position as a "physical scientist with the FBI" (id.). Finally, another Lifecodes employee, Kevin McElfresh, "who holds a PhD in molecular and population genetics and is the assistant manager and laboratory supervisor at Lifecodes," agreed with Giusti that there was a VNTR match and added that "the statistical probabilit[y] of such a match being incorrect was one in thirteen million" (id.).

To counter this array of experts from academia and Lifecodes, the defense produced a lone expert, "John T. Castle, owner and operator of Castle Forensic Laboratories in Dallas" (id.). Castle, as the court delicately put it, "holds a Bachelor's degree in chemistry and was certified to teach life and earth sciences in public schools" (id.). Among other things, Castle asserted that "radioactive technology," which had been introduced into medicine in the late 1930s, "was too new to be generally accepted in the scientific community."[9]

Again, the only difficult question for the appellate courts was whether to apply the *Frye* standard or a reliability-relevance standard that will be discussed more fully later. Applying the latter standard to the record of the hearing, the court of appeals and later Texas's highest court determined that the DNA match and the "one in thirteen million" statistic were admissible.

The courts in cases like *Andrews, Cobey,* and *Kelly* can hardly be faulted for responding to the lopsided records before them by finding sufficient general acceptance or scientific validity to justify the admission of the DNA profiling, but we soon shall see that aspects of the panegyric for DNA evidence were open to question. Although advocates in later cases can try to challenge these precedents, the early cases can have a snowball effect. Other courts cite the initial opinions as establishing general acceptance or scientific validity. The snowball grows until it becomes increasingly difficult to stop even if there are serious grounds to question the scientific technique. Indeed, it has been said that this is currently the case with dermatoglyphic fingerprinting (Faigman et al. 2006). Fingerprints obviously are highly variable, but the validity and reliability of analysts working with latent prints of varying quality have not been rigorously studied. Yet the technique is well ensconced, and courts are reluctant to overturn what is now approaching a century of judicial acceptance. The problem, of course, is that judicial acceptance in a more forgiving era cannot necessarily be equated to contemporary scientific acceptance or scientific validity.

With DNA evidence, however, this hardening of the legal arteries did not occur. The first wave of DNA cases, from 1986 to 1988, was defined by exuberant and essentially uncontested claims from the prosecution, exemplified by *Andrews, Cobey,* and *Kelly.* Some of the force of this first wave was arrested by a rare combination of scientific and legal advocacy. The details of the laboratory procedures soon were questioned, and limitations were identified in the statistical and population-genetics models underlying vanishingly small estimates of the frequencies of VNTR genotypes. In a second wave of cases, from 1989 to 1990, doubts raised by defendants about the analytical procedures at particular laboratories led some courts to reject DNA evidence, or aspects of it, in individual cases. The first case to deflate the more hyperbolic claims of the forensic DNA laboratories—and thereby to undermine judicial confidence in DNA evidence—was *People v. Castro.*

The Double Murder and the Handyman

Vilma Ponce was a twenty-two-year-old woman who lived in an apartment in the Bronx with her common-law husband and their two-year-old daughter, Natasha. Vilma, seven months pregnant, and Natasha had been stabbed to death in the apartment. The little girl had been stabbed sixteen times, and Vilma's body had been perforated nearly sixty times.[10] The husband, returning to the apartment on that fateful February day in 1987, had seen an "almost Shakespearean apparition," a man leaving the building, his face, arms, and sneakers smeared with blood (Levy 1996, 37).

Without DNA evidence, the state did not have a compelling case. The man accused of the murders was Joseph Castro, a thirty-five-year-old janitor or handyman who lived nearby and did odd jobs in the neighborhood. In January 1987, working with the building superintendent's nephew, he had helped install a lock on the apartment door. Police found that the lock did not work because it was improperly installed. They theorized that Castro might have used this defect to get into the apartment easily. Castro came to their attention because the detective working the case returned to the area the next day and spotted a man who fit the husband's description of the bloodstained man. The detective interviewed and photographed this man—Castro—and put together a simple photo spread of several Hispanic men. The husband was unable to make an identification (Sheindlin 1996, 16–17). On the second day after the murders, the husband recognized a man on the street as the bloody apparition and spoke to him. In an almost manic state, he rushed to the police. Now he was sure about the detective's picture of Castro. There also was hearsay testimony that Castro had repeatedly made suggestive remarks to Vilma. Finally, Castro's statements to the police were fishy. For instance, he had denied being in the building on the day of the murder, but several tenants had seen him there. A good prosecutor could do a lot with these facts, but would jurors agree that they amounted to proof beyond a reasonable doubt—especially when, by his own admission, Vilma's husband had once hit his wife and broken her jaw? He might have altered the lock and set Castro up as the villain (Levy 1996, 35).

What sewed up the state's case was a dark stain on a watch that Castro was wearing when he finally was arrested. The husband had mentioned seeing blood on a watch worn by the man who had passed him. At the police station, a detective told Castro that it looked like he had blood on his watch. Castro shrugged it off: "If it's blood, it's my blood" (Sheindlin 1996, 30). The detective removed the watch from Castro's wrist. Serologic tests indicated that the blood spots were not Castro's and could have been Vilma's. But they could have been from a lot of other people as well.

The lead prosecutor, Risa Sugarman, having read about the DNA testing in the *Andrews* case, obtained a court order for a sample of Castro's blood. It went to Lifecodes' lab in nearby Westchester along with samples from Vilma and Natasha. The prosecutor herself retrieved the watch from the medical examiner's office, where it had been stored after the serological tests, and personally delivered it to Lifecodes (ibid., 31–32). The laboratory concluded that the stain on the watch was Vilma Ponce's blood and estimated that the multiple-locus VNTR profile had a frequency of 1 in 189 million in the Hispanic population.[11]

The Pretrial Hearing

When the case appeared on the docket of Bronx County judge Gerald Sheindlin, no appellate court in New York had ruled on DNA evidence. In the absence of binding precedent, it was Judge Sheindlin's responsibility to evaluate the defendant's objection that DNA profiling, as practiced by Lifecodes, had yet to be generally accepted. He granted the defendant a pretrial hearing that dragged on for more than three months and produced a transcript of 5,000 pages.

The state's expert witnesses were three research geneticists in New York State and Baird and Giusti, Lifecodes' employees. The star of this strong group was Richard Roberts, the assistant director of the renowned Cold Spring Harbor Laboratory. His 1977 discovery of the introns mentioned in Chapter 3 would earn him a share of the 1993 Nobel Prize. Roberts explained the nature of RFLP-VNTR profiling and testified that the system could not produce false-positive results (Levy 1996, 43). Testifying for the defense were an assistant professor of neurogenetics at Columbia University's College of Physicians and Surgeons, the director of a kidney-disease institute, geneticists at Washington University and the University of Edinburgh, and a mathematician-turned-geneticist with positions at Harvard and MIT.

Several members of the cast of characters in this "first serious challenge to DNA 'fingerprinting'" (McFadden 1989, B1; Schmeck 1989, B1) were to play major roles in later cases and developments. Judge Sheindlin did not appreciate what would be unleashed when he granted Castro's court-appointed counsel's request "to bring in a couple lawyers for the hearing" (Sheindlin 1996, 60). The lawyers were Barry Scheck and Peter Neufeld, who were at the forefront of the defense bar's efforts to catch up with the prosecutorial establishment when it came to DNA evidence. Scheck and Neufeld would make their most lasting contribution to the legal system by founding and staffing the Innocence Project at Yeshiva University's Cardozo School of Law, where Scheck teaches as a clinical professor. Their project would be instrumental in freeing scores of falsely convicted prisoners, in promoting reforms in the justice system, and in rekindling debate over capital punishment. Indeed, the project's very public success in harnessing the power of DNA evidence to exonerate the wrongfully convicted also contributed symbiotically to the acceptance of the evidence for the prosecution. But that was in the future. In 1989, Scheck and Neufeld were determined to show that DNA evidence was not ready for the courtroom. Their personalities were suited for the task. "[H]ighly intelligent [and] extremely argumentative," they exuded an "arrogant certitude" (Levy 1996, 44). And they were well paired. The much taller Neufeld had "the somewhat distracted but deliberate manner of a professor," while the more diminutive Scheck was "an irascible, aggressive street fighter" with an "in-your-face style" (Sheindlin 1996, 60–61). In *Castro* itself, Scheck's antics included "raising his arms over his head, like a football referee, and mouth[ing] 'Touchdown!' when he felt he had scored a point in his cross-examination of Baird—as, indeed, he had" (ibid., 67).

On the science side, the dominant personality was Eric Lander, who would act as a court-appointed expert in later cases and would champion a much-maligned "ceiling principle" for estimating VNTR allele frequencies in criminal cases (a topic that I take up in Chapters 6 and 7). *Time* magazine, in its 2004 profiles of the world's 100 most influential people, singled out Lander as a moving force in the Human Genome Project and described him as "a big teddy bear of a man" (Elmer-DeWitt 2004). Having immersed himself in mathematics at Princeton and then at Oxford as a Rhodes scholar, Lander started his teaching career at Harvard Business School. Bored by economics, he "picked up biology on the street corner," auditing a class at Harvard and reading what interested him. Before long, he obtained a fellowship at MIT's Whitehead Institute for Biomedical Research (Hopkin 2007). Scheck found him to be "right up our alley," a "fast-talking New York Jew from Brooklyn who loved to argue and knew a lot of law" (Parloff 1989, 53). From the bench of his Bronx courtroom, Judge Sheindlin perceived Lander as "a brilliant man" but "strident and officious—bordering on arrogant" and "relishing his role as a kind of surrogate attorney" for Neufeld and Scheck (Sheindlin 1996, 72). At one point, listening to a particularly absurd explanation of an anomalous result in the control lane of an autorad (the lab records, it later turned out, had misidentified the source of the control DNA as a male when it came from a female), "Lander . . . laughed silently and mouthed the word 'Bullshit'" (ibid., 80). He was definitely not a teddy bear.

This unusual, if not unorthodox, defense team's challenges fell into two categories. Much of the critique went to the nitty-gritty of how Lifecodes had conducted and documented its tests and resolved ambiguities or anomalies in the autoradiographs. The judge was shocked at the revelations that arose during the cross-examination of Baird: "The X-rays were mislabeled. They showed the wrong tests performed, the wrong dates, and a whole cascade of sloppy procedures. I sat on the bench and watched with dismay and disbelief. . . . I figured that laboratory conditions were above all this. Was I ever wrong!" (Sheindlin 1996, 66–67). Another problem for Lifecodes and the prosecution was the autoradiograph for the DXYS14 locus. This is an unusual locus in that more than two polymorphic fragments can be present.[12] Three bands appeared in the lane with Vilma Ponce's DNA, but five bands were present in the watch lane. On its face, this different pattern seems to establish that the watch did not have Vilma's DNA on it. This was Lander's conclusion—he testified that the autorad exonerated Castro. Lifecodes sought to explain away the extra two bands as the result of bacterial contamination, but the lab had failed to conduct tests that might have settled the question.

In addition to such case-specific objections, the defense experts raised more generic concerns about the procedures—or the lack of them— followed by forensic DNA laboratories generally. For example, Lander identified problems in Lifecodes' computation of the 1-in-189-million random-match probability. These more generally applicable objections were better fodder for a *Frye* hearing. The *Frye* standard, after all, requires only that a scientific theory or technique be generally accepted as valid and reliable—not that it be executed perfectly or, for that matter, competently on a given occasion. *Frye* itself concerned the general acceptance of two propositions—that a rise in systolic blood pressure is diagnostic of conscious deception, and that the instrumentation for monitoring these changes is accurate. These questions cut across cases. Whether the machine is running properly and whether the readings are recorded correctly are more appropriately determined on a case-by-case basis regardless of whether the generic *Frye* issues are present. Case-specific departures from an appropriate protocol for testing sometimes should result in the exclusion of the results, but not because of *Frye*. Instead, the issue should be handled under the normal doctrine of relevance. Are the errors in the application of a generally accepted technology so grave as to make the probative value of the evidence too slight in comparison with the dangers of misunderstanding and undue consumption of time? If so, the judge should exclude the evidence. If not, then the case-specific problems can be left to the jury. Traditionally and logically, case-specific mistakes in applying a generally accepted technique are not part of a *Frye* inquiry (Kaye, Bernstein, and Mnookin 2004, 281–282).

In *Castro*, however, the court decided to add a "third prong" to the normal *Frye* inquiry. "Prong I" of the court's inquiry asked whether there was "a theory, which is generally accepted in the scientific community, which supports the conclusion that DNA forensic testing can produce reliable results." "Prong II" was the existence of "techniques or experiments that . . . are capable of producing reliable results in DNA identification and which are generally accepted in the scientific community." The new "Prong III" was whether the laboratory performed "the accepted scientific techniques in analyzing the forensic samples in this particular case" (545 N.Y.S.2d at 987). "[I]n this complex area of DNA identification," the court announced, "[t]he focus of this controversy must be shifted. It must be centered around the resolution of the third prong" (id.).

The court was correct to evaluate the relevance of the evidence produced for the particular case before the trial rather than automatically leaving it to the jury to weigh the evidence despite possibly fatal flaws in the laboratory's application of accepted theories and procedures. Analytically, however, it seems better to regard this as an independent inquiry than as a necessary feature for *Frye* hearings. It can be efficient to address the case-specific question in the course of a *Frye* hearing, but a pretrial hearing on the application of a complex but generally accepted technique also might be desirable even when the *Frye* issue of general acceptance has been fully resolved in previous cases. Also, whatever procedure is followed in deciding whether the laboratory's implementation of an accepted and valid technology is good enough to be put before a jury, the burden of proof should fall where it normally does—on the opponent of the evidence (Kaye, Bernstein, and Mnookin 2004, 294–297).

To the chagrin and professed amazement of Neufeld and Sheck, the *Castro* court found the first two prongs of its expanded *Frye* test satisfied. It rejected the opinion expressed by Lander and one other expert that it would be "approximately six months before DNA identification will be in a position to generate reliable results." It found their reasoning perplexing because neither expert felt that any new scientific breakthrough was needed. "Accordingly," the court concluded that "to breathe any meaning into the opinion of these highly respected and rather brilliant scientists, one must conclude that the test is presently reliable and will remain so for the next six months" (545 N.Y.S.2d at 995).

The third prong was another story. When scientists of the caliber of the experts in *Castro* differ, courts normally are at a loss to decide who is right, and the decision can go either way. Here, however, an extraordinary turn of events forced the court's hand. Lander was attending a scientific meeting at the Banbury Center of the Cold Spring Harbor Laboratory organized by academic and forensic scientists.[13] He gave Roberts a copy of his fifty-page report on the case. "I hadn't really seen the evidence in great detail before, and I quickly became rather concerned," Roberts said later (R. Lewin 1989; Levy 1996, 47). Roberts proposed that all the outside experts confer privately. They met for a morning in a borrowed law office in Manhattan (R. Lewin 1989; Lander 1992, 201).[14] This unchaperoned tête-à-tête resulted in a joint statement concluding that "the DNA data in this case are not scientifically reliable enough to support the assertion that the samples do or do not match" (Parloff 1989).

Although the strict rules devised for presenting evidence to jurors at trials are not compulsory in hearings to determine the admissibility of evidence, the court rebuffed Scheck's proffer of the two-page report on the ground that it was "rank hearsay" (Sheindlin 1996, 76). Taking the witness stand again, various expert witnesses subscribed to the conclusions expressed in the report. Roberts told the court that he could not say whether the DNA in the case matched (Levy 1996, 47). Lander repudiated his earlier testimony that the extra alleles at the DXYS14 locus exonerated Castro in favor of the consensus view that the testing was inconclusive—as the report put it, "If these data were submitted to a peer-reviewed journal in support of a conclusion, they would not be accepted. Further experimentation would be required." Only Baird continued to insist that the autoradiographs demonstrated a match to the victim, and even he would not "bet [the] ranch on it" (Sheindlin 1996, 75). Because Lifecodes' testing had consumed the DNA on the watch, the unanimous consensus of the outside scientists was the kiss of death to the prosecution's evidence. The court ruled that the tests were admissible to prove that the blood was not Castro's—every expert agreed on this point. But on the crucial issue of the match to Vilma Ponce, the court ruled that the evidence was inadmissible: "In a piercing attack upon each molecule of evidence presented, the defense was successful in demonstrating to this court that the testing laboratory failed in its responsibility to perform the accepted scientific techniques and experiments in several major respects" (*Castro*, 545 N.Y.S.2d at 996).[15] The judge later expressed his frustration with Lifecodes less politely: "To tell you the truth, the lab's incompetence infuriated me. How dare they come into a courtroom in a double-murder case and offer such careless work? It was an insult to invoke the name of science for such junk" (Sheindlin 1996, 81).

The Aftermath

The evidentiary ruling gave Joseph Castro a bargaining chip. The prosecution prepared for a trial that still would include DNA evidence. Although Lifecodes' work could not be used to pinpoint that the blood on the watch was Ponce's, it still could be admitted to establish that Castro was walking around with someone else's blood on his watch. As a jury was about to be selected, Castro's counsel advised the court that Castro wanted to plead guilty if he could be assured of a sentence of fifteen years to life. The judge and the prosecutor were opposed: "I can't give a guy fifteen years for two murders. That would be ridiculous" (Sheindlin 1996, 84). The deal that emerged was twenty years to life. At the guilty-plea hearing, the court asked Castro, "Did you have blood from the deceased on your watch?" With "his mouth twisted in a smirk," Castro answered, "Yes" (ibid.).

The events in the Bronx courthouse had profound national repercussions. Courts could no longer accept bland assurances that false positives were impossible. (In fairness, what these experts probably were trying to say is that outside contamination of samples and deficiencies in the chemicals and gels used in the testing could not produce false matches. Compare Haflon 1998, 816–817.) Neufeld and Scheck emerged as the defense bar's preeminent specialists on DNA evidence, and they continued to attack both the general acceptance of the forensic technology and its applications, most prominently in the murder trial of football hero O. J. Simpson (Chapter 7). Eric Lander wrote a series of articles and letters in scientific journals publicizing his critiques of Lifecodes' work and spreading his doubts about the reliability of forensic laboratories and the statistical implications of matches. Roberts complained about the adversarial system, observing that "[w]e all did so much better when we sat down without the lawyers and had a reasoned discussion. Perhaps it's time the system changed" (R. Lewin 1989, 1033). The scientists' joint report called for a committee of the National Academy of Sciences to review the forensic use of DNA tests and opined that "[t]he *Frye* hearing is not the appropriate time to begin the process of peer review of the data."

One astute legal historian (Mnookin 2006, 227–228; see also Jasanoff 1995, 56–57) counters Roberts's distaste for the adversarial system by observing that DNA evidence received such microscopic scrutiny from some of the nation's leading biologists only because of the adversarial system that drew them into the "piercing attack upon each molecule of evidence" (*Castro,* 545 N.Y.S.2d at 985). *Castro* certainly focused an uncomfortable spotlight on Lifecodes' work, but one should not misconstrue Mnookin's observations as arguing that criminal adjudication ordinarily is a satisfactory quality-control mechanism. Typically, defense lawyers do not have the expertise, time, and resources to scrutinize the state's scientific evidence. Even today, many potential problems with DNA evidence in specific cases probably go unnoticed (Thompson et al. 2003; W. C. Thompson 2006). Adversarial proceedings cannot be relied on as the sole or perhaps even the dominant mechanism for detecting laboratory errors. Quality-assurance standards are vital to prevent mistakes, and procedures such as court-appointed experts to remove some the excesses of adversarial expertise merit consideration.

Similarly, one must be wary of post hoc, ergo propter hoc reasoning. It seems likely that the broader criticisms developed in *Castro* of Lifecodes' procedures for declaring a match and for estimating the frequency would have surfaced even if the case had never existed. They were in the air, so to speak. Some had been voiced by experts in connection with protein electrophoresis (Chapters 1 and 2). Others were raised in the earlier DNA case *People v. Wesley.* Indeed, the defendants' attack on the frequency estimates in *Wesley* persuaded the court to shave these numbers by a factor of ten. The National Academies already were seeking funding for a study of the new forensic science. At a Banbury conference held before *Castro,* participants also called for "designated quality standards" and "periodic inspections" of "government and private laboratories doing DNA analysis for forensic use" (Westin 1989, 34), and for "[l]aboratory performance standards" (Caskey and Hammond 1989, 130). Barry Scheck told the assembled scientists in Banbury that "because the *Frye* test is stupid in the context of novel scientific techniques," there should be an eighteen-month moratorium on prosecutorial DNA testing to allow "blind trial testing, looking at . . . different kinds of probes in different contexts, seeing whether our allele frequency tables are good," and additional measures to verify that VNTR testing was being performed and interpreted correctly in different laboratories (Ballantyne, Sensabaugh, and Witkowski 1989, 97–98). Lander also attended the conference, presenting a paper challenging the statistical basis for estimating random-match probabilities (Lander 1989b) and sharply questioning Baird's ability to declare that the bands in adjacent lanes matched (Ballantyne, Sensabaugh, and Witowski 1989, 183–190). In fact, it was Lander's already-critical stance at the conference that led Neufeld and Scheck to seek his assistance in *Castro* (Levy 1996, 43–44).

This is not to say that *Castro* was unimportant. Headlines and news reports across the nation began to ask whether DNA evidence was reliable (Mnookin 2006; Aronson 2007, 76). The district attorney's office brazenly declared that it had achieved a "'victory of national importance' for the criminal-justice system and the prosecutor" because of the court's findings of general acceptance of the principles of DNA profiling (McFadden 1989). Scheck, on the other hand, insisted that "[t]here are good grounds for reopening [many] cases" because "[t]hey are all flawed in the same way" (ibid.). Judge Sheindlin found such headlines "maddening" and Scheck's call for reopening cases "chilling" (Sheindlin 1996, 86).

In the new atmosphere, it became easier for defendants to find experts to testify on their behalf. Indeed, a cottage industry of defense-oriented experts sprang up; a few geneticists even traded their ivory-tower offices or laboratories for the more lucrative profession of experts for hire (Chapter 7; Coleman and Swenson 1994, 103). In addition, the stark portrayal of Lifecodes' departures from its own protocol for declaring a match and its other troubling conduct may have made judges more willing to grant defense requests for discovery and for the appointment of experts to assist indigent defendants.

From a doctrinal perspective, however, *Castro*'s significance was limited. In a way, Neufeld and Scheck had been too successful. By persuading the court that Lifecodes had shortchanged the prosecution with its poor record keeping, contaminated probes, and overstated results, the defense made it possible for the court to resolve the case without addressing the questions of statistics and population genetics that cut across many cases.

Castro's Progeny

After *Castro,* some attacks on specific features of gel electrophoresis and the analysis of autoradiographs proved effective. In Portland, Maine, for instance, the prosecution's case against Kenneth McLeod for sexually assaulting a five-year-old girl evaporated when the defense uncovered irregularities such as a mislabeled autorad. Michael Baird was once again in the uncomfortable position of apparently playing fast and loose with Lifecodes' match criteria (Norman 1989).[16] The prosecutor angrily wrote Lifecodes not to send any further bills because "[t]he state will not suffer further by paying for Lifecodes' incompetence" (Sheindlin 1996, 97).

Lifecodes was not the only laboratory to come under a cloud. Cellmark Diagnostics received criticism in cases such as *State v. Schwartz* (1989). Thomas Schwartz was indicted for stabbing Carrie Coonrod to death in Minneapolis. The physical evidence included a pair of bloodstained blue jeans from his residence and a bloodstained shirt said to belong to Schwartz that was found in the vicinity of the murder. The state laboratory found that the blood groups matched those of the victim, and samples were sent to Cellmark Diagnostics for DNA testing. Cellmark reported that "[a]ll bands in the DNA banding pattern obtained from the blood of Carrie Coonrod are contained in the DNA banding pattern obtained from the stain removed from the plaid shirt. The frequency of this DNA banding pattern in the Caucasian population is approximately 1 in 33 billion." These statements notwithstanding, the report added that "no definitive conclusion can be reached," but it concluded that "it is the opinion of the undersigned that the DNA banding patterns obtained from the stain removed from the blue jeans and the blood of Carrie Coonrod are from the same individual" (447 N.W.2d at 424). After a pretrial hearing that featured conflicting testimony from twelve expert witnesses, the trial judge ruled that this conclusion was admissible, but asked the appellate courts to confirm this ruling before proceeding to trial.

The Supreme Court of Minnesota held that the evidence could not be admitted. Although the court determined that "DNA typing is generally acceptable" (id. at 426), it questioned the "reliability" of Cellmark's results. It was "troubled by the fact that Cellmark admitted having 'falsely identified two samples as coming from the same subject' during a proficiency test performed by the California Association of Crime Laboratory Directors." It was disturbed by the reported matches to the victim when "the banding patterns did not fit [Cellmark's] match criteria." It was concerned that "[t]he director of Cellmark's Research and Development Laboratory, Dr. Robin Cotton, admitted that . . . Cellmark did not meet all the minimum guidelines [of the FBI and the California Association of Crime Laboratory Directors for validating a testing protocol], such as formal methodology validation and published results of experimental studies in peer review journals" (id. at 426–427). Finally, it was perturbed by Cellmark's refusal to comply with certain discovery requests. Although "Cellmark disclosed to the defense its . . . protocol, laboratory notes from the testing in this case, the autoradiographs produced during RFLP analysis and statistical frequency tables," a "request for more specific information regarding its methodology and population data base was denied by Cellmark" (id. at 427), and "Cellmark has not yet published data regarding its methodology and its probes are only selectively available" (id. at 428). Under these circumstances, the supreme court concluded that although "forensic DNA typing has gained general acceptance in the scientific community, . . . the laboratory in this case did not comport with . . . guidelines" involving "appropriate standards and controls, and the availability of testing data and results." *Schwartz,* more than *Castro,* came close to holding that forensic DNA testing was not ready for prime time.

Yet even in this second wave of cases, defendants rarely denied that the basic principles of forensic DNA testing were generally accepted and scientifically sound. Also, the courts continued to agree that every individual's genome is unique (with the exception of identical twins) and that gel electrophoresis of VNTR RFLPs can discern some of the differences in these individual genomes. Thus exclusion of DNA evidence was atypical, and it appeared that better adherence to standards, freer disclosure of data, and additional publications were just around the corner. Significantly, the FBI was entering the fray and promoting standardization. The bureau had started to conduct its own studies in 1986. By the end of 1988, it had established an advisory group to develop a standard set of probes and procedures that could be adopted by all laboratories.[17]

The first opinion of a federal court, in *United States v. Jakobetz,* seemed to vindicate these efforts. It came at about the same time at which the National Association of Criminal Defense Lawyers (NACDL) established a DNA Task Force. (Prosecutors had their own groups and meetings for training their members in presenting the evidence and countering defense challenges.) The NACDL task force was headed by Neufeld and Scheck, who declared that DNA evidence might be a case of "the emperor having no clothes" (Sheindlin 1996, 98). The federal district court in *Jakobetz* did not agree. A young woman from Burlington, Vermont, was driving south on a vacation. She stopped at a rest area along Interstate 91 in Westminster to make a telephone call and to use the restroom. On her way out of the restroom, she was grabbed from behind, thrown to the floor, and hand-cuffed, her mouth was stuffed with paper towels, and her head was covered with a pillowcase. She was forced into the back of a tractor-trailer, which traveled for about half an hour. A man entered the back of the trailer and proceeded to brutally and repeatedly assault her. After driving for another four hours, he left her on the side of the road in the Bronx, New York. Even then, her ordeal was not over. Three in the morning on June 14, 1989, was not the best time to find help. After half an hour, a passing motorist called the police. They took her to a hospital. A semen sample, taken from a vaginal swab, was sent to the FBI laboratory for DNA analysis.

The investigation quickly pointed to Jakobetz. The first clue came from the records for the pay phones at the Westminster rest area. Just before the woman had placed her telephone call there, someone else had made a forty-six-minute, collect call to a number listed under the name of P. Zanon. P. Zanon was married to Randolph Jakobetz. Jakobetz drove a truck for a construction company in St. Albans Bay, Vermont. The company's records showed him driving a rig from St. Albans Bay to Westbury, Long Island, on June 13, 1989, the day of the kidnapping. His route took him on I-91 through Westminster, and the company had a toll receipt that he had submitted for reimbursement for travel across the Throgs Neck Bridge, which connects the Bronx with Queens—the most direct route to Westbury. The time stamped on the receipt was 3:28 A.M., June 14, 1989, just the right amount of time for driving from where the woman was released to the Throgs Neck Bridge.

Additional evidence left little doubt about Jakobetz's involvement. A search of the trailer hauled by Jakobetz revealed samples of head hair and pubic hair that were similar to those of the young woman. Armed with warrants, officers found handcuff keys, two knives, and two green pillow-cases similar to those described by the victim in Jakobetz's tractor-trailer cab. In Jakobetz's home, they found a set of handcuffs. In a shed behind his residence, they found nine rolls of undeveloped 35mm film. One of the rolls of film contained pictures taken during a fishing trip by the victim's father and his friends. (The victim's father had lent her his camera containing partially exposed film. The camera and several other possessions were missing when she finally returned to her car at the rest stop.) In addition, when she viewed a photo spread, she positively identified Jakobetz as the man who had abducted and raped her. Because she had been driven across state lines, a federal crime had been committed, and Jakobetz was charged with this kidnapping.

The DNA evidence confirmed what already was clear. The court of appeals, which ultimately upheld the thirty-year sentence that Jakobetz received, described it as establishing "one chance in 300 million that the DNA from the semen sample could have come from someone in the Caucasian population other than Jakobetz" (955 F.2d at 789). The astute reader will detect the transposition fallacy noted in Chapter 2. One in 300 million would be the chance, as computed by the FBI, that if the semen came from someone unrelated to Jakobetz, Jakobetz would happen to have the same VNTR alleles. It is computed, as we shall see, by assuming that the semen came from an unrelated individual. The chance that this assumption is true—that, given a match to Jakobetz, the semen actually came from someone else—is a different animal.

In any event, the defense was determined to keep DNA out of the trial. The district court held an eight-day hearing featuring the testimony of "imminent [sic] and highly regarded" experts—five for the prosecution and four for the defense (747 F. Supp. at 258). It correctly distinguished between the chemical techniques for deciding whether two samples match and the statistical exercise of estimating the frequency of such a match in a given population. As for the former, the court credited the government witnesses' testimony that RFLP-VNTR analysis as practiced by the FBI was generally accepted by molecular biologists and highly reliable. The court recognized that "the publications by no means unequivocally endorse DNA profiling"—indeed, its opinion cited the highly critical commentary from Eric Lander (1989a) inspired by his duel with Lifecodes in *Castro* and published in *Nature*. However, the FBI dodged this bullet when the opinion rhapsodically described the "rigorous standards [and] protocols established by the FBI [that] preclude abuse of the technique [in an] RFLP matching procedure [that] is exceptionally 'fail-safe'" (747 F. Supp. at 259). In addition, the court did not need to find general scientific acceptance. The federal Court of Appeals for the Second Circuit anticipated the Supreme Court's approach in *Daubert*. In a case involving "voiceprints," a much more dubious form of identification than DNA, the Second Circuit had abandoned the *Frye* standard in favor of a "reliability" or "relevancy-plus" standard (Kaye, Bernstein, and Mnookin 2004, 192–193). Hence the district court in *Jakobetz* was unperturbed by one or two critical articles in the scientific literature: "neither unanimity of scientific opinion nor a strong majority is a prerequisite to finding a scientific technique reliable."[18]

The approval of the FBI's work in *Jakobetz* indicated that if defendants were to succeed in keeping DNA evidence out of court, they would need to find more universal and lasting defects in the evidence than those that had been uncovered in *People v. Castro* and *State v. Schwartz*. For this purpose, they turned to the fields of statistics and population genetics. In *Castro*, the defense had found problems with the 1-in-189-million random-match probability. Lander had published these criticisms in *Nature*. Could this or other challenges persuade the courts to preclude such numbers in all cases, to exclude opinions like the one in *Schwartz* about the origin of a DNA stain, and perhaps even to block testimony about a match? The battle lines were drawn.

CHAPTER FIVE

The Intensifying Debate over Probability and Population Genetics

IF THE DNA TYPES of everyone in the world were recorded in a super-computer, it would be a trivial matter to determine the frequency of any DNA genotype. Indeed, we would not even need this number, because the police could investigate the suspect list spit out by the computer to identify the likely culprit. In the absence of a universal database, however, it is necessary to explain the probative value of a match. Prosecutors thus sought to impress jurors with vanishingly small numbers for the probability of plucking out someone from the general population who just happened to have a matching DNA type. The DNA laboratories obliged by providing expert witnesses to attest to figures of one in many millions or even billions.

In a third wave of DNA cases from 1989 to 1992 or so, objections to such vanishingly small estimates became common. Many defendants attacked the manner in which these probabilities were computed. As in *Castro* and *Jakobetz* (Chapter 4), they challenged the scientific validity or general acceptance of the estimation procedure. Some defendants also maintained that the figures were simply too impressive—that even if these probabilities were accurate, they must be excluded lest they seduce or overwhelm a jury. In the language of the federal and most state rules of evidence, they argued that the prejudicial impact of evidence substantially outweighed its probative value. Each objection requires its own analysis.

The Aversion to Numbers

The prejudice argument was widely rejected in connection with blood-serum proteins and enzymes (Chapter 2). Although the new DNA cases could be distinguished on the ground that the DNA frequencies were many orders of magnitude smaller, this was not seen as a reason to change the rules. In a Florida rape case decided on the heels of *Andrews,* Kevin McElfresh, the laboratory supervisor at Lifecodes Corporation, "explained the significance of the match of DNA patterns" in the following way:

> Q. And what would be the answer to that question as far as the likelihood of finding another individual whose bands would match up in the same fashion as this?
> A. The final number was that you would expect to find only one individual in 234 billion that would have the same banding pattern that we found in this case.

Not content to stop here, the prosecutor in *Martinez v. State* continued:

> Q. What is the total earth population, if you know?
> A. Five billion.
> Q. This is in excess of the number of people today?
> A. Yes. Basically that's what that number ultimately means is that that pattern is unique within the population of this planet.

Just in case the jurors missed something, the prosecutor asked:

> Q. Is that consistent with your opinion earlier that the semen involved in this case came from Fernando Martinez?
> A. That is correct. (549 So. 2d at 695).

The defendant argued that the trial court should have excluded this testimony because "a figure 47 times larger than the world's current total population was 'nonsensical'; and it was so overwhelming as to deprive the jury of its function in fairly appraising all of the evidence" (id.). The Florida District Court of Appeal was unmoved. It recognized that it had not considered this argument about prejudice in *Andrews* and that the *Carlson* line of cases in Minnesota supported the objection. Nonetheless, it dismissed these cases as aberrations. Citing various other opinions on the statistics associated with immunogenetic markers, it concluded that "where statistical probability testimony is scientifically and reliably grounded, it is admissible, however high the probabilities may be" (id. at 696).

But Minnesota stuck to its guns (for a time). In *State v. Schwartz* (see Chapter 4), prosecutors argued that because "[p]robability calculations are integral to DNA typing," the court should relax "the *Kim* limitation, at least regarding DNA evidence" (447 N.W.2d at 428). The Minnesota Supreme Court conceded that "other jurisdictions admit this type of evidence," but it declined to alter its unique exclusionary rule. The state legislature responded with a statute making population-frequency statistics admissible, but the Minnesota courts largely ignored it, apparently regarding it as an unconstitutional encroachment on judicial power.[1] The law changed in 1994, however, when the Minnesota Supreme Court created a "DNA exception" to *Kim*. In *State v. Bloom* (1994), the court held that *Kim* should not be applied to an "extremely conservative" random-match probability for a multiple locus DNA profile. (This conservative "interim ceiling" method is described later in this chapter.)

The *Bloom* court did not rely on the argument from *Schwartz* that numbers were "integral to DNA typing." After all, probabilities are no less important in interpreting a match with protein markers. This time, the state cleverly presented a new argument through an unusual affidavit from a population geneticist, Daniel Hartl. Hartl, who then was at the Washington University Medical School in St. Louis, was well known as a critic of DNA evidence. He later joined the Harvard faculty after collaborating with Harvard's Richard Lewontin on a controversial article that branded the standard procedure for estimating population frequencies "unjustified and generally unreliable" (Lewontin and Hartl 1991, 1750). In *Bloom*, however, he swore that

> the *Kim* rule was correct at a time when genetic typing could be performed only with blood groups and other types of genetic systems that are not highly polymorphic; the effect of the *Kim* rule is to prevent a series of genetic matches at different loci, none individually uncommon, from being combined by a long chain of multiplication of probabilities to yield a small, and unjustified, match probability for the whole set of genes. (516 N.W.2d at 166)

Having defended the Minnesota rule against multiplication, Hartl proceeded to use a form of multiplication with DNA loci to report "a 1 in 634,687 chance of a random match across the five loci" in *Bloom* (id. at 160 n.2).

At first blush, it seems hard to reconcile these two positions. A "long chain of multiplication" is no more justifiable when the alleles are uncommon than when they are common. If multiplication is good for the goose of VNTR loci, why is not the same form of multiplication good for the gander of the immunogenetic loci? Presumably, Hartl was concerned only with being "conservative" in the multiplication process and would have approved of the conservative version of the multiplication for protein markers, just as he did for VNTR alleles.

In *Bloom,* however, the court seized on the affidavit to create a DNA-only exception to the Minnesota rule against multiplication. Because the distinction is unfounded, the court would have done better simply to overrule the *Kim* rule as applied to any genetic system in which the probability is validly computed.

Furthermore, *Bloom* approved of proposed testimony by Hartl that the "match constituted 'overwhelming evidence that, to a reasonable degree of scientific certainty, the DNA from the victim's vaginal swab came from [the defendant], to the exclusion of all others' " (id. at 160 n.2). This portion of the affidavit illustrates how science is repackaged in the courtroom. The reasonable-degree-of-scientific-certainty language almost certainly was drafted by the lawyers. Scientists have no use for this phrase (outside the courtroom). Indeed, "a reasonable degree of scientific certainty" is not a defined concept in scientific disciplines or even in law (Kaye, Bernstein, and Mnookin 2004, 363). It is legal mumbo jumbo derived from archaic cases in which lawyers discovered that if a medical doctor did not utter the incantation "to a reasonable degree of medical certainty," his testimony might be excluded because doctors were not supposed to talk about mere probabilities (Hassman 1977; J. L. Lewin 1998). Modern cases usually recognize that suitably explained information about less-than-certain possibilities can be helpful in various circumstances, but experts want to (or are induced to) incant not only "medical certainty" but also "clinical certainty," "psychological certainty," "psychiatric certainty," "engineering certainty," "architectural certainty," "ballistic certainty," "professional certainty," and even "forensic certainty" and "legal certainty."

In any event, cases like *Martinez* and (eventually) *Bloom* made it abundantly clear that a probability is not inadmissible just because it is small. The objection that small probabilities are too prejudicial carries no more weight with DNA loci than it did with immunogenetic markers. This is not to say that there is no possibility of prejudice from a statistic. There is a risk that judges or jurors will misconstrue a random-match probability as a statement of the probability of innocence. This is the transposition fallacy that the prosecutor in *People v. Collins* made famous to generations of law students decades before the introduction of DNA evidence (Chapter 2).

Although counsel should beware of naive transposition and perhaps juries should be cautioned against it, the mere possibility of transposition hardly justifies automatic exclusion of a properly computed and presented probability.[2] Even in cases in which the expert or the prosecutor mistakenly transposes the conditional probability, the error can be harmless because the proper transposition would not be very different or would be even more damaging to the defendant. This will be the case when the other evidence of guilt is substantial or the random-match probability and the chance that the DNA came from a close relative are extremely small. Indeed, if the non-DNA evidence in the case establishes at least a fifty-fifty chance of guilt, then naive transposition favors the defendant.[3]

Another argument holds that the random-match probability is prejudicial because it fails to account for the chance of laboratory error. Philip Reilly, a physician and lawyer with a long-standing interest in genetics and the law, explained to one reporter that "when crime laboratories issued odds like one in a million or one in a billion, scientific experts were 'offended' because they knew such odds were unreasonable. In most laboratories, the ordinary rate of error due to human sloppiness alone would severely undermine purported odds of a billion to one" (Kolata 1992b). Hartl's expert report in one hotly contested federal case and a later article in *Science* condemned the random-match probability as "simply meaningless" (Brief of Defendants-Appellants, *United States v. Bonds,* 1992, 52) and "terribly misleading" (Lewontin and Hartl 1991, 1749) in failing to incorporate the probability of laboratory error.

The validity of this objection turns on the ability of jurors to appreciate the fact that a tiny probability of a random person's having the same DNA type as that in the sample from the crime scene addresses only one issue—could the samples actually match because of an unfortunate coincidence? Other explanations for a reported match also are possible (Koehler 1993). Courts generally recognize that the random-match probability is not an index of guilt—after all, the DNA might have been planted, it might have come from an identical twin or other close relative, or the laboratory might have made a mistake in labeling or handling the samples (Kaye, Bernstein, and Mnookin 2004, 449). These possibilities are not hard to grasp, and it should be obvious that the random-match probability does not account for them. The evidentiary issue is whether the possibility that presenting the random-match probability without a parallel figure for the chance of a false positive due to laboratory error (or a single figure that incorporates the risk of a false-positive laboratory error) is unduly likely to mislead the jury. Despite concerns expressed by defendants and academics (e.g., Koehler, Chia, and Lindsey 1995), in no reported case has the admission of DNA evidence ever been reversed on this ground. The courts reason that the defendant can attack the weight of the DNA evidence by bringing up these possibilities (Kaye and Sensabaugh 2006).

The Validity of the Numbers

The nearly universal conclusion that properly computed DNA probabilities are not necessarily prejudicial does not imply that numbers such as those we have encountered—1 in 234 billion *(Martinez)*, 1 in 33 billion *(Schwartz)*, 1 in 1.4 billion *(Wesley)*, 1 in 189 million *(Castro)*, and 1 in 300 million *(Jakobetz)*—are correctly computed. Thus the defense in *Jako-betz* not only maintained that the small random-match probability was prejudicial, but it also questioned the "scientific technique used to calculate the frequency with which a particular pattern of alleles occurs in the relevant population." To examine the validity of the FBI's computation, we must consider (1) the rule for declaring matches, (2) the source and extent of the data on allele frequencies, and (3) the model for combining the individual frequencies to find the joint frequency of the combination of alleles—the genotype. *Jakobetz* illustrates how these issues were the subject of expert disagreement in the court cases.

The Match Window

Gel electrophoresis of VNTR alleles is messy. When samples containing identical copies of a DNA fragment are run on the same gel, they do not always produce bands in precisely the same positions. The fragments are the same lengths, but the experimental apparatus is incapable of measuring the VNTR fragment lengths exactly. As with using an ordinary yardstick or measuring tape to determine the dimensions of a room, measurement error is present. DNA analysts were willing to declare that two fragments matched if the bands matched visually, and if they fell within a specified distance of one another. The FBI laboratory declared matches within a 5% "match window."

The concept of a "match" here is somewhat artificial and has engendered considerable confusion. To clarify the issue, a few numbers and symbols will be helpful. Suppose that a particular fragment has migrated a distance that corresponds to some number of base pairs X as indicated by known molecular-weight markers run simultaneously in the gel. Because of slight variations in the gels and the other experimental conditions, such measurements are almost never exact. The true length could well be somewhat different. A range of possible values for this value can be constructed by adding and subtracting a percentage of the observed length. The FBI essentially said that the true length is not exactly X, but an interval $X \pm (2.5\% \times X)$. The FBI then could declare a match between two RFLPs when the two interval estimates overlapped. For example, if an allele in a crime-scene sample appeared to be 2,000 base pairs (bp) long, 2.5% would be 50 bp, and the interval would go from 1,950 to 2,050 bp. If an allele from the suspect appeared to be 2,400 bp long, 2.5% would be 60 bp, and the corresponding interval would go from be 2,340 to 2,460 bp. Because these two intervals do not overlap, a match could not be declared. Conversely, similar arithmetic shows that any allele from a suspect with a measured length between 1,903 and 2,103 bp could be considered to match the band weighing in at 2,000 bp. The corresponding interval estimates now overlap the interval for the crime-scene allele. As this example shows, because each of the two bands being compared has an uncertainty of ±2.5% of its measured values, the match window is about ±5% (see National Research Council Committee on DNA Forensic Science: An Update 1996, 44, 140 [hereinafter cited as NRC 1996]).

The FBI arrived at the 2.5% figure for the uncertainty in each measurement by experiments involving pairs of measurements of the same DNA sequences. It found that the match windows using this figure were wide enough to encompass all the differences seen in the calibration experiments. In *Jakobetz,* Joseph Nadeau, then a staff scientist at the Jackson Laboratory in Maine (he is now chairman of the Genetics Department at Case Western Reserve University School of Medicine) contended that this window was unacceptable because it was not derived by considering the statistical properties of the match rule. Because two alleles separated by as much as 5% of their mean value could be reported to match, and because the standard error of measurement was about 1%, a gap of some five standard errors would not preclude a match. To some statisticians, this seemed incredibly generous. In the *Journal of the American Statistical Association,* Seymour Geisser, who testified for the defense in roughly 100 DNA cases and "was renowned in the statistical world" (Berry 2005), advised his colleagues that "a tolerance of about 4 standard deviations of the difference . . . is an extraordinarily wide net to declare a match" (Geisser 1992, 609). Daniel Hartl expressed a similar opinion in *United States v. Yee,* in which another all-out attack on the FBI's methods was under way. In *Yee* (discussed more fully later in this chapter), Hartl opined that "too many false matches are declared" and reported that the FBI's window "should be scaled down by approximately one half" to be consistent with accepted scientific practice (W. C. Thompson 1993, 49 n.123).

In advancing this criticism in *Jakobetz,* however, the defense was engaging in what prosecutors like to call "blowing smoke." Nadeau believed that the interval estimates should have been about half as wide (1% rather than 2.5%) (*see* 747 F. Supp. at 257). As the district court pointed out, however, "all sixteen band matches (eight alleles from each the victim and the suspect on four different autorads) were within plus or minus 1%" (id.). That the FBI protocol also would have allowed an analyst to declare a match with more widely separated bands was irrelevant in this case.

In other cases, of course, a narrower window might make a difference, but the objection was never persuasive to the courts, and for good reason. From a statistical standpoint, the argument makes little sense. The choice of any particular match window is arbitrary. What matters is how it operates in classifying pairs of samples of DNA. The operating characteristics depend entirely on how wide the window is open and not at all on how one chooses to open it this far. Narrow windows are undesirable because they produce too many false exclusions. A false exclusion can arise from a discrepancy at a single allele. Broad windows reduce the chance of these false exclusions, but they also could produce more matches to people who are not the source of the crime-scene DNA. For this to happen, however, every allele must match even though the defendant is not the source of the crime-scene sample. How often this occurs with unrelated people who just happen to have the same VNTRs should be given by the random-match probability. A window that is open so wide as to let nearly everyone in the population through it will have an unacceptably high random-match probability. Thus the laboratory needs a window that keeps the probabilities of both types of errors reasonably small.

A choice is not improper just because a different number of standard errors might be used as a decision criterion in a different situation. Studies indicated that repeated measurements of the same fragments were roughly consistent with the normal distribution (Geisser 1992, 608). This distribution—the famous bell-shaped curve—applies to many phenomena. Repeated measurements tend to cluster about a true value, with extreme deviations being much less likely than modest ones. The chance of a large departure from the true value is given by the area in the tails of the normal curve, and the magnitude of the departure is conveniently expressed in units of standard deviations. Consider Geisser's claim that "a tolerance of about 4 standard deviations . . . is an extraordinarily wide net." Because 99.9937% of the area under a normal curve lies within ±4 standard deviations, the chance of a false exclusion with this net is about 0.0063% for a single allele. But remember that it takes only one band to exclude a suspect. With four heterozygous loci, eight comparisons are made. If every allele is independent, then the chance of a false exclusion (at least one pair of alleles is declared not to match) is $1 - 0.999937^8 = 0.05\%$. In other words, in about

1 case out of every 200, the "extraordinarily wide net" classifies two bands from the same source as nonmatching—it falsely exonerates a suspect. This calculation is an oversimplification,[4] but the point is that if a match window is going to be used to make comparisons across several loci, it needs to be quite wide for a single allele.

The FBI's match window therefore was reasonable because it seemed to keep the probabilities of false inclusions and false exclusions small. Also, because the jury could use an accurate estimate of the rarity of the multiple locus VNTR type—as determined with this window—in assessing the probative value of the match, the debate over the width of the match window was a distraction (Kaye 1995c). The more serious concern, and the one that caused the greatest consternation for the FBI, was whether the estimates of the proportion of people who would have matching VNTR types in the general population—ascertained with whatever match window one likes—were sound.

Allele Frequencies

Once a match window is clearly defined and a pair of DNA samples is known to match at all the corresponding alleles, the probability of this match occurring in the general population (or some other pertinent group) provides an indication of its probative value. The procedure used by the FBI and other laboratories involved examining DNA from a small sample of the population, determining the relevant VNTR allele frequencies in the sample, and then combining the individual frequencies according to a population-genetics model.

In *Jakobetz,* the FBI relied on a "Caucasian data base . . . derived from blood samples of approximately 225 FBI agents from throughout the United States" (747 F. Supp. at 253). It analyzed these samples to obtain a list of measurements for the lengths of each allele at each VNTR locus. The most straightforward procedure to estimate the frequency of a given allele then would be to count the number of alleles within the match window surrounding the measured value for the crime-scene sample. Suppose that the measured length of an allele in the crime-scene sample is 2,000 bp. Earlier we saw that the FBI could declare a match to suspects whose alleles ranged from about 1,900 to 2,100 bp (the ±5% window). The proportion of bands in the database falling into this range provides an estimate of the frequency of the 2,000 bp allele in the Caucasian population.

Cellmark Diagnostics and Lifecodes Corporation used this "floating-bin" procedure, but the FBI adopted a more complicated "fixed-bin" method.[5] Rather than a bin being centered on the band in question, the bins are created in advance by dividing the entire range of allele lengths into fixed intervals. A certain proportion of the alleles in the database will fall into each bin. One can then determine which bins the measured alleles belong to and use these bin frequencies as estimates of the allele frequencies in the larger population. Complications arise, however, when some bins are empty, or nearly so, and when the match window around a defendant's band includes several bins. Bruce Budowle, a geneticist at the FBI laboratory who led the FBI's DNA efforts, and Thomas Caskey, a distinguished geneticist then at the Baylor School of Medicine, testified in *Jakobetz* that the FBI used "very conservative" adjustments to handle these matters.[6]

One might well wonder whether a convenience sample of 225 FBI recruits is sufficient to represent the U.S. Caucasian population, and the defense in *Jakobetz* did not overlook this objection (747 F. Supp. at 261). The government countered with testimony from Kenneth Kidd, a leading population geneticist at Yale Medical School, who would later receive the Justice Department's Profiles in DNA Courage Award. Kidd insisted that (in the district court's words) "once it is determined that the alleles are randomly occurring throughout a targeted population, sample size can be decreased to as little as one hundred individuals" and that "the FBI data base . . . provides an adequate representation of Caucasians in the United States for purposes of VNTR frequencies" (id.). Impressed with Kidd's testimony and the intuitions of Budowle and Caskey that the FBI's binning procedure was conservative enough to compensate for any weaknesses in the database, the court held that the allele-frequency estimates were admissible.

Certainly, these estimates rested on more extensive data than the so-called random sample of twenty students that Jeffreys used to buttress the claim that his multilocus probes supplied "individual-specific" identification (Chapter 3). Nevertheless, a number of respected statisticians and other expert witnesses continued to call for truly random samples. Geisser (1992, 608), for instance, complained that "the sampling of populations is not random, and not even haphazard—rather, it is catch as catch can, with the result that sometimes the catch should be canned." Lander (1992, 204) tartly observed that "the sampling schemes used by testing laboratories leave much to be desired." However, the courts almost always rebuffed defendants' arguments that general acceptance or scientific validity required larger, truly random samples (Kaye 1993).

The Basic Product Rule for Combining
Allele Frequencies

The last step in estimating a VNTR genotype frequency is to combine the various allele frequencies into a single number—for example, the 1 in 300 million frequency for the occurrence of the VNTRs in *Jakobetz*. A common misconception is that one just multiplies the individual frequencies according to the "product rule" discussed in cases such as *People v. Collins*, which involved a robbery by an interracial couple who drove away in a partly yellow automobile (Chapter 2).[7] Multiplication *is* crucial, but the FBI's procedure (which I shall call the "basic product rule" despite the fact that it is a bit more complicated than the multiplication in *Collins*) produces larger frequencies than the mere product of the allele frequencies.

An example shows why. Suppose that a profile from DNA found at a crime scene consists of four loci, with one allele at the first locus and two alleles at each of the other three. Let us focus on just the first two loci, which can be designated A and B. This part of the full profile can be designated A_1–B_1B_2, the dash indicating the single band at the A locus. The allele frequencies in the database of Caucasians are, let us say, $p_{A1} = 3.5\%$, $p_{B1} = 2.9\%$, and $p_{B2} = 6.8\%$. The basic product rule first estimates the frequency of the alleles at each locus (the "single-locus genotypes") and then combines these figures to estimate the frequency of the combination of single-locus genotypes (the "multiple single-locus" or "multilocus genotype"). Here is how each step works.

Single-Locus Genotypes: Hardy-Weinberg Equilibrium

The presence of two distinct alleles at the B locus means that whoever provided the DNA is heterozygous there—one allele is on a chromosome inherited from the individual's mother, and the other is on the paternal chromosome. As will be explained more fully in the next section, in a large, "randomly mating" population the fraction of those who will have the allele B_1 on the paternal chromosome and the allele B_2 on the maternal chromosome is approximately the product of the two allele frequencies, 0.029×0.068. About the same fraction will have the B_1 on the maternal chromosome and the B_2 on the paternal one. The laboratory cannot say which of these two, equally likely possibilities pertains to the evidence sample, so it estimates the relative frequency in the white population with either arrangement to be $2 \times 0.029 \times 0.068 = 0.003944$. More generally, the formula for the frequency P of a heterozygous single-locus genotype in a large, randomly mating population is $P = 2p_1p_2$, where p_1 is the frequency of one allele, and p_2 is the frequency of the other.

At the A locus, however, only the allele A_1 is seen. This might be because the source of the DNA taken from the crime scene has two copies of allele A_1—one inherited from each parent. The frequency of the homozygous single-locus genotype in a randomly mating white population is $0.035 \times 0.035 = 0.001225$, about 1 in 1,000. If we generalize from this example, in a large, randomly mating population, homozygote frequencies are simply p^2. When a population is in what is known as "Hardy-Weinberg equilibrium," the equations $P = 2\ p_1p_2$ for heterozygotes and $P = p^2$ for homozygotes describe the relationship between single-locus genotype frequencies and allele frequencies.[8]

But the single band A_1 might really be two bands that are close together, or there might be a second band that is relatively small and has migrated to the edge of the gel during the electrophoresis. In these circumstances, only one band would show up on the autoradiograph. The FBI therefore makes a "conservative" assumption. It acts as if there is a second, unseen band. Not knowing the frequency of this band, it assumes that it occurs 100% of the time. With this modification, the genotype frequency $P = p^2$ for apparent homozygotes becomes $P = 2p$. Thus the genotype frequency at the A locus then is estimated to be $2p = 2 \times 0.035 = 0.070$—almost sixty times larger than $p^2 = 0.001225$.

Multilocus Genotypes: Linkage Equilibrium

The basic product rule now calls for the two single-locus frequencies to be multiplied to yield a two-locus profile frequency of

$$2p_{A1} \times 2p_{B1}p_{B2} = 4p_{A1}p_{B1}p_{B2} = 0.000276$$

The four-locus genotype frequency would be even smaller, because we would multiply by two more single-locus frequencies to obtain the product

$$2p_{A1} \times 2p_{B1}p_{B2} \times 2p_{C1}p_{C2} \times 2p_{D1}p_{D2} = 2^4 p_{A1}p_{B1}p_{B2}p_{C1}p_{C2}p_{D1}p_{D2}.$$

When the frequencies of multilocus genotypes are the simple products of the frequencies of single-locus genotypes, the population is said to be in "linkage equilibrium." Both linkage equilibrium *and* Hardy-Weinberg equilibrium are required for statistical independence, a point that eluded some courts (Kaye 1997a).

Threats to Independence

The basic product rule treats the alleles as being statistically independent, permitting them to be multiplied. What justifies this assumption? Why should we believe that the idealized situations of Hardy-Weinberg and linkage equilibria apply to real populations? Throughout science and statistics, simplified models are used to represent a more complex reality. Physicists speak of frictionless surfaces, perfectly elastic collisions, and the gravitational interactions of only two bodies in a universe populated with inconceivably many bodies. If the assumptions used to construct the model are reasonably close to the actual situation, then they are justified. ("Reasonably close" means that realistic departures from the assumptions will produce similar results.) The forensic and scientific controversy over the basic product rule thus had to do with whether there were good reasons—based on theory and data—to believe that the conditions that lead to Hardy-Weinberg and linkage equilibria apply in the U.S. population and whether any departures from equilibria are so substantial as to render random-match probabilities too inaccurate to be admissible.

What are these conditions? For simplicity, we continue to focus on a biallelic polymorphism that is not sex linked. That is, there only two alleles (A_1 and A_2) at a locus, and the locus is not on (or influenced by anything on) the sex chromosomes (X and Y). At this locus, a proportion p_1 of the sperm and ova are of type A_1, and a proportion p_2 are A_2. If people mate independently of these alleles, if the number of surviving offspring is also independent of genotype, if the alleles assort independently, and if there are no mutations, then we easily can compute the expected proportions of each single-locus genotype in the next generation. Random mating is equivalent to combining gametes at random, and the chance of picking a sperm and an egg of the same type is p_1^2 (for an A_1A_1 offspring) and p_2^2 (for an A_2A_2 offspring); the chance of picking a sperm and an egg of different types is $p_1p_2 + p_2p_1 = 2p_1p_2$. These are the Hardy-Weinberg proportions, and they will be established (within the limits of chance fluctuations) in a single generation of random mating.

Here, "random mating" is a term of art. It does not mean that individuals choose their mates at random, but only that the selections are unrelated to the alleles in question. If this condition holds, then linkage equilibrium also follows—asymptotically. With unlinked loci (those that are not inherited together), each generation cuts the departure from equilibrium in half (NRC 1996, 106).

In *Jakobetz*, there was no dispute about most of these conditions for statistical independence. For example, the VNTR loci used for identification lie on different chromosomes (or far apart on the same chromosome) and hence are not physically linked. Random mating seems assured because forensic VNTRs do not produce any external traits, let alone ones that might make people who possess certain alleles more sexually attractive. In contrast to visible and socially significant traits like skin color, how could mate selection be influenced by unseen and unexpressed DNA sequences?

Jakobetz's answer was "population structure." Studies had shown that VNTR allele frequencies vary somewhat across the major racial or ethnic groups in the United States. Of course, these categories are largely cultural and are not sharply defined genetically, but the variations across these "races" are the remnants of differences among ancestral populations. To the extent that people mate endogamously—within their own self-recognized racial groups—the population has some genetic stratification, or "structure." This affects genotype frequencies because it means that mate selection is not independent of VNTR alleles. Selecting partners now depends on race, which is associated with VNTR allele frequencies. Hence mating is no longer random with respect to VNTRs.

The forensic DNA testers handled this type of structure in the U.S. population as a whole by stratifying the databases according to race and testifying to the estimated genotype frequencies among Caucasians, African Americans, Hispanics, and other groups. If the DNA profiling was accurate and all these numbers were infinitesimal, then the possibility that the defendant was unrelated to the true perpetrator and just had the misfortune of sharing his DNA profile could be dismissed. However, why assume that population structure stops (or is best described) at the gross level of "races"? Among whites, for example, there might be endogamously mating ethnic or religious subgroups with different VNTR allele frequencies. There are Swedish Americans and Italian Americans, Korean Americans and Japanese Americans, and sub-Saharan African Americans and northern African Americans. A vast number of such subgroups within the groups represented by the primitive reference databases might have somewhat different allele-frequency distributions and might mate primarily among themselves.

Concerns about population structure had been voiced in a few of the cases on serological evidence, but the courts were not receptive (Chapter 2). No prominent scientists were drawn into this seeming backwater of the law. There was no conclave at Cold Spring Harbor to examine a newly emerging technology and no news reports in *Science* and *Nature* on conflicting testimony in court cases. DNA evidence, in contrast, fascinated the public and the scientific community. The *Castro* case and especially Eric

Lander had made challenges to the emerging technology scientifically respectable (Chapter 4), and prosecutors and defenders across the country were recruiting geneticists and statisticians to support and attack the evidence. In *Jakobetz,* the defense enlisted the services of Richard Lewontin, the Alexander Agassiz Professor of Zoology and Professor of Biology at Harvard University. An evolutionary biologist, geneticist, and social commentator,[9] Lewontin pioneered the idea of applying molecular biology to questions of genetic variation and evolution, and he did seminal work on the mathematical basis of population genetics and evolutionary theory. He has been a powerful critic of the notion that there is a meaningful genetic basis for dividing people into racial categories (e.g., Lewontin 2005, 2006), and his 1972 study of worldwide variations in protein polymorphisms (see Chapter 6) was instrumental in deconstructing the idea that the traditional human "races" have a single, objective genetic foundation (e.g., Brown and Armelagos 2001).

In *Jakobetz,* Lewontin "did discredit the government's experts who casually concluded that VNTRs must randomly occur throughout the population because individuals do not consciously consider VNTRs when they chose their mates" (747 F. Supp. at 260). Even so, the court rejected Lewontin's view that because of possible population structure, "it is entirely inappropriate to use one data base for all Caucasians and to use the product rule to calculate an allele pattern's frequency" (id.). As with the use of a small convenience sample, the district court found that "to the extent that substructure might exist for VNTRs within Caucasians, the FBI has sufficiently proven that it has compensated for this possibility by using conservative binning procedures" (id.). In this respect, the court credited the testimony of Kidd and Budowle. "Dr. Kidd testified that he has looked at data from many subgroups, including Italians, Swedes, Irish, Amish, and mixed Europeans, and all have 'very small differences' in allele frequencies" (id.). "[T]he differences in subgroups," he insisted, "are 'absolutely insignificant'" to basic-product-rule computations. Likewise, Budowle testified to the "amazing" similarity in VNTR frequencies in German, Dutch, French, Lebanese, and Israeli samples. He "stated, in complete exasperation, that 'there comes a point where you got to say, how many populations do we have to do to convince?'" (id.). Just what Budowle meant by "doing populations" is not clear from the opinion. As with Stolorow's validation studies of the thin-gel multisystem for electrophoresis (Chapter 1), this work was as yet unpublished and apparently unreplicated. The court recognized that "neither Dr. Kidd nor Dr. Budowle cite to any published studies to support these conclusions" (id. at 261), but this did not prevent the judge from concluding that the basic product rule, used with large fixed bins, was scientifically valid. Neither did the debate between Lewontin, the professor from Harvard, and Kidd, the professor from Yale, stop the court of appeals from adding that the procedure was generally accepted.

Relying on work that has not been published or otherwise subjected to the scrutiny of other scientists is risky. The U.S. Supreme Court recognized as much when it listed "peer review and publication" as one important factor to be considered in deciding whether a theory or finding qualifies as "good science" (*Daubert* 1993, 509 U.S. at 593). Of course, "[s]ome propositions . . . are too particular, too new, or of too limited interest to be published" (id.), but that was not the case here; moreover, during the third wave of DNA cases, the FBI also refused to make its database public (W. C. Thompson 1993, 78; cf. Abbott 1992; Giannelli 1997a, 420–421). Nonetheless, the early impressions of the prosecution's experts proved to be correct. In time, statistical studies aplenty would confirm that population structure was not a significant problem in estimating random-match probabilities in broad, racial populations.

At this point, however, courts struggled with opposing testimony like that in *Jakobetz*. In *People v. Castro,* Eric Lander had caught Lifecodes using a floating-bin width larger than its match window, and he maintained that the population represented in the company's Hispanic database exhibited population structure (Lander 1989a, 504). In *Caldwell v. State,* Jung Choi, a molecular biologist from the Georgia Institute of Technology, disputed the testimony of Wyatt W. Anderson, the University of Georgia's Alumni Foundation Distinguished Professor of Genetics, that the assumptions underlying the basic product rule were "not unreasonable" (393 S.E.2d at 443). In response, the Supreme Court of Georgia concluded that Lifecodes' "straight binning method" was satisfactory but balked at admitting Lifecodes' basic-product-rule frequency of 1 in 24 million.[10]

These occasional defense victories notwithstanding, most courts continued to find general acceptance or scientific validity. Neufeld, Scheck, and other defense lawyers lost an intense battle against the FBI's match-binning procedures in *United States v. Yee.* Three members of the Hell's Angels motorcycle gang in Cleveland were accused of killing David Hartlaub a record-store clerk, who they thought was a member of a rival gang. Hartlaub was shot inside his van fourteen times with a silenced machine gun (Sheindlin 1996, 98). Some of the blood in the van was not Hartlaub's, and VNTR testing by the FBI indicated that it came from one of the three Hell's Angels, Johnny Ray Bonds. The estimated random-match probability was a modest 1/35,000. When the defendant objected that the FBI's methods of VNTR analysis and interpretation were not generally accepted, Magistrate Judge James Carr conducted a six-week hearing that bloodied the FBI laboratory. One defense expert, working with the data from some of the FBI's validation studies, "raise[d] troublesome questions about the quality of the Bureau's work" (134 F.R.D. at 207). The magistrate was perturbed that the FBI's chief scientist, "Dr. Budowle[,] did not respond persuasively to [the] criticisms, and he refused to acknowledge the potential significance or merit of a competent scientist's critique and to consider the desirability for further experimentation and confirmation" (id.). Despite "the unfortunate, and to some extent unjustifiable flaws" (id. at 206) and "the remarkably poor quality of the F.B.I.'s work and infidelity to important scientific principles" (id. at 210), however, the magistrate concluded that the FBI was "able to declare matches accurately" (id. at 206). The FBI also survived criticisms from Lander, Lewontin, Hartl, and others of its database and match-binning probability. These scientists did persuade the magistrate that "the potential effect of substructure cannot be known or even estimated in any given case, and there is no factor that rationally and indisputably will compensate for its presence" (id. at 211), but he found the testimony and explanations of Caskey, Kidd, and Patrick Michael Conneally, a distinguished professor of medical genetics and neurology at the Indiana University School of Medicine, on the limited effect of population structure and related matters to represent better the views of the relevant scientific community. The district court adopted the magistrate's conclusion that match binning and the basic product rule could "provide a scientifically acceptable estimate of the resulting probabilities" (id. at 206), and the U.S. Court of Appeals for the Sixth Circuit affirmed the district court.

The Lewontin-Hartl Article

The disagreements over the basic product rule in cases like *Jakobetz* and *Yee* reached the flash point when Lewontin and Hartl did what academics often do—they published. They "reworked their expert witness reports" in *Yee* (Derksen 2000, 821; cf. Brief of Defendants-Appellants, *United States v. Bonds,* 1992, 26 n.9) and submitted "their losing courtroom criticism" (Levy 1996, 111) to the journal *Science.* The manuscript advanced the view that population structure made the basic-product-rule calculations invalid and unreliable. It reported that, at least for serum enzymes and proteins, "there is, on average, one-third more genetic variation among Irish, Spanish, Italians, Slavs, Swedes, and other subpopulations, than there is, on the average, between Europeans, Asians, Africans, Amerinds, and Oceanians" (Lewontin and Hartl 1991, 1747). Pointing out that VNTR allele frequencies could vary substantially between groups such as Israelis and French, the article concluded:

> Both the theory of population genetics and the available data imply that the probability of a random match of a given VNTR phenotype cannot be estimated reliably for "Caucasians," probably not for "blacks," and certainly not for "Hispanics," if the present method of calculation and the databases presently available are used. As currently calculated, the estimates may be in error, possibly by two or more orders of magnitude. The error may be in favor of the defendant, or against the defendant, and the magnitude and direction of the error is [sic] impossible to evaluate in any particular case without additional data. (Lewontin and Hartl 1991, 1749)

After suggesting some alternatives to the basic product rule (but overlooking others; see Chapter 6), the article ended by suggesting that even these alternatives were not yet feasible "[w]ithout the type of subpopulation studies already carried out for blood groups and enzymes" and reiterated that "estimates of the probability of a matching DNA profile based on VNTR data, as currently calculated, are unjustified and generally unreliable" (ibid.).

The publication process did not go smoothly. Hartl and Lewontin (1994) charged that after "our original article [was] peer-reviewed and already in galley proof," Daniel Koshland, the editor of *Science,* demanded that it "be substantially altered and weakened," and that he then took "the unprecedented step of soliciting a non-peer-reviewed attack . . . that was published in the same issue." Koshland's response was scathing. He recounted that "a distinguished population geneticist" who peer-reviewed the original article "found that the authors drew conclusions far beyond their data" (Koshland 1994, 202). (The next chapter discusses whether their conclusions were well supported.) Therefore, "I offered Lewontin and Hartl two options: they could make their conclusions more consonant with their data . . . or they could publish their more extreme statements in an opinion piece, rather than as a validated scientific article" (ibid.). When Lewontin threatened to raise "the biggest stink [Koshland] had ever heard" (Roberts 1991, 1722), Koshland allowed the original article to be published without modification.[11] However, he "asked other population geneticists to write a Perspective, so that the public would at least know that the subject was controversial" (Koshland 1994, 202–203). In Koshland's view, if his decision was extraordinary, it was only because the situation was extraordinary. "I would do the same," he wrote, "if we received a paper about global warming that extrapolated to a conclusion that we will all burn up in 10 years if we do not stop using fossil fuels." Lewontin and

Hartl had not written a typical article that would "affect only a few scientists" and for which "differing views can follow later in a classical tradition." They had crafted a critique with "erroneous" statements, and it would "cause headlines and confusion if a debatable point [were] exalted, even temporarily, into a fact" (ibid., 203).

In retrospect, this fear seems exaggerated. The article was bound to cause headlines and confusion regardless of what else appeared in the same issue. Also, *Science* is a weekly journal. It is unlikely that a gap of a few weeks in publishing a rebuttal would have made much difference. Just as defendants would trumpet the Lewontin-Hartl article, prosecutors would soon be apprising courts of opposing articles. Moreover, Koshland interceded only after two of the prosecution's experts in *Yee,* Kidd and Caskey, had warned that the article would have an unfortunate and unwarranted legal impact.[12] A prepublication manuscript had been circulating among prosecutors and DNA experts who worried that because of the prestige and visibility of *Science* magazine and the scientific reputations of Lewontin and Hartl, "[t]he impending publication of their article threatened the admissibility of DNA analysis in a way that their courtroom testimony never could" (Levy 1996, 111). Indeed, when James Wooley, the federal prosecutor in *Yee,* read the article, he telephoned Hartl to express his concern. Hartl felt that he was being pressured to withdraw the article, and Lewontin accused Wooley of trying "to suppress scientific evidence"— charges that Wooley denied (Levy 1996, 113–115).[13]

Ironically, although prosecutors saw the article as "potentially deadly ammunition" to be used against them (ibid., 112), the effort to secure a contemporaneous rebuttal may have backfired. The charges and denials of government and editorial meddling, along with the stark contrast between the article and the rebuttal, became the subject of a news story in *Science* (Roberts 1991). It was not long before court opinions were quoting the headline, "Fight Erupts over DNA Fingerprinting," the subtitle, "A Bitter Debate Is Raging," and the reporter's description of Lewontin and Hartl as "two of the leading lights of population genetics" who had "the support of numerous colleagues" (ibid., 1721). It was hard for courts looking for evidence of general acceptance of the basic product rule to find it in the face of such an account. As Lewontin (1992, 39) asked, "If professors from Harvard disagree with professors from Yale (as is the case), what is a judge to do?" Lewontin's answer was that the judge should exclude basic-product-rule calculations. Another answer, developed in later chapters, is that the judge should examine the scope of and the reasons for the disagreement to discern whether the scientific controversy actually represents a potent barrier to the admissibility of the evidence.

The National Research Council Speaks

In a great many cases from the time of *Castro* through that of *Yee,* experts bickered and battled over laboratory procedures and probability and statistics. Egged on by counsel and perhaps professional pride, they became increasingly polarized. Some scientists thought that there was a better way. The consensus report of the experts on the DNA testing in *Castro* called on the National Academy of Science to convene a committee to study the emerging technology (Chapter 4). In December 1989, after an initial setback,[14] the scientists who had subscribed to the report got their wish. The National Research Council (NRC)—the operating arm of the National Academy of Sciences, the National Academy of Engineering, and the Institute of Medicine—initiated a sweeping review of forensic DNA technology. Victor McKusick, a legendary physician-scientist at Johns Hopkins University, widely acknowledged as the "father of genetic medicine" (Downer 2002; see also Hendricks 2000), was tapped to chair a committee of geneticists and forensic scientists, as well as a sprinkling of bioethicists, an engineer, a law and sociology professor, and a federal judge (NRC Committee on DNA Technology in Forensic Science 1992, 173–176 [hereafter cited as NRC 1992]).

The report was expected within a year (Norman 1989, 1558), but this was not to be. The committee had to engage a diverse array of issues, from laboratory technique to law. Its nontechnical members needed to learn about the genetic science and technology. The geneticists needed to learn about forensic science and law. As the months passed, the committee became entangled in the statistical debate over the basic product rule. Only two of the committee's original fourteen members mentioned population genetics as an area of expertise in their short biographies—Eric Lander and Mary-Claire King. (King, who had worked with Alan Wilson in the 1970s to compare sequences in humans and chimpanzees, pioneered the search for the *BRCA1* cancer gene, and used genetic tests to reunite children with their grandparents after their parents had "disappeared" during Argentina's "dirty war," was then a professor of epidemiology and genetics at the University of California at Berkeley.) No committee members were statisticians. Lander came to the committee convinced that population structure was a major unresolved problem, but studies were appearing questioning his claim of "spectacular departures from Hardy-Weinberg equilibrium" (Lander, 1989a, 504), and a growing number of statistical studies appeared to support the use of the basic product rule (Chapter 6). Richard Lempert, the sociologist-lawyer on the committee, described the changing landscape before the committee: "[Y]ou would have a blockbuster article come out . . . and then three months later, you would have another article disputing that. . . . What was the good science? It couldn't be the latest science, because the latest science would have been peer reviewed in the process of publication, but not . . . necessarily by all the right people" (Derksen 2000, 822).

The committee, hoping to advance and improve the use of a powerful crime-fighting tool, desperately desired a unanimous report. Courts looking for general acceptance would not find it in a house divided against itself. On some matters, agreement was not extraordinarily difficult. The report recommended that laboratories follow rigorous protocols for quality control and assurance; that they adopt objective rules for declaring that VNTR bands matched; and that they reserve some of the crime-scene sample, whenever possible, for independent testing by the defense. The committee advised expert witnesses or prosecutors to avoid asserting that VNTR profiles are unique[15] and to present a false-positive error rate as determined by blind proficiency testing along with the random-match probability. At the structural level, the report proposed that laboratory accreditation and proficiency testing be made mandatory. To ease the daunting task of judging the validity of emerging technologies in the courtroom, the committee called on Congress to commission a new, blue-ribbon panel of scientists, ethicists, and lawyers—independent of the FBI—to evaluate further developments in DNA technology. This would cut back on the need for massive, *Castro*-like pretrial hearings. To streamline any such hearings, the report (NRC 1992, 23, 133) also stated that courts "should take judicial notice of [the] underpinnings of DNA typing . . . even in the context of a *Frye* hearing." (Judicial notice is a doctrine that allows a court to dispense with case-specific proof of a fact that is not subject to reasonable debate.) In addition, the report implored courts to fund expert witnesses for the defense and to allow full discovery of all data and laboratory records.

Even today, not all of these recommendations have been adopted—and these were the easiest issues for the committee to agree on. The goal of unanimity was almost unattainable with respect to the product rule. Eric Lander "took the cautious view," and Thomas Caskey, "the pragmatic" (Roberts 1992a, 300). Philip Reilly, a physician and lawyer on the committee, told a reporter for *Science* that initially "the Lander camp held sway, and early drafts of the statistics chapter were very conservative" (ibid., 301). An October 15, 1991, draft chapter that was introduced by a defendant in a Seattle murder case stated that the basic product rule was "scientifically invalid and should not be introduced into evidence" (Sherman 1992, 1). This draft adopted the position of Lewontin and Hartl (1991,

1749) that random-match probabilities should be obtained either by "counting" the number of occurrences of the entire DNA profile in a database or by a "ceiling principle" applied to studies (that did not yet exist) of VNTR allele frequencies in ethnic subpopulations (ibid.). Lewontin and Hartl's counting rule was to tally the number of occurrences (x) of the full DNA genotype in the population database; if this number was zero, as it almost always would be, then the laboratory would report this fact and estimate the random-match probability as $1/N$, where N is the size of the database. For example, if there were 225 samples in the database, and if none of them matched the defendant's profile, then the laboratory would report that the random-match probability was $1/225$. This is simple enough to do, but one could argue that a better estimate of the population frequency is o/N—the proportion of matching genotypes actually observed in the database. If one regards the database as, in effect, augmented by the observation of one profile of the defendant's genotype, then Lewontin and Hartl's numerator makes sense, but the denominator would be $N + 1$ instead of N.[16] Although Lander (1992, 204) apparently favored the $(x + 1)/N$ approach and the draft chapter may have used it (Sherman 1992, 46), the chapter that finally appeared in print substituted a more complicated upper 95% confidence limit on x/N. For example, if there were 225 samples in the database, and if none of them matched the defendant's profile, then the laboratory would report that the random-match probability was between 0 and 0.013.[17]

Both versions of "counting" grossly understate the force of the evidence. The Lewontin-Hartl $(x + 1)/N$ statistic is tantamount to playing 225 hands of poker, never observing a royal flush, and then concluding that the probability of a royal flush is less than $1/225$, or 0.4%. The estimate is true as far it goes, but smaller estimates also would be pretty safe. Coming in at 1.3%, the committee's 95% confidence limit on x/N is even larger in this instance.

Two committee members "were so disgruntled that they leaked an early draft of the statistics chapter to FBI scientist Bruce Budowle, prompting John Hicks, the director of the FBI's crime laboratory, to send angry letters to the National Academy. Having Lander coordinate that chapter is like having 'the fox guarding the hen house,' Budowle complained to *Science*" (Roberts 1992a, 301).[18] After a threatened minority opinion, "the committee hammered out a compromise between the warring camps in the dispute over the statistical method to interpret a match" (ibid., 300). This compromise had two parts—"the ceiling principle" presented in Lander (1991b, 902) and endorsed in Lewontin and Hartl (1991, 1749), and an "interim ceiling principle" reminiscent of an approach that had been used at least once in the days of serology testing (Chapter 2, 64).[19]

Lander (1991b, 902) had called the ceiling principle "empirical yet practical." The NRC report described it as "practical and sound" (NRC 1992, 82). Lander had written about the need to acquire "perhaps a dozen or so well-separated ethnic population samples." The NRC committee "strongly" recommended "[r]andom samples of 100 persons . . . drawn from each of 15–20 populations, each representing a group relatively homogeneous genetically" (ibid., 83). The "ceiling frequency" for an allele would be the largest frequency encountered in any of these samples. Moreover, to be particularly cautious, if this largest observed frequency was less than 5%, the ceiling frequency would be rounded up to that number. These allele ceiling frequencies then would be multiplied (along with the factors of two for heterozygous loci) to provide what was believed to be an upper bound for the genotype frequency. Because the allele frequencies would come from many different populations, the profile frequency would not be the best estimate of the frequency in any real population, structured or otherwise. It would be biased in favor of defendants in the statistical sense that its expected value would be larger than the true value. We can call this approach the product rule with ceiling frequencies (or, more simply, the ceiling product rule) to distinguish it from the product rule with broad population frequencies (the basic product rule).

Standing alone, the ceiling product rule was not acceptable to the full committee. The random samples of populations, such as English, German, Italian, Russian, Navaho, Puerto Rican, Chinese, Japanese, Vietnamese, and West Africans (NRC 1992, 84), did not exist.[20] McKusick estimated that collecting them would take about a year and cost $1 million (Roberts 1992a). To circumvent this delay, the report recommended a temporary expedient—using the highest frequency in any of the major population groups in the United States (Caucasians, African Americans, Hispanics, Asians, and Native Americans), or 10%, whichever was higher.[21] This combination of the interim ceiling product rule followed by the full ceiling product rule satisfied Lander: "It is sufficiently conservative, yet sufficiently usable. I don't think anyone would fight it" (ibid., 301). How wrong he was! The committee's "practical and sound" compromise soon ignited a firestorm of scientific critiques, legal briefs, and conflicting court opinions.

The 1992 report was misunderstood or misrepresented from the day it was published. Indeed, two days before the report was scheduled for release to the public, the *New York Times* ran a front-page story under the headline "U.S. Panel Seeking Restriction on Use of DNA in Courts: Labs' Standards Faulted, Judges Are Asked to Bar Genetic 'Fingerprinting' Until Basis in Science Is Stronger" (Kolata 1992c).[22] Yet the report itself said no such thing. Rather, the *Times* inferred that the committee must have been seeking a moratorium because it called for blind proficiency testing, accreditation, and other quality-control and assurance standards not yet in place and then wrote that "so long as the safeguards that we discuss in this report are followed, admissibility of DNA typing should be encouraged" (NRC 1992, 145). The argument involves what lawyers call a negative pregnant. The proposition is in the positive form, "If A, then B," and it is transformed into the negative form, "If not-A, then not-B." But this inference is not a logical necessity. "If it rains, I shall carry my umbrella" is logically consistent with "If it does not rain, I shall carry my umbrella." For instance, I might want the umbrella to screen me from the sun. Thus the argument is about language, which is to say, the intent of the speaker. Did the committee intend its call for improved standards and other changes to bar the admission of evidence in the period before those reforms were put into operation? The report was not always explicit on this point, but whenever the committee did address what to do in the interim, it came down in favor of admitting the evidence. This was the obvious raison d'être of the interim ceiling principle, and the committee explicitly contemplated another "interim" method for admitting test results from unaccredited laboratories, namely, that "these laboratories demonstrate that they are effectively in compliance with the requirements for accreditation" (NRC 1992, 107). Still, someone determined to read the report as implicitly demanding a moratorium could argue that the reason the committee was explicit in some instances and not others was that, in general, the committee saw its recommendations as prerequisites for admissibility.

In response to the ambiguity in the report, the NRC acted swiftly to squash the interpretation of the *Times*. The chair of the committee, Victor McKusick, held a press conference the same day and maintained that the "story 'seriously misrepresents our findings' and gives a 'misleading' impression" (Davidson 1992, B4). "We did not say . . . that courts should cease to admit this evidence" (Ezzell 1992). "We think that DNA can be used in court without interruption" (Kolata, 1992a, A19). This was not the first time that the reporter, Gina Kolata, had been accused of misrepresenting the views of scientists in order to suggest that DNA typing was not generally accepted. In January 1990, in another front-page article, she had claimed that in "numerous interviews," "[l]eading molecular biologists [said that VNTR-RFLP testing] is too unreliable to be used in court" (Kolata 1990, A1). A majority of the seven biologists interviewed sent the *Times* a letter denouncing the article as a "misrepresentation" of the value of DNA typing, but the newspaper did not print this response (Levy 1996, 50).

The *Times* could hardly ignore the NRC's news conference. The next day it included on the front page an article forthrightly subtitled "Times Account in Error." Kolata's confession of error was more grudging. In a second article, she suggested that her only error lay "in saying that the panel called *directly* for a moratorium on the use of DNA typing" (Kolata 1992a, A1, emphasis added). She reported that two committee members, Lander and Thomas Marr, had told her that although "the committee did not mean to call for a moratorium, . . . an effective moratorium was implied," and that "[s]everal law professors and defense lawyers" agreed that the report implicitly called for a moratorium (ibid., A19). However, it appears that the "legal experts" were simply reacting over the telephone to the reporter's description of the report and to snippets that she read to them. The report had yet to be issued, and most of them had had no chance to study it and offer a considered judgment.[23] Defense attorneys would continue to press this argument, but Lander and Marr signed on to a "Statement by the Committee" inserted into the bound version of the report that accused the *Times* of a "serious . . . misrepresentation" and specified that "[w]e recommend that the use of DNA for forensic purposes, including the resolution of both civil and criminal cases, be continued while improvements and changes suggested in this report are being made. There is no need for a general moratorium" (NRC 1992, x).[24]

Watching from the sidelines, Lewontin (1992, 39–40) defended "the original interpretation" from "one of [the *Times*'] most experienced and sophisticated science reporters." On "the critical issue of population comparisons," he found a sentence in the report that "should be copied in large letters and hung on the door of every public defender in the United States," for it used the legally loaded terms "reliable or valid."[25] Lewontin conceded that the committee might not actually have intended to bar the use of DNA evidence or of "the one-in-a-million claims that prosecutors have relied on to dazzle juries." Nonetheless, the "effective content" of the report was unmistakable: "Whether by ineptitude or design, the NRC Committee has produced a document rather more resistant to spin than some may have hoped."

But what Lewontin regarded as "spin" and "scientific politics" prevailed. The courts did not treat the 1992 report as recommending or necessitating a general moratorium. However, they did see it as confirming what already was apparent—that some highly respected population geneticists and statisticians were at loggerheads over how a random-match probability should be computed. The result was a fourth wave of cases in which some appellate courts accepted the defense argument that because the basic product rule was not generally accepted or scientifically valid, these estimates rendered DNA evidence inadmissible. Assessing these cases requires us to look more deeply at the scientific and statistical arguments over the product rule. The next chapter therefore examines the expanding scientific debate and the judicial opinions that tried to make sense of it.

CHAPTER SIX

The Initial Reaction to the 1992 NRC Report

To say that the NRC report failed to cool the heated debate over the impact of population structure on basic-product-rule probabilities would be an understatement. If anything, its endorsement of the ceiling product rule inflamed the controversy (Roeder 1994, 223). It is possible that with the imprimatur of the National Academy of Sciences, the ceiling approach ultimately would have become dominant in the courtroom, but in the short run it held little appeal for prosecutors and defense attorneys alike. Prosecutors presenting DNA matches could have lived with a few decimal points shaved off the random-match probabilities, but what about the cases already on appeal in which rapists and murderers had been convicted with the help of the basic product rule? Prosecutors wanted these convictions to stick. Arguing that the error in the trials was harmless or retrying old cases with more conservative statistics was not always an attractive option. Defendants had even less incentive to embrace ceiling computations. From their perspective, it would be far better for scientific disputes about DNA evidence to continue. As for the scientific community, rallying around what might strike lawyers as a reasonable solution was no substitute for showing how much of a kludge the committee had come up with. Also, something more than the normal one-upmanship was at work. Forensic scientists in DNA-testing and other laboratories see their work as integral to solving crimes and contributing to justice. Other scientists drawn into high-stakes litigation by the government may have seen their opinions as a necessary antidote to the extreme positions adopted by defense counsel. In this highly charged atmosphere, the committee's report was a lightning rod for criticism. Any perception that the committee was watering down extremely powerful evidence was certain to be expressed in academic and judicial settings.

The Initial Reception of the NRC
Report in the Courts

If there was any doubt left in early 1992 that prominent scientists were divided over the adequacy of basic-product-rule computations, the NRC report laid it to rest. Two opinions of the California Court of Appeals show the trend that worried prosecutors and the FBI. The 1991 case *People v. Axell* was another third-wave case like *Jakobetz* and *Yee*. In the face of powerful criticism from defense experts, the court still managed to find general acceptance. In February 1988, a cook coming to work in the morning at the Top Hat Burger restaurant in Ventura, California, saw someone with dark, shoulder-length hair leaving the restaurant. Inside, she found the body of a fellow worker in a pool of blood. Lynda Axell, a woman who worked the graveyard shift at the nearby Party Place, emerged as a suspect. A criminalist found that long hairs in the area and on the body were physically similar to Axell's and unlike those of other suspects.[1] Cellmark Diagnostics found that Axell's VNTR profile matched that in fifteen hair roots found at the scene of the murder. But would Cellmark's damning DNA analysis be admitted into evidence?

At a pretrial hearing on general acceptance, the state produced an impressive array of experts, including Richard Roberts and Kenneth Kidd, to confirm that Cellmark's procedures were generally accepted. Kidd was satisfied with the basic product rule used with Cellmark's Hispanic database (derived from samples provided by the Los Angeles Red Cross Blood Bank). The defense responded with a large number of experts whose names also were becoming familiar in courtroom circles. On the issues of population genetics and statistics, Diane Lavett (the geneticist who, under the name of Diane Juricek, had incurred the ire of the forensic serology community for her testimony and writing in the protein wars of Chapter 1), Seymour Geisser (introduced in Chapter 5), Laurence Mueller (a population geneticist at the University of California at Irvine), and Charles Taylor (a biology professor at UCLA) collectively questioned the size and representativeness of Cellmark's Hispanic database, the width of Cellmark's bins, and the independence assumptions. Geisser, for example, wanted a sample of 5,000 to 10,000 people and more studies to prove statistical independence of alleles. The state countered with rebuttal testimony from Michael Conneally (see Chapter 5) and Lisa Forman, a geneticist at Cellmark who then was studying the Cellmark database. The trial court concluded that the existing science established that even if Cellmark's figure of 1 in 6 billion for the population frequency was not definitive, it was in the ballpark. It admitted Cellmark's DNA findings. A jury found Axell guilty, and she appealed. The California Court of Appeal agreed with the trial court's evaluation of the scientific dispute and affirmed the conviction.[2] When the California Supreme Court denied review, it looked like smooth sailing for prosecutors in California.

Then came the Lewontin-Hartl article and the NRC report.[3] In *People v. Barney* and *People v. Howard*—two consolidated cases of kidnapping, robbery, attempted rape, and murder—another division of California's intermediate court reevaluated *Axell*. The "heart of these appeals" was the "fundamental disagreement among population geneticists concerning the determination of the statistical significance of a match of DNA patterns" (10 Cal. Rptr. 2d at 740). The court's opinion described these newer materials and quoted liberally from the reports of science journalists about them. Citing language in the NRC report about a "substantial controversy," the court departed from *Axell*. It held that given the "palpable" lack of general acceptance, the estimates obtained with the basic product rule should not have been admitted at trial (id. at 743).

Judge Ming Chin, the author of the *Barney* opinion (and today a justice of the California Supreme Court), was correct in seeing the NRC report as recognizing a schism in the scientific community. Some judges, however, read much more into the report. In *State v. Anderson*, the New Mexico Court of Appeals wrote in 1993: "The report discusses the debate over the need for subpopulation databases, and concludes that they indeed are necessary. This report is indicative of the absence of general acceptance. There is not just one author trying to make a point, but rather a group of people that has reached a consensus in rejecting one aspect of the current methods of forensic use of DNA evidence" (853 P.2d at 146). The claim that the NRC committee "reached a consensus in rejecting" the basic product rule flies in the face of the report itself and news accounts of the committee and its deliberations. The *Barney* court was far closer to the truth when it acknowledged that "[t]he report does not, however, choose sides in the debate"; it merely "assume[s] for the sake of discussion that population substructure may exist" (10 Cal. Rptr. 2d at 741). Unable to agree that population structure leads to dramatic errors in estimating VNTR genotype frequencies, the committee "decided to assume that population substructure might exist" and to propound the two ceiling methods as a solution to this possible problem (NRC 1992, 80, 94).

Despite the committee's agnosticism on the scientific need for the ceiling methods, other courts cited its endorsement of ceiling frequencies as sufficiently cautious as grounds for excluding basic-product-rule calculations. *Commonwealth v. Lanigan* is one of the earliest of these cases. There, the Supreme Judicial Court of Massachusetts noted that the prosecution's failure to follow the "considered, conservative" prescription of the NRC committee "underscored the wisdom of . . . excluding the test evidence" (596 N.E.2d at 316).

The Arizona Supreme Court reached the same result in *State v. Bible*. A nine-year-old girl bicycling to a ranch in Flagstaff disappeared. Thee weeks later, her battered body was found hidden in the woods. The day she disappeared, police saw Richard Lynn Bible driving a stolen car that, other than being repainted, matched the description of a car driving away from the vicinity in which the girl last was seen. A high-speed car chase ensued. Bible fled from the car. Police tracked him with dogs. They found him hiding under a ledge and covered with twigs, leaves, and branches. Numerous items in the stolen car—vodka bottles, cigars, hot chocolate, rubber bands, and even metal from the steering column—matched items later found strewn about the girl's decomposing corpse. Fibers of hair on the defendant's clothing and wallet and in the car matched the girl's hair. Hair fibers near the little girl's corpse matched the defendant's hair. Blood was spattered across the defendant's boots, pants, and shirt. Bloodstains on the shirt contained the same enzyme as the girl's blood—an enzyme found in less than 3% of the population. The pièce de resistance in this surfeit of circumstantial evidence came when Cellmark Diagnostics reported that DNA in the blood on Bible's shirt matched the girl's DNA and that this particular DNA was extremely rare in the Caucasian population, occurring with a relative frequency between 1 in 60 million and 1 in 14 billion.

Before trial, the defense objected to expert testimony about this estimate. The court held a pretrial hearing and concluded that Cellmark's methods of DNA identification were generally accepted in the scientific community. It admitted the laboratory findings and calculations, as well as opposing testimony from Laurence Mueller, who "explained in detail to the jury the lack of Hardy-Weinberg equilibrium, [the existence of] linkage disequilibrium, the possible unreliability of the database, and the consequent misleading end result of the product rule" (858 P.2d at 1191).

The Arizona Supreme Court held that the introduction of the DNA match was error. Although the opposing articles in *Science* at the end of 1991 largely recapitulated the debate between the population geneticists in *Yee,* the Arizona court asserted that "the debate that erupted in *Science* . . . changes the scientific landscape considerably" (id. at 1188). It noted the shift in California from *Axell* to *Barney* and quoted the NRC report's acknowledgment that the product rule "has provoked considerable debate among population geneticists" (id.). In essence, the court concluded that the principles and methods employed by Cellmark to declare a DNA match were scientifically accepted, but that the basic product rule used with the allele frequencies in Cellmark's database was not.[4]

Despite the perceived error in admitting DNA evidence, the state supreme court did not reverse the conviction. In light of the other overwhelming circumstantial evidence incriminating Bible and the testimony of Mueller advising the jury that Cellmark's probability calculation was wrong, the court deemed the error harmless and allowed the conviction and sentence to stand.[5]

Although the *Barney* and *Bible* courts nixed the basic product rule, they did not consider whether the interim ceiling version of the product rule could produce admissible numbers.[6] Indeed, the Arizona Supreme Court's opinion left open other possibilities as well. The court explicitly reserved judgment on the question whether "the prosecution could introduce evidence of a match, accompanied by evidence that a match means that it is possible or probable that the two samples came from the same individual," noting that "[w]e need not and do not decide the propriety of such trial strategy because it is not before us" (858 P.2d at 1190). Once the court had indicated why the expert testimony was erroneously admitted, any definition of testimony that would have been admissible could only have constituted dicta—nonbinding judicial commentary unnecessary to the decision in the case. Nevertheless, this "careful and cautious approach" (id. at 1193) left the lower courts adrift. It remained unclear whether an expert could testify to the bare fact of a match or could go further to offer an opinion about the significance of the match based on the ceiling product rule.

One reason that the computation of random-match probabilities proved to be such an issue in the courtroom was the common rejection of the possibility noted in *Bible* that providing these numbers was not truly essential to the admissibility of DNA evidence. The courts in *Axell* and *Barney,* for example, treated the presentation of a random-match probability as an integral and apparently indispensable part of the expert testimony. The Washington Supreme Court was even clearer about the need for numbers. In *State v. Cauthron,* Cellmark's expert, Robin Cotton, avoided any numbers by testifying that defendant's DNA matched that in semen samples taken from rape victims and that "the DNA could not have come from anyone else on earth" (846 P.2d 516). On appeal, citing opinions in Alabama and Massachusetts, the Washington Supreme Court reversed the conviction on the ground that no "probability statistics" and "background probability information" accompanied Cotton's expert opinion (id.). According to these courts, without a generally accepted method of quantifying the probability of a coincidental match, an expert could not testify to the existence of the match. The basic product rule thus became the Achilles' heel of DNA evidence.

This demand for quantification was a new development for forensic science and medicine. Fingerprint examiners, tool-mark analysts, and medical examiners reporting on autopsies almost never couch their findings in terms of numerical probabilities. They might speak of "unusual," "rare," or even "unique" events, but, by and large, they lack systematic data and mathematical models with which to generate probabilities. The NRC report, however, supported a numerical imperative by stating that "[t]o say that two patterns match, without providing any scientifically valid estimate (or, at least, an upper bound) of the frequency with which such matches might occur by chance, is meaningless" (NRC 1992, 74). The *Cauthron* opinion quoted this demand for numbers as vindicating "the conclusions reached in the courts" (846 P.2d at 516).

The notion that science requires experts to give numbers in every DNA case, however, is misguided. The issue is not one of scientific practice but of legal policy. As a legal matter, a completely unexplained statement of a "match" should be inadmissible because it is too cryptic to be weighed fairly by the jury. The *Yee* court made this point nicely: "Without the probability assessment, the jury does not know what to make of the fact that the patterns match: the jury does not know whether the [matching] patterns are as common as pictures with two eyes, or as unique as the Mona Lisa" (134 F.R.D. at 181). At the other extreme, an overblown characterization of a match at a handful of loci as unique should be excluded as unfairly prejudicial. Conclusions like these follow from the policies behind the pertinent rules of evidence. They are not scientific judgments. Yet the NRC report prompted some courts to think that the admissibility of nonnumerical characterizations must be resolved by reference to scientific norms. In *Commonwealth v. Daggett*, a plurality of the justices of Massachusetts's highest court disparaged expert testimony that used phrases like "highly likely" (622 N.E.2d at 277) but did not include numbers. The plurality noted that the state "cited no authorities and presented no testimony . . . that the use of such terms is generally accepted by the scientific community in evaluating the significance of a match" (id. at 275 n.4). "The point is," these judges insisted, "not that this court should require a numerical frequency, but that the scientific community clearly does" (id.).

This view is mistaken. The general-acceptance standard addresses the validity and reliability of the methodology that produces evidence of identity. The fact of a match is scientifically valid evidence of identity as long as it can be shown from theory and data that the genotype has been ascertained in a valid manner and is not ubiquitous in the relevant population. But how to present to a jury valid scientific evidence of a match is a legal rather than a scientific issue falling far outside the domain of the general-acceptance test and the fields of statistics and population genetics. It would not be "meaningless" to inform the jury that two samples match and that this match makes it more probable, in an amount that is not precisely known, that the DNA in the samples comes from the same person (cf. National Research Council Committee on Scientific Assessment of Bullet Lead Elemental Composition Comparison 2004, 107). Nor, when all estimates of the frequency are in the millionths or billionths, would it be meaningless to inform the jury that there is a match that is known to be extremely rare in the general population. Courts may reach differing results on the legal propriety of qualitative as opposed to quantitative assessments, but they only fool themselves when they act as if scientific opinion automatically dictates the correct answer.

In the wake of the NRC report, a few courts did opine on the admissibility of the interim ceiling product rule as an alternative to the basic product rule. The Supreme Court of Washington was so "encouraged" by the NRC report that it commented in *State v. Cauthron* that "[a]lthough we lack the scientific expertise to either assess or explain the methodology [of the ceiling principle], its adoption by the Committee indicates sufficient acceptance within the scientific community" (846 P.2d at 517). On remand, the trial court in *Cauthron* decided that the issue of general acceptance of the interim ceiling principle was settled. Consequently, it ordered a hearing on whether Cellmark adhered to this procedure, and it ruled that "the statistical evidence in a DNA case is no longer subject to a *Frye* hearing if it is founded on the NRC approved Interim Ceiling Principle" (Kaye 1995b, 102). Likewise, applying a scientific-validity standard similar to that in *Daubert v. Merrell Dow Pharmaceuticals,* the Supreme Court of Wyoming in *Springfield v. State* gave its approval to interim-ceiling calculations in the rape and robbery prosecution of a man "of three-fourths Native American ancestry from the Crow tribe and one-fourth from the black race" (860 P.2d at 437).[7] In short, in the fourth wave of cases battering the legal shoreline, the 1992 report seemed to be undermining the admissibility of the standard estimates of population frequencies and encouraging "interim-ceiling" ones (id.).

Hitting the Ceiling

At the same time at which the NRC report was entering the legal arena, it was proving to be a hot potato in the scientific world. A few geneticists gave the ceiling compromise positive marks.[8] Francisco J. Ayala was one. A University Professor of Biology and Philosophy at the University of California at Irvine, Ayala is a past president of the American Association for the Advancement of Science and a recipient of the National Medal of Science. His research focuses on population and evolutionary genetics, including the origin of species, genetic diversity of populations, the origin of malaria, the population structure of parasitic protozoa, and the molecular clock of evolution. In the late 1980s, as chairman of the NRC's Board of Biology, he worked to persuade federal agencies to fund the NRC Committee on DNA Technology in Forensic Science (Ayala 1993, 55–56). When the committee's report was completed, he described the "ceiling principle" as "an ingenious solution" to the "problem of ascertaining genetic frequencies in the relevant subpopulations" (ibid., 58; Ayala 1992, 275).

Other population geneticists were less appreciative. Lewontin initially described the interim product rule as "just totally irrational," with the 10% minimum allele frequency having been pulled "out of the air" (Aldhous 1993, 755). Likewise, Lewontin provided a letter to a public defender in California stating that the interim ceiling rule "has no rational basis and has been chosen by entirely arbitrary means." The letter suggested that because of "the lack of the necessary statistical information," the rule was not "a valid estimate of matching probabilities" (Lewontin 1993b). Indeed, if, as Lewontin, Hartl, Lander, and others had argued, the "conservative" features of the FBI's match binning were not sufficient to overcome the unknown effect of population structure, how could one know that the interim ceiling product rule would be adequate? How much conservatism is enough? In 1991, Lander had insisted that "[o]ne cannot compensate for a bias without knowing how large it is" (Lander 1991a, 821). A year later, he was advocating the interim ceiling product rule as a mechanism for compensating for the alleged and still unknown "bias" in the basic product rule.[9] Likewise, Lewontin and Hartl seemed persuaded that the interim rule would work as advertised. In a letter to *Science* continuing to press for studies of VNTR allele frequencies across the globe, they wrote that although "the lower bound of 10% used for allele frequencies is arbitrary," "[e]veryone agrees that it is conservative. . . . Whether or not it is excessively conservative is a matter that can [only] be resolved empirically by ethnic group studies" (Hartl and Lewontin 1993, 474).

However, not quite everyone agreed that even the ceiling rule always underestimated genotype frequencies. Joel Cohen, the mathematical biologist and MacArthur fellow at Rockefeller University who had exposed some of the statistical flaws in Jeffreys's early articles (Chapter 3), dashed off a letter to the *American Journal of Human Genetics* showing that "the ceiling principle is not always conservative in assigning genotype frequencies for forensic DNA testing" (Cohen 1992). Most population geneticists, however, regarded as remote the conditions that would produce this counterintuitive result (Crow and Denniston 1993; NRC 1996, 158).

Thus the key question that divided the defenders and critics of the ceiling rule was, as Hartl and Lewontin (1993, 474, emphasis added) framed it, "[w]hether or not it is *excessively* conservative." Of course, this was not a strictly empirical question. Empirical studies could only show how strongly and how often the rule overestimated genotype frequencies in certain groups. Whether such overestimation was too high a price to pay to ensure that the state never presented an unduly impressive genotype frequency was a question of policy and judgment. Lander, for instance, was content to sacrifice accuracy to security. The population geneticists who usually testified for the prosecution were not as risk-averse. Those already on record as thinking that the independence assumptions with large bins and the $2p$ rule for homozygotes were amply conservative were not about to concede that there was any reason to be *more* conservative. Other voices from across the globe joined them in a chorus of criticism of the NRC committee's compromise.

One leading statistical geneticist, Newton Morton of the University of Southampton in England, emphasized the arbitrariness of the ceiling method and its inability to supply a true upper bound on the probability. Morton, Collins, and Balasz (1993, 1896) observed that

> the culprit might be assumed to be a Lapp for one allele and a Hottentot for another. However extensively our species might be sampled, an ethnic group could always be found with an apparently higher gene frequency, through drift, mistyping, or sampling error. If the suspect has a wooden leg and evidence were presented that the culprit had a wooden leg, it would be immaterial if there is a Cambodian village where wooden legs are common. Any juror can understand that the same principle applies to DNA, and so a court in Kansas or Cornwall will be reluctant to assign guilt on the basis of gene frequencies in Cambodia or simply invented [as prescribed by the interim ceiling product rule].

The entire approach was, in their view, "absurdly conservative." A trio of statisticians and geneticists then at Yale—including Neil Risch, who later would be called "*the* statistical geneticist of our time" (Gitschier 2005)—made similar points. Furthermore, they criticized the NRC committee's proposed plan to draw samples of size 100 from many ethnic groups as pointless and certain to "exaggerate maximum allele frequencies" because of "tremendous" sampling error (Devlin, Risch, and Roeder 1993b, 749). Elizabeth Thompson, the chair of the University of Washington's Department of Statistics, submitted a caustic affidavit in a murder case branding the ceiling principle "data-driven, interest-ridden, pseudo-statistical, ad hoc methodology, to which no statistician or scientist should be a party" (W. C. Thompson 1993, 79 n.272).[10]

Although Lander continued to describe the ceiling principle as a reasonable accommodation (Lander 1993a), some committee members distanced themselves from it. Testifying for the prosecution in a California case, *People v. Marlow,* Mary-Claire King advised the court that although "[s]he 'was actually the person responsible for writing' the section in the NRC report dealing with the variation of the product rule known as the 'ceiling principle,'" the procedure was "a really unbelievably conservative approach, very, very pro-defense approach" that yields a "vastly too conservative estimate." (41 Cal. Rptr. 2d at 22), Richard Lempert (1993a, 43) wrote that his enthusiasm for the method had waned by the time the committee signed off on the report because "a number of studies had appeared which suggested that the kinds of convenience samples used to establish DNA frequencies for forensic purposes were considerably more robust to the threat of population substructure than one might have supposed."[11]

What, then, was the situation when the committee endorsed the ceiling procedures to circumvent the debate over population structure? Was the ceiling principle an overreaction to the concerns that various experts expressed about the extent of population subdivision and its impact on basic-product-rule probabilities? The remainder of this chapter argues that it was, and that the scientific debate and the legal response to it suffered from a failure to focus on the population to which a random-match probability was supposed to apply.

Evaluating the Population-Genetics Debate

The three main proponents of the theory that population structure rendered the basic product rule invalid were Lander, Lewontin, and Hartl. To be sure, other experts, such as Laurence Mueller, expostulated similarly in courtrooms, but these individuals published little of their own to support the population-genetics argument. Lander, Lewontin, and Hartl maintained that there were strong reasons to doubt the accuracy of the simple model of random mating in the major racial or ethnic categories (the "population groups" or "populations," as I shall call them) of African Americans, Caucasians, Hispanic Americans, and Asian Americans. At the same time, they did not question the usefulness of the model for estimating genotype frequencies within more homogeneous "subpopulations" such as Italian Americans or Swedish Americans. The data and arguments of these "population structuralists" were varied, and I shall examine them in turn. I shall suggest that these scientists retreated from an initial position that theory and evidence supported the claim that the major populations were so highly structured that estimates of VNTR genotype frequencies in these populations were unreliable to the defensive position that the claim had not been directly disproved. In addition, I shall explain how the scientific and legal debate suffered from the failure to distinguish between two questions: How does population structure affect the estimate of the genotype frequency in a population, and how different is the genotype frequency across the subpopulations in each broad population group? The population structuralists tended to emphasize the latter question even in cases where only the former was relevant to determining the probative force of a DNA match in a particular case.[12]

Excess Homozygosity and Other Tests
for Independence

Lander deserves credit for first raising the possibility that correlations in VNTR allele frequencies could arise because of population structure. In 1989, for example, he wrote that "there is no reason to expect that the loci are in Hardy-Weinberg equilibrium in such heterogeneous groups as Caucasians, Blacks, and Hispanics. Indeed, some early studies are finding statistically significant deviations from Hardy-Weinberg equilibrium, indicating the presence of genetically distinct subgroups" (Lander 1989b, 149). In *People v. Castro,* Lander reported "spectacular deviations from Hardy-Weinberg equilibrium . . . indicating, perhaps not surprisingly, the presence of genetically distinct subgroups within the Hispanic sample" (Lander 1989a, 504).

Hardy-Weinberg equilibrium, it will be recalled, simply means that the alleles at a single locus are statistically independent. As explained in Chapter 5, this implies that if the proportion of all the alleles in the sex cells in a population that are of type i is p_i, then the proportion of homozygotes—people with the allele i on both chromosomes—is p_i^2. Looking over

Lifecodes' small database of Hispanics, Lander discovered proportions greater than the expected numbers p_i^2. The population-structure explanation of this discrepancy goes something like this. Suppose that two subpopulations of equal size—say, Puerto Rican and Cuban—constitute the Hispanic population reflected in the database. Suppose further that the proportion of an allele A is 1/5 in Puerto Ricans and 1/10 in Cubans. Assume that there is no interaction between the groups—Puerto Ricans have sexual relations only with Puerto Ricans, and Cubans have congress only with Cubans. Finally, assume that within each subpopulation, mating is random—it has nothing to do with whether a partner has the allele in question or a different allele. Then in the Puerto Rican subpopulation, the proportion of homozygotes AA will be about $(1/5)^2$. The expected proportion of Cuban homozygotes AA is $(1/10)^2$. Weighting these homozygote proportions by the fraction of the Hispanic population that is Puerto Rican and Cuban gives the homozygote proportion in the composite Hispanic population. This proportion is approximately $(1/2)(1/5)^2 + (1/2)(1/10)^2 = 250/10,000$. This figure is a weighted average of the product of the allele frequencies in each subpopulation, where the weights are the fraction of the population represented by each subpopulation.

If mating were random in the entire Hispanic population—if mixed Puerto Rican–Cuban couples were as likely to exist as any other combination—then the proportion of AA homozygotes would be the square of the overall proportion for the allele A. The frequency of A in the total population is just $(1/2)(1/5) + (1/2)(1/10)$, and the square of this quantity is $225/10,000$. This figure is the product of the weighted average of the allele frequencies in each subpopulation. In this example, there is an excess of 25 homozygotes per 10,000 Hispanics (0.25%) due to the population structure. Mathematically, it comes about because one averages and then multiplies when one should be multiplying and then averaging. The two operations are not commutative.

The example illustrates how population structure can lead to excess homozygosity. Inbreeding or population structure generally decreases the proportion of heterozygotes and increases the proportion of homozygotes because relatives have more alleles in common than unrelated individuals. Therefore, the offspring of relatives have a greater chance of receiving the same alleles from both parents, that is, of being homozygotes, and a smaller chance of inheriting dissimilar alleles, that is, of being heterozygotes. But inbreeding and population structure are not the only causes of an apparent excess of homozygotes, and many population geneticists believed that the "spectacular deviations" in Lifecodes' small Hispanic database were too spectacular to be explained in this fashion. Lander was reporting excesses of 9% and 13% at two loci. A population would have to be incredibly stratified to produce such huge excesses (Roeder 1994, 230). Ranajit Chakraborty (1991, 896) suggested that the excesses reported by Lander would have required "more than 20–30 subpopulations each of which should not have exchanged any gene among them for more than 40,000 years since their divergence from a common ancestry"—a state of affairs "clearly contrary to the origin and demography of the U.S. populations, where even among the orthodox religious populations the gene migration has been rather substantial (at least of the order of 10%/generation during the last century. . . .)."[13]

Consequently, it was not long before alternative explanations were proposed. Chakraborty pointed to the well-known phenomenon of small, more mobile fragments piling up at the end of the gel. When this occurs, only the larger allele from a heterozygote will show up on the autoradiograph, giving a false impression of homozygosity. Therefore, he maintained that "covert nondetectability of extreme-size alleles is a much *simpler* explanation of heterozygote deficiency of binned allele data" (ibid., 895; emphasis in original).[14] The Yale group suggested that Lander was counting many "close heterozygotes" as homozygotes because gel electrophoresis could not distinguish accurately between two VNTR fragments of nearly equal length (Devlin, Risch, and Roeder 1990, 1417). When they corrected for this "coalescence," they found "no obvious violation" of Hardy-Weinberg equilibrium in other data sets (ibid., 1418; see also Devlin and Risch 1993). A rather technical exchange followed in which Lander (1991b, 901) and Green and Lander (1991) defended Lander's computations in *Castro,* while Devlin, Risch, and Roeder (1991) insisted that their procedures were much more appropriate and realistic. In addition, Devlin and Risch (1992b, 552) explored the effect of undetected single bands in the FBI databases and concluded that "when a reasonable estimate of the frequency of null alleles can be made, there is no evidence that these VNTR loci violate independence assumptions."

Judges could not have found the details of this debate easy to follow. Indeed, some courts misconstrued even relatively simple discussions of Hardy-Weinberg equilibrium. The opinion of the Arizona Supreme Court in *State v. Bible,* introduced earlier in this chapter, reveals how the failure of a court to understand a scientist's testimony can lead it astray. The *Bible* court did not squarely hold that basic-product-rule estimates were never admissible. Instead, it concluded that the basic-product-rule estimate in this particular case was not produced by a generally accepted method because "the database relied on is not in Hardy-Weinberg equilibrium" (858 P.2d at 1189). As with Lifecodes' database in *Castro,* an analysis of

Cellmark's database in *Bible* showed an excess of homozygotes. Even "[s]tate expert Dr. [Lisa] Forman [of Cellmark] conceded that the 1988 Caucasian database used by Cellmark in this case was not in Hardy-Weinberg equilibrium" (858 P.2d at 1187). Or so the court thought.

At first, Forman's testimony seems to support the court's characterization:

> Q. Does the data base that was used in 1988 meet Hardy-Weinberg equilibrium?
> A. In the way that it was analyzed, with the gel system that was used, you could not say that it meets Hardy-Weinberg expectations.

However, Forman immediately explained that the apparent excess in the database was not a manifestation of a departure from Hardy-Weinberg equilibrium in the Caucasian population:

> Q. Have you done further testing on the database?
> A. Certainly.
> Q. And what have the results of your further investigation been?
> A. The results of the further investigation show that the reason we appeared to have too many individuals with one band, too many homozygotes than would be predicted from this model, was because we were only looking at a small window. We were only looking at the top of the gel down to where one of these molecular weight markers had migrated 20 centimeters. In that case we were cutting off all bands that were smaller than the molecular weight marker band.
> When we left those bands on, when we ran the gel for a shorter period of time, . . . we found the bands that were missing were there. We had just cut them off in the data base, but they were still there. . . .
> Q. What has your decision been in terms of how to evaluate that statistically?
> A. What we do is the most conservative approach and that is to say everybody has two bands, even if you can only see one, so instead of doing a p^2 . . . we treat everyone as though they are a heterozygote.
> Q. So in essence, does that drive your numbers down, make them—
> A. It makes the appearance of that DNA band seem more common. It is more generous to the person who has that band. . . .
> State v. Bible, No. 14105 (Ariz. Super. Ct. Mar. 28, 1990) (testimony of Lisa Forman, 13 Transcript 66–69).

Had the Arizona court understood that population structure pertains to the pattern of mating and the distribution of alleles across subpopulations, perhaps it would not have misconstrued this testimony as a concession of an unresolved problem with estimating the multilocus frequency from Cellmark's database. As one reads beyond the first question and answer, the testimony clearly makes two points that were supported by the scientific literature. First, it asserts that the database is consistent with a population in Hardy-Weinberg equilibrium; appearances to the contrary are artifacts due to small alleles running off the gel. This is the explanation for the excessive number of single bands in Lifecodes' Hispanic database that Chakraborty (1991) proffered and Lander (1991b) rejected. If one small allele in a pair of DNA fragments is drawn to the end of the gel, only the larger allele will show up on the autoradiograph. These single bands from heterozygotes will be counted as homozygotes. The database derived from these autoradiographs will show an excess of homozygotes, but this excess will not reflect any excess in the population. Thus the deviations in the database from Hardy-Weinberg expectations do not support the claim that the Caucasian population is so structured as to undermine estimates that assume Hardy-Weinberg or linkage equilibrium. As Forman had explained, when the electrophoresis lasted a shorter time so that the smaller fragments did not migrate all the way to the end of the gel, "we found the bands that were missing were there. We had just cut them off in the data base, but they were still there." She denied rather than "conceded" that there was an excessive number of homozygotes in the population.

Second, her testimony asserts that the use of this database instead of one in which all single-locus profiles were true homozygotes was generous to the defendant. Had a longer gel been used and the second allele detected, the single-locus frequency would have been estimated as $2p_1p_2$, where p_2 is the frequency of the second allele that was not visible on the autoradiograph from the shorter gel. This frequency p_2 has to be less than 1 (100%). Given the number of VNTR alleles at each locus, it probably is closer to 0 than 1, but it is excluded from the multiplication because the laboratory uses the larger single-locus-frequency estimate of $2p_1$ (see Chapter 5). Consequently, in using the database from the shorter gel together with the $2p$ rule for apparent homozygotes, Cellmark was erring on the side of the defendant. Employed in this fashion, the database is not objectionable. Misunderstanding the testimony and the scientific debate about excess homozygosity, however, the state supreme court characterized Cellmark's database as "flawed" for "*Frye* purposes."

In addition to examining independence at each locus (Hardy-Weinberg equilibrium), researchers also tested for violations of the assumption of independence across loci (linkage equilibrium). Bruce Weir, the William Neal Reynolds Professor of Statistics and Genetics at North Carolina State

University (now chairman of the Department of Biostatistics at the University of Washington), pored over databases supplied by the FBI and Lifecodes, searching for correlations among pairs of alleles both at and across the VNTR loci. He found few departures from the values expected if all the alleles were statistically independent. He concluded that any differences would not have a large impact on multilocus genotype frequencies (Weir 1992a, 1992b). Risch and Devlin (1992a, 1992b) formed all possible pairs of the profiles in the databases to determine how often different people would match at a single locus, at two loci, and more. For the millions of pairs, they found a "close correspondence between the number of matches observed and the number expected under independence" (Risch and Devlin 1992b, 719; see also Herrin 1993).

In response to such studies, the proponents of the population-structure theory shifted their rhetorical ground. Lander dismissed the negative results of statistical tests for excess homozygosity as "virtually meaningless because the tests have such low statistical power to detect substructure even if is present" Lander (1991a, 821). Lewontin and Hartl (1991, 1747) also maintained that "statistical tests for [Hardy-Weinberg equilibrium] are virtually useless as indicators of population substructure because, even for large genetic differences between subgroups, the resulting deviations from HWE are generally so small as to be undetectable by statistical tests." In fact, they decried, "[s]tatistical tests for HWE are so lacking in power that they are probably the worst way to look for genetic differentiation between subgroups in a population." They insisted that "[t]he proper approach is the straightforward one of sampling the individual subgroups and examining the differences in the genotype frequencies among them." Attributing the position to "most observers," Lander (1991a, 821) said the same thing: "[T]he right way to settle the question of population homogeneity is to sample ethnically distinct populations and to observe the actual degree of genetic differentiation."[15]

Thus by 1993, the claims that excess homozygosity indicated extreme population structure were largely abandoned in scientific journals. This leg of the argument for population stratification had more or less collapsed. The consensus seemed to be that the flurry of studies of homozygosity in particular databases was not such a fruitful exercise in ascertaining whether genotype frequencies were reasonably accurate.[16] Population structuralists like Lewontin and Hartl basically were saying, "Show us it isn't so!" In part, however, they took this skeptical position because their expectation was that there would be substantial heterogeneity due to subpopulations in the roughly defined populations of "Caucasians," "Hispanics," and the like. But how likely was this?

Variations in Frequencies within
and across Populations

At first blush, the population structuralists' view might seem implausible. One might expect that the difference in allele frequencies would be greater across population groups like Caucasian Americans and African Americans than among subgroups within these categories. If gene flow is more extensive among the subgroups (French and German, for example) than across major groups, then subpopulations should have frequencies that are closer to one another. Lewontin and Hartl (1991) believed that this would not be the case. They alluded to "a considerable body of evidence indicating genetic substructure within what are called the 'Caucasian,' 'black,' and 'Hispanic' populations" (ibid., 1746). Their evidence consisted of differences in the frequencies of blood groups and serum proteins and enzymes in people living in different countries. Specifically, the article relied on a Lewontin's 1972 study of seventeen polymorphisms that showed that although "most human genetic diversity reflects differences among individuals within local populations (85.4% when averaged over all 17 genes)," the "remaining 14.6% of human genetic diversity broke down as follows: 8.3% among local ethnic and linguistic subpopulations within races, and 6.3% between the major races" (ibid., 1747). Although the blood-group and protein loci in this classic study, being subject to the force of natural selection, would be expected to be different from essentially neutral VNTR sites, Lewontin and Hartl insisted that the seventeen genes were "typical," and they emphasized that "for these genes, there is, on average, one-third more genetic variation among Irish, Spanish, Italians, Slavs, Swedes, and other subpopulations, than there is, on the average, between Europeans, Asians, Africans, Amerinds, and Oceanians." They also cited sociological and historical studies to show that "there has been very little time for mixing of genes from diverse populations of origin" and that "[t]he notion of an American 'melting pot' is [certainly not] true . . . for marriage, which is strongly affected by religion and ethnicity" (ibid., 1748). Likewise, Lander (1991a, 821) attacked the intuition that within-population variance was considerably greater than across-population variance, writing that the choice of forensic laboratories to compile databases of "broad groups—Caucasians, blacks, and Hispanics . . . rather than on narrower ethnic subgroups . . . seemed to be based on a notion that there was more genetic variability among races than within each race, although population geneticists showed decades ago that the reverse was true." The only proof that Lander cited for the "reverse" view was Lewontin's 1972 survey.

Some population geneticists were aghast at these statements. Conneally was so disturbed that he advised the federal district court in *Jakobetz* and a California superior court that Lewontin and Hartl were "fruit fly geneticists" who "are not necessarily as qualified to talk about human populations" (People v. Marlow, 21, 31 n.52). Morton, Collins, and Balasz (1993, 1896) pointed to numerical errors and suggested that "[n]o human geneticist would" accept some of Lewontin and Hartl's claims. Chakraborty and Kidd (1991, 1737) accused Lewontin and Hartl of overlooking "the extensive mobility and mixing of groups in the general U.S. population that occurred during and following World War II." They observed that "[b]oth the proportion of marriages of mixed ethnicity (20%) and that of marriages outside the 10-mile radius (67.6%) per generation are high" and noted that "population genetic theory shows that even a small amount of gene migration across ethnic and religious boundaries will quickly homogenize populations." They readily conceded that "[s]ubstructuring exists everywhere, no matter how the population as a mating unit is defined. However, as the component subpopulations are genetically similar, because of gene exchange almost since the beginning of their evolution, the net effect of substructuring is trivial" (ibid., 1738). In their eyes, "the vast literature on blood groups and protein markers has demonstrated that the existing subdivisions within populations do not produce any appreciable departures of single or multiloci genotypic frequencies from the ones predicted with the Hardy-Weinberg and multiplication rules" (ibid., 1737). Morton, Collins, and Balasz (1993, 1892) also denigrated the statements of Lewontin, Hartl, and Lander about population structure as devoid of "theoretical basis or empirical support." They accused Lewontin and Hartl of picking atypical, incorrect, and outdated allele frequencies that exaggerated the extent of subpopulation heterogeneity. Morton and his colleagues concluded that "less than two percent of the diversity selected by [Lewontin and Hartl] is due to the national kinship to which they attribute it, little of which persists in regional forensic samples" (ibid., 1896; see also Morton 1995a). Still smaller estimates of subpopulation heterogeneity were juxtaposed to Lewontin and Hartl's 8.3%. Devlin, Risch, and Roeder (1993b) maintained that "the estimate of diversity based on variance of allele frequencies among subpopulations is usually quite small—approximately 0.1%." Chakraborty (1993) reiterated that "the extent of regional difference within a racial group is far less than that between races" and that "analysis of hypervariable DNA loci [demonstrates that] the mean kinship within race is 0.4%," which is "less by an order of magnitude . . . than for blood groups and isoenzymes." Referring to work done in the 1980s, Devlin and Roeder (1997, 727) wrote that "a large body of consistent results" using "standard methods for partitioning variance" contradicted "Lewontin's surprising result" and established that "ethnic diversity dominates subpopulation diversity." "Even a study . . . using Lewontin's methods, as well as similar populations and loci, but with substantially larger samples, found that 5.6% of the diversity was attributable to differences between subpopulations and 10.4% to differences among ethnic groups." Even this figure was not "completely applicable" to the United States because small, isolated populations (such as tribes of Pygmies) were given as much weight as "the large, open subpopulations that have populated the U.S." (ibid., 728). Devlin and Roeder suggested that a "more applicable" result was a 1982 study that attributed 10% of the genetic diversity to variation across the major population group and "only 0.5% or less . . . to differences in English, Germans, and Italians" (ibid; see also Devlin, Risch, and Roeder 1993a).

Oddly, Lewontin and Hartl (1991) did not acknowledge the existence of any studies that might have undermined their dramatic claim of "one-third more genetic variation among Irish, Spanish, Italians, Slavs, Swedes, and other subpopulations, than there is, on the average, between Europeans, Asians, Africans, Amerinds, and Oceanians." When confronted with some of these studies, Hartl and Lewontin (1993, 474) modified and averaged them along with Lewtontin's classic 1972 findings to reverse their original statement—now there was nearly one-third more variation across the racial or geographic divides than there was among the subpopulations.[17] They merrily presented this result as if it fully confirmed their original claim, writing that "[w]e reiterate the conclusion that there is approximately as much genetic variation among ethnic groups within major races as there is among the races." Lander (1991b, 901) apparently also abandoned the original claim of one-third less variation across populations. His fallback position was that it did not matter that Lewontin presented "outlier examples to make his point" because "the key issue is to define the possible range of outliers."

In short, in early 1992, the affirmative arguments in the scientific literature for population structure were running on fumes. Lander, Lewontin, and Hartl correctly raised the possibility that extreme population structure might exist with regard to the VNTR loci. But their evidence for its existence—the worldwide studies of genetic diversity at the loci for blood and serum proteins and the claims of excess homozygosity in some population databases—did not show the severe stratification that would vitiate basic-product-rule calculations. In effect, the population structuralists and the defenders of independence had adopted the same line—"Show us it isn't so!" The former expected to see lots of population structure and hence violations of Hardy-Weinberg and linkage equilibria for VNTRs; the latter were confident that the simple model of random mating provided satisfactory estimates of genotype frequencies in the crude populations defined by the forensic scientists.

Thus, contrary to the interpretations or intimations of several observers (Lander 1991b, 901; W. C. Thompson 1993, 70–71 n.215; Redmayne 1997, 1050–1051), it was not a rigorous commitment to empiricism versus a naive infatuation with theory that divided the two camps. Instead, it was a set of prior beliefs about the nature of populations and the type of empirical evidence that would be needed to overcome those beliefs. As we have seen, the population structuralists seemed to accept—apparently, on theoretical grounds—the equilibrium assumptions within subpopulations. Yet for broader populations, they were willing to accept only direct evidence in the form of studies of variations in the VNTR allele frequencies in diverse locales around the globe. Lander (1989b, 149) scoffed that "there seems to be a serious misperception that Hardy-Weinberg equilibrium is a law of physics that must apply to a population." He perceived "no reason to expect that the loci are in Hardy-Weinberg equilibrium in such heterogeneous groups as Caucasians, Blacks, and Hispanics." In contrast, the defenders of independence perceived ample empirical evidence of adequate homogeneity in the form of studies of other polymorphisms[18] and in the failure to detect departures from the population genotype frequencies predicted with the independence assumptions. Thus both sides claimed the empirical high ground. On the one hand, according to Hartl and his colleagues, "[t]he validity of the conventional product rule has been challenged because analysis of blood group and enzyme-coding loci have given evidence of substructuring within the human races" (Krane et al. 1992, 10583). On the other hand, Devlin and Risch (1992b, 552) inveighed that "claims of violation of HWE should be viewed with caution, as they run counter to both theoretically and empirically derived expectations." Morton (1992, 2556) ruthlessly criticized "opponents" of "the validity of DNA-based identification" (specifically, Cohen and Lander) for citing "general principles that would apply with equal force were the suspect a *Drosophila* or other organism whose population structure is little known, neglecting the large body of evidence in humans."

In addition, the defenders of independence were reporting new results that confounded the structuralists' expectations. To discover whether the true population genotype frequencies are much larger than those computed ignoring population structure, one can simulate highly structured populations, much as Chakraborty and Kidd (1991) did in response to Lewontin and Hartl's effort to contrast blood-group frequencies for

Poles and Italians (see discussion later in this chapter). If VNTR allele data on two subpopulations are available, the single-locus and multilocus genotype frequencies validly can be estimated under the independence assumptions in each subpopulation. (The possible structure in the general population is eliminated or reduced by focusing on the subpopulations.) Next, a database mixing these subgroups can be constructed, simulating a highly structured population. Using this simulated database, one can estimate allele frequencies and compute genotype frequencies as if the population were homogeneous. If the artificial population frequencies are close to the true frequencies in the simulated population, one must conclude that even the exaggerated substructuring does not produce much error.[19]

Although detailed data on fully homogeneous subpopulations were not available, suggestive analyses along these lines were performed by mixing various other groups: southeastern and southwestern Hispanic Americans, Caucasian Americans, Afro-Caribbeans, Asians, and Caucasians living in England. When artificially mixed populations of these groups were formed and the basic-product-rule estimates ignoring population structure were compared with the estimates of population genotype frequencies taking structure into account, the differences were minor (Evett and Gill 1991, 229–230; Devlin and Risch 1992a, 544–546; cf. Evett, Scranage, and Pinchin 1993). The robustness of the basic product rule was tested in an even more extreme way by using the wrong reference group to arrive at random-match probabilities or likelihood ratios—for example, by analyzing a match between two Caucasians as if it were a match between two Afro-Caribbeans. Again, the effects attributable to differences in the databases were small (Evett and Pinchin 1991, 271; Berry, Evett, and Pinchin 1992; Evett 1992; Monson and Budowle 1993, 1044–1049).

In sum, as of 1992, no population geneticist disputed the proposition that allele frequencies varied across both populations and subpopulations. By this time, it also was widely recognized that when a population is composed of such separately mating subpopulations, it is not strictly correct to use the basic product rule with allele frequencies produced by sampling the population as a whole. The population structuralists insisted that these facts vitiated any reliance on the basic product rule. They began by observing "spectacular deviations" from Hardy-Weinberg equilibrium and claiming that genetic variation was greater across subpopulations than it was across the major population groups. The defenders of independence responded by refuting these claims. The two camps clashed over the details of the rebuttal, but it soon became clear that the defenders of independence had largely demolished the structuralists' affirmative case.

Unbowed, the structuralists argued that it was up to the defenders to prove that Hardy-Weinberg and linkage equilibria existed despite some degree of population structure. The defenders offered statistical demonstrations that such departures from equilibria were of no practical importance. Ian Evett, a statistician with the British Forensic Science Service, wrote that on the basis of his simulations designed to evaluate the impact using the wrong database to estimate the frequency of three-locus genotype frequencies in the London area, DNA evidence would produce no more than one "wrongful conviction on average about once every 250 years" (Evett 1992, 145). Such simulations did not appease the population structuralists, who demanded direct inspection of allele frequencies in a large number of subpopulations.

To the uninitiated, the demand for this one form of proof might seem dogmatic. One of the nation's most distinguished population geneticists, James Crow, advised a reporter for *Science* that Lewontin and Hartl were "perfectionists" (Roberts 1991, 1723). Another geneticist who requested anonymity described the structuralists' stance as "a religious argument. We are talking about matters of faith that are not likely to be settled by reason" (ibid., 1721).

The Problem of the "Reference Population"

In tossing their 1991 bombshell from *Yee* to *Science* magazine, Lewontin and Hartl (1991, 1746) posed a fundamental question: "What is the reference population from which the randomly chosen individual is to be taken?" More specifically, we should ask: Is the pertinent frequency to be found from a sample drawn from the general population? From a particular geographic area? From people resembling or related to the defendant? These questions are neither new nor special to DNA evidence. One simple principle supplies the answers. The relevant population consists of all people who might have been the source of the evidence sample. This point is acknowledged by virtually all commentators—lawyers (Lempert 1991, 1993c; Kaye 1993, 2004, 2007a), scientists (e.g., Budowle, Monson, and Wooley 1992; Lewontin 1993c, 1997), and statisticians (e.g., Buckleton, Walsh, and Evett 1991; Evett and Weir 1991, 1992; Berry 1992; Weir 1992c, 1993; Weir and Evett 1993; Roeder 1994).

The only reason to introduce a random-match probability is to quantify the probative value of the match with respect to the hypothesis that someone other than the defendant is the source of the DNA that incriminates the defendant. Who this "someone" might be depends on the facts of the case. If a white tourist dies from a wound inflicted by a blowgun deep in the Amazon jungle, the appropriate reference population will be the Indian denizens of that region (and any other whites in the locale). If a tourist from Senegal is mugged in a dark alley in a part of New York City that is frequented by whites, blacks, Hispanics, and Asians, there is no reason to assume that the assailant comes from any particular one of these populations, let alone any special subpopulation within them. In general, the appropriate reference population is not a function of the victim's ancestry. In some cases, the class of plausible perpetrators will be limited to individuals who share the defendant's ancestry, but the suspect population can be quite different. It can be a narrow subpopulation, as in the Amazon basin case; a broad "racial" group, as when a woman reports that a white man raped her at a rest stop on an interstate highway; or a whole series of major population groups, as in the New York City tourist case. We can designate the first class of cases "subpopulation cases" and denominate the remaining classes "population cases."

This distinction is crucial to ascertaining whether and when the population-structure objection is well taken. The early scientific literature blurred it. As Lempert (1993c, 3 n.7) observed, "[I]mportant articles at that time [from 1989 to 1991, by Lewontin and Hartl, Cohen, and Lander] implicitly assumed that the proper population database consisted of a sample of defendant's coethnics." This defendant's-ancestry fallacy was evident in several court opinions. In *People v. Mohit*, for example, a court in Westchester County, New York, was concerned that the race and ethnicity of the dentist accused of raping his anesthetized patient was not represented in the FBI's database. It noted:

> The issue of inbreeding is of particular importance in this case. The defendant, Dr. Mohit, was born in the Iranian town of Shushtar. His ancestors over at least the past five generations were of Persian descent, all from the same town or a town close by. They are all of the Shiite Muslim religion. Dr. Mohit claimed that for religious reasons, and as a matter of tradition, inbreeding was very common in his family. He indicated that his maternal grandmother was the daughter of his father's great-grandparents. Marriage among first cousins was common in his town. (579 N.Y.S.2d at 997)

The issue, however, is not the frequency of matching DNA patterns for inbred families of Shiite Muslims from Shushtar, Iran, but their frequency in the vicinity of Westchester County, New York, or, more precisely, their frequency among people other than Dr. Mohit who might have left their semen on the patient. Unless this group consists largely of Dr. Mohit's relatives and coethnics, there is no need to estimate the frequency among people of his racial and ethnic background. The frequency among broadly defined racial and ethnic groups—among populations, not subpopulations—is the apposite figure.

Given the difference between the pertinent probability in population cases and subpopulation cases, we need to consider the effect of population structure in each situation. In a statistical sleight of hand, Lewontin and Hartl (1991) substituted one situation for the other. Initially, they used allele frequencies for six blood groups in Poles and Italians to find that "Poles are about 250 times more likely to match for these [six-locus] genotypes than are Italians." This would be a concern if an Italian database of allele frequencies were used to compute a Polish genotype frequency. The relevant question in a general-population case, however, is not whether one subpopulation's allele frequencies are mistaken for another. None of the subpopulation frequencies are even known. Yet Lewontin and Hartl (1991, 1749) apparently transformed the wrong-subpopulation figure of 250 into a measure of the error in using population-wide allele frequencies to estimate population-wide genotype frequencies. They concluded that basic-product-rule estimates for the populations of Caucasians, blacks, and Hispanics "may be in error, possibly by two or more orders of magnitude" and that "[t]he error may be in favor of the defendant, or against the defendant."

Population cases. The population-structure objection is not without force, but it does not have the full impact that Lewontin and Hartl attributed to it. In fact, Chakraborty and Kidd (1991, 1736–1737) used a population database composed of equal numbers of Italians and Poles to estimate the average genotype frequency in this composite population. They showed that the population-frequency estimate is off by a factor of 3, not 250. The difference in the estimate that took population structure into account and the estimate that ignored it was the difference between "12 versus 37 in 1 million." The error, such as it is, in using a random-mating model in a structured population flows from a simple algebraic fact (also demonstrated in the numerical example in Chapter 5)—the product of a series of weighted averages is not necessarily equal to the average of a series of weighted products. The former is the calculation that uses population-wide frequencies with the basic product rule; the latter is the computation that, ideally, should be done but cannot be performed because the subpopulation allele frequencies are not known. Nevertheless, all this mathematics means that (1) on average, the error due to population structure inures to the defendant's benefit, and (2) the differences between the computed and the true population-wide genotype frequencies will rarely be large (Chakraborty et al. 1992; Kaye 1993; Weeks, Young, and Li 1995). As a result, the population-structure objection is weak when the relevant population in which to estimate the frequency is a collection of subpopulations having different VNTR frequencies.[20]

In these general-population cases, even courts that required general scientific acceptance could have admitted basic-product-rule estimates, the arguments of population structuralists notwithstanding. No population geneticist or statistician denied that the relevant population in which to estimate a population genotype frequency consists of all the people who might have committed the crime. Neither did any population geneticist or statistician dispute the mathematical truism that the basic product rule, when used with allele frequencies for a structured population in which the independence assumptions hold within each subpopulation, tends to overstate the frequencies of VNTR genotypes in that structured population (for an example, see Kaye 1993, 141). Not surprisingly, the judicial opinions stopped with the recognition of a prominent debate about population genetics and rarely dissected the details of the scientific exchanges. The existence of the controversy was hard to miss, but the subtleties of the debate were less visible. In *Commonwealth v. Lanigan*, for instance, the Massachusetts Supreme Judicial Court correctly stated that the population-structure argument concerned "the possibility that using allele frequencies of larger population groups might produce an inaccurate frequency estimate *for members of substructure groups*" (596 N.E.2d at 315, emphasis added). Noting "the lively and still very current dispute" (id. at 316) over that issue, the court upheld the exclusion of the evidence of a match between Lanigan's DNA and a semen stain on the clothing of a child whom Lanigan was accused of raping. However, there was no apparent reason to estimate the genotype frequency in any particular subpopulation. Consequently, the debate over what to do for "members of substructure groups" did not justify exclusion of the evidence. The same is true of other general-population cases such as *Barney* and *Bible*.

Subpopulation cases. The concern with subpopulations has more traction when it actually is important to report the frequency in a subpopulation like Poles or Italians. When the group of people who might have left the crime sample is a narrow and possibly insular subpopulation, it is more likely that the genotype frequencies in this subpopulation will be seriously misjudged by using allele frequency from the overall population. The situation is yet more troublesome when the subpopulation is not even part of the standard reference databases, as in *People v. Atoigue*, a sexual-assault case that arose in Guam. Also, when the defendant and the rest of the suspect population are from the same subpopulation, population-frequency estimates will tend to overstate the import of the match (Krane et al. 1992). In all these subpopulation cases, the difference of opinion between the population structuralists and the defenders of independence could not be resolved by identifying the specific points of agreement and disagreement. Although it might be rare to find a subpopulation in which one allele frequency after another is on the high side, so that the multilocus genotype frequency derived from the overarching population database would be too low (and thereby detrimental to the defendant), some method of indicating the possible error in the standard calculations was called for.[21]

Strangely, the NRC report also failed to focus on the validity of the basic product rule in population cases. The closest that the report came to acknowledging this approach was the following passage:

> Some legal commentators have pointed out that frequencies should properly be based on the population of possible perpetrators, rather than on the population to which a particular suspect belongs. Although that argument is formally correct, practicalities often preclude use of that approach. (NRC 1992, 85)

The report nowhere identified its practical objections to defining the reference population in the only logically acceptable manner imaginable. It also did not defend the implicit assumption that it is normally appropriate to seek some estimate of the frequency in the suspect's subpopulation as opposed to the broader population. The committee's desire to arrive at a single number that might apply to all populations and subpopulations was understandable, but there were better alternatives. The committee chose not to analyze them and thus created a gap in an otherwise comprehensive report.

The long-awaited report, in its treatment of the population-genetics issues, came closer to endorsing the views of the population structuralists than those of the defenders of independence. Certainly, it chose not to repudiate the argument that population structure was a major problem in population and subpopulation cases alike, rendering invalid all basic-product-rule estimates of random-match probabilities. Unable to achieve consensus on the adequacy of the basic product rule, and without critically examining the arguments against independence, the committee endorsed variations of the "ceiling method" previously proposed by Lander, Lewontin, and Hartl. Although the report was influential in the courts, the scientific community greeted it with shrill calls to establish a second committee with more representation from population geneticists and statisticians. Having published a critique of the NRC report in a law review, and blissfully ignorant of the growing angst over the report within the National Academy of Sciences, I soon found myself drawn into the effort to try again.

CHAPTER SEVEN

Ending the Debate over Population Genetics

THE NRC COMMITTEE'S treatment of population structure was not the only aspect of the report that met with hostility. In addition to the dismay among the defenders of independence over the peculiarities and deficiencies in the formulation of the committee's ceiling approaches (e.g., Weir 1993), there were complaints about the report's miscalculations and misinterpretations of elementary statistical concepts (Weir 1992c, 1993), its timidity in not considering alternatives to match binning (Berry 1994; Weir 1992c, 1993; Devlin, Risch, and Roeder 1993b; Evett, Scranage, and Pinchin 1993; Kaye 1993; Roeder 1994), and its apparent ignorance of the well-established "affinal model" (discussed later) for accounting for inbreeding or population structure (Morton et al. 1993, 1895).[1]

Of course, a major reason for the rough edges and lacunae was the desire of a polarized committee to get a report out the door (Chapter 5). The committee regarded its task as the production of a practical document for the legal system, not an academic tome. In addition, although Eric Lander and Mary-Claire King were the designated specialists in population genetics or statistics, there was a perception that the committee lacked the intellectual resources to tackle the issues of statistics and population genetics. Committee member Richard Lempert (1994, 258) referred to "practical concerns" and the "committee's makeup" as justifications for skirting alternatives to the conventional presentation of a match-binning probability. From outside the committee, Neil Risch asserted that "[t]he major problem is that there was no population geneticist on that panel." Even the committee's chairman, Victor McKusick, diplomatically conceded that "[w]e probably could have done with more representation in that respect" (Aldous 1993, 755).

Given the very vocal discontent with these parts of the report among some population geneticists and the risk that the courts would misconstrue a controversy among scientists over the policy question of how conservative forensic laboratories should be as if it were a debate over the scientific claim that the ceiling methods generated estimates that were on the high side (Chapter 6), it was not long before a second committee was empaneled to "update" the 1992 report. By the time this committee completed its work, however, the legal landscape had shifted as the result of continued research into the population-genetics issues and strategic publications in key journals. This chapter examines these developments, starting with the politics of convening a second committee.

Lobbying the Academy

In addition to publishing critical reviews of the 1992 NRC report, several vocal scientists importuned the National Academy of Sciences (NAS) to try again. From England, Newton Morton (1993) wrote the president of the academy in near-apocalytic terms. Warning that "[a]ll reviews will be critical of the NRC report on the grounds that the committee was unqualified in statistics and population genetics," Morton's letter maintained that "decisions of courts . . . to rule expert DNA evidence inadmissible on the basis of the NRC report . . . elevate what formerly was a scientific controversy into a confrontation between science and authority that has rarely occurred since the Inquisition attacked Galileo." In his view, the report had become "a bulwark of authority against science," making "[n]eglect of this issue pusillanimous and immoral."

In a memorandum to the NRC's Commission on Life Sciences, which had directed the study, Eric Lander (1993b) dismissed such pleas as sour grapes on the part of only "three authors who argued to the committee that there was no need for a conservative approach" and whose views were rejected after "careful" deliberation.[2] This assessment seems more measured than Morton's, but the implicit claim that the critics did not support "conservative" approaches is false. The population structuralists and the defenders of independence both supported conservative measures, albeit very different ones. The defenders of independence contended that the FBI's approach already was conservative in using large bins for estimating allele frequencies, in using $2p$ instead of p^2 for single bands, and in a few other particulars. The additional overestimation achieved by the ceiling versions of the product rule was, they maintained, excessive (see Chapter 6). Morton (1992, 2560) opposed not conservatism as such, but a method that he regarded as "absurdly conservative." He proposed an approach that would "allow a court to be conservative without being absurd." Moreover, the "current 'controversy'" over the adequacy of the report involved more than three critics of the ceiling approaches to multiplication. Morton (1993) had contended that somewhere between "all" and "most competent scientists reject the NRC report," which he derided as "a laughingstock" (Morton 1995b, 142).

As for "the current round of articles," though, Lander saw "nothing to change my mind" because "[t]hey present additional data of precisely the sort already submitted to the committee." He opposed reopening "the substance of the committee's decision" because "[i]t would send a clear message to the courts that there is *not* general acceptance." He thought that "[t]he NRC report has had the salutary effect of easing the admissibility of DNA evidence for the FBI, since courts have deferred to the NRC report as proof of the legal requirement of 'general acceptability.'" However, he did favor "a 1–2 page reaffirmation and clarification" to block the efforts of "[s]ome lawyers . . . to twist a few sentences in the NRC report to find loopholes in the committee's recommended 'ceiling principle' approach" (Lander 1993b).

While Lander was suggesting that the FBI regarded "the NRC committee's recommendation to be . . . completely workable" and that the FBI was content with it as "a *settled* rule," the FBI was presenting a different face to the NAS. It complained that since the report had appeared, "eleven of thirty appellate decisions . . . question[ed] the admissibility of DNA typing evidence," sometimes citing the NRC report (National Research Council Commission on Life Sciences 1993; emphasis in original). Yet the FBI's legal analysis was no more convincing than Lander's. Courts have been known to cite almost anything to support a conclusion, and the fact that some fraction of the opinions expressing misgivings about DNA evidence referred to the NRC report, often along with the Lewontin-Hartl article or other evidence of a scientific dispute, tells us very little about the impact of the report per se. Nevertheless, the Department of Justice was worried enough to offer to help fund a new study to address concerns expressed "in some quarters of the scientific and legal communities about . . . the recommendations for estimating gene frequencies in light of the genetic structure of human populations" (ibid.).

On May 14, 1993, a little over a year after the release of the 1992 report, a triumvirate of NRC commissions (National Research Council

Commission on Life Sciences, Commission on Behavioral and Social Sciences and Education, and Commission on Physical Sciences, Mathematics, and Applications 1993) distributed to interested individuals a memorandum citing the opinions expressed by "[s]ome scientists" that "statistical errors exist, that the authoring committee lacked sufficient expertise in population genetics and statistics, and that the committee relied heavily on an unreliable scientific paper." The memorandum described a plan to "form a committee of approximately 12 members consisting of experts in molecular biology, statistics, and population genetics, forensic science, and law" to revisit "the issue of population substructure" and other sources of "uncertainty" in DNA evidence. The idea was to hold three meetings and produce a report of 30 to 50 pages in six months. In fact, the actual report took closer to two years to complete and filled 254 pages.

Upon receiving the memorandum, one of the lawyers on the first committee, Richard Lempert, wrote the academy's president to convey his distress—not at the prospect of a second panel, but at the manner of its creation and implementation. Lempert (1993b) acknowledged a "glaringly apparent" and "inexplicable lack of specialists in population genetics and statistics on the [first] panel,"[3] but he warned that "the decision to 'fast track' the next report . . . will appear consistent with the charge of a 'bought and paid for' report, and, as importantly, means that there is little time for Committee deliberation." To counter the FBI's claim of a crisis in the courts, Lempert suggested that criminals were not being and would not be set free because "in most rape prosecutions DNA evidence is simply icing on the cake." As he saw it, the appellate courts were affirming convictions by treating the perceived errors in using random-match probabilities computed with the basic product rule as harmless, and in those cases that were being remanded for further proceedings involving "modified ceiling principle probabilities . . . it is quite likely that these will be sufficient to sustain conviction." To avoid "problems of appearance" and to give a crop of new committee members time to be "brought up to speed on the technology before Committee decision making begins," he urged a more deliberate committee procedure and funding from agencies other than the FBI.

In June, the NRC assembled a "planning group" chaired by James Crow to "identify the key major issues that must be addressed to resolve quickly the current debate over statistical evaluation of DNA evidence" (NRC Commission on Life Sciences 1993). The purpose of the group, according to the project's director, was to "design the study," which already had been approved (Fischer 1993).

Richard Lewontin was not pleased. Bruce Alberts, a cell biologist celebrated for his work on the protein complexes that allow chromosomes to be replicated, had just been appointed the new president of the NAS. Lewontin wrote Alberts about the "extremely unfortunate situation . . . bequeathed to you by Frank Press as one of his last Presidential acts" (Lewontin 1993a). According to Lewontin, the FBI had repudiated the 1992 report, "a moderate and judicious summary of the situation," and had "taken unprecedented steps in contravention of the National Academy's own procedures to get a new DNA report more to its liking." He reported to Alberts that the NRC's Commission on Life Sciences had supported Lander's idea of "a clarifying statement from the original panel," but that the FBI director "personally approached Frank Press," who then went "over the heads of the Life Sciences Commission" to obtain approval from the NAS's Executive Committee. As if this were not enough, Lewontin complained, Press called together an "ad hoc committee" that included Bruce Weir, "a principal consultant for the FBI in this matter," to confirm this course of action.[4] Lewontin's prognosis for the viability of a new report was bleak:

> If Weir and his allies indeed dominate the new Report Committee, then it will be greeted by cynicism (well deserved, in my opinion) as nothing but a deliberately rigged attempt to engineer a politically desired end. If, on the other hand, others (like me) who have consistently opposed the FBI's policy are put on the Committee to "balance" it, there will ensue divisiveness and struggle. . . . *There is no way that the NAS/NRC can come out of this affair undamaged if it persists in Frank Press's ill-advised course.* (Ibid., emphasis in original)

At the "ad hoc" meeting of the planning group, Eric Lander had questioned the need to reexamine any major issues, and Lewontin endorsed what he said was "a perfectly reasonable course of action" proposed by Lander, to form an "Oversight Committee" to assess, every six months or so, the progress of the forensic science community in creating "independent quality control mechanisms and . . . adequate representative databases."

The NAS did not heed Lander's or Lewontin's advice. The planning committee recommended a study to update the 1992 report,[5] and the NAS announced the plan for the fast-track study, contingent on additional funding.[6] Defense lawyer Peter Neufeld was furious. "It's offensive that law enforcement can be dictating to the independent scientific community how they should examine problems," he told the *National Law Journal* (Sherman 1993, 3). As far as he could tell, there was "no confusion in the courts. The FBI has lost every appeal in the nation where the issue of statistics has been preserved."[7] A year later, Neufeld would be arguing that the empaneling of a second committee was not so much a sign of the FBI's control over the scientific establishment, but rather "the most powerful evidence that could exist of the continuing nature of the scientific controversy over DNA statistics" (*People v. Simpson,* Defendant's Memorandum of Points and Authorities in Support of Defendant's Motion to Exclude DNA Evidence, Oct. 5, 1994, *28 [hereinafter cited as Defendant's Memorandum]). Although Lander saw the FBI as willing to live with the interim ceiling product rule, Neufeld charged that the FBI "can't live with the report, so they want it changed" (Sherman 1993, 3).

The NRC Speaks Again

At first, the NRC seemed to move fairly quickly. By the end of August, James Crow was asked to chair the new committee. As with the first committee, however, funding came slowly. As a result, a year slipped by before the members were named. Along with Crow, the resident population geneticists were Masatoshi Nei of Pennsylvania State University and Thomas Nagylaki of the University of Chicago. Their forte was theoretical population genetics. Two statistics professors from Stanford and Chicago added to the mathematical acumen of the group. Two medical doctors and geneticists (one had served on the previous committee) and one forensic scientist (also a member of the first committee) completed the complement of full-time scientists. Two law professors specializing in evidence law (including the author) and a professor of law and psychology supplied additional perspectives.

It took only five months or so for the group to collect public input, to review the literature, to agree on the main recommendations, and to begin the writing, but it was not until May 1996 that editing, revising, external reviewing, and printing were completed (NRC 1996, vi). Perhaps the six-month goal would have worked had the committee confined itself to the single issue of population structure, but its mandate extended at least to all the "statistical and population genetics issues in the use of DNA evidence."[8] The result was a comprehensive report that endorsed some key recommendations of the 1992 report for quality control and assurance, for sample splitting (to allow defendants the opportunity to obtain independent tests whenever feasible), and for the creation of an independent, standing government committee to evaluate new developments in forensic DNA technology.

Also, the report provided a more complete analysis of many legal and statistical issues. For example, the statistical and legal literature contained cogent arguments for (and against) presenting alternatives to random-match probabilities for DNA and many forms of trace evidence (see Kaye,

Bernstein, and Mnookin 2004, 475–505). Yet the first committee had made no effort "to assess the relative merits of Bayesian and frequentist approaches." Instead, it disposed of the issue with the observation that "outside the field of paternity testing, no forensic laboratory in this country has, to our knowledge, used Bayesian methods to interpret the implications of DNA matches in criminal cases" (NRC 1992, 85).[9] The second report went further by describing the actual case law and arguments on this topic, but the committee ultimately took no position on which mode of presentation the courts should allow. Furthermore, the report rejected the Bayesian perspective on the impact of the emerging practice of searching a database of DNA genotypes of convicted offenders to identify a suspect in a rape or other case in favor of the first committee's view that trawling a large database greatly degrades the probative value of a DNA match (NRC 1996, 161, 163–165; for discussion, see Kaye 2009a). Finally, the report recommended additional research into enhancing juror assessments of probabilistic evidence—a recommendation that did not please everyone at the FBI (ibid., 203–204).[10]

On the issue of population structure, which was the raison d'être for the new committee, the 1996 report saw no reason to compromise. It sidestepped the question whether the interim ceiling product rule ever was necessary by analyzing newer data acquired by the FBI and other researchers. This report concluded that "sufficient data have been gathered that neither ceiling principle is needed," especially in view of "alternative procedures, all of which are conservative to some degree" (ibid., 158). "With such procedures available," the committee reiterated, "the interim ceiling principle is not needed and can be abandoned" (ibid., 159).

These alternative procedures emphasized the distinction between general-population and subpopulation cases. When "the race of the person who left the evidence-sample DNA is known," the committee suggested that "the database for the person's race should be used, [but] if the race is not known, calculations for all the racial groups to which possible suspects belong should be made" using the basic product rule (ibid., 122).[11] To this extent, the report supported the standard practice of the forensic laboratories to give figures for random-match probabilities within the major racial groups such as whites, blacks, Hispanics, and East Asians, at least in general-population cases. In subpopulation cases—for example, when "the suspect and other possible sources of the sample [are] all members of an isolated village," the committee again endorsed the basic product rule, but with "the allele frequencies for the specific subgroup" (ibid.). Because such detailed data rarely were available, however, this recommendation was none too practical. As a second-best substitute in these situations, the committee proposed adjusting the allele frequencies from the population containing the subgroup according to "population-structure equations" and then multiplying them according to the product rule (ibid.). This was the "affinal model" urged by Morton and others.[12] It tends to boost the random-match probability upward, but its effect is less drastic than those of the ceiling methods, and it is less prone to manipulation.[13] It is useful in cases of crimes on Indian reservations, but it can be used in all cases, and courts promptly determined that it could produce admissible random-match probabilities.[14] Ironically, the magistrate judge in *Yee* had declared that "the potential effect of substructure cannot be known or even estimated in any given case, and there is no factor that rationally and indisputably will compensate for its presence" (134 F.R.D. at 211). If the second committee was correct, it had accomplished this feat (at least if the word "clearly" could be substituted for "indisputably").

The report was not front-page news. *Science* reported that the "fierce debate" of four years ago had been replaced by "surprisingly little public comment" (Marshall 1996, 803). Lewontin shrugged off "the old population genetics question" as "not at the center" of the debate, although he could not resist charging that "the conclusions of the report were 'bought' by the Department of Justice" (ibid.).[15] The *New York Times,* having erred in interpreting the 1992 report as demanding a moratorium on DNA testing, went to the other extreme with a headline proclaiming "Expert Panel Calls Evidence from DNA Fully Reliable" (Leary 1996). This time, no one contradicted the *Times.*

Several factors might explain the lack of widespread outcry over the 1996 report. For one, its rejection of the ceiling approach was widely expected. For another, it arrived too late to be a dramatic event. By 1996, the defenders of independence had won their war in most courtrooms, and the admissibility of DNA evidence was all but assured. This is not to say that the report was unimportant. In courtrooms across the country, it reinforced the disposition of courts to consign the ceiling principle to the dustbin of history, and it short-circuited several defense arguments about whether and how the estimated frequencies or probabilities should be presented in court. The remainder of this chapter describes some of the litigation over DNA probabilities that transpired while the forensic science and legal communities awaited the promised report and some cases that emerged soon afterward. Later chapters then turn to more recent cases and scientific developments.

Simpson Meets His Match

The murder trial of former football hero O. J. Simpson was an international sensation and a turning point in the history of DNA evidence. Whether it was the "trial of the century," as it was called at the time, or just the "media trial of the century" (Lee and Labriola 2001, 223), the case riveted the attention of the nation with its mix of extreme violence, spousal abuse, racial tensions, Hollywood life, and DNA science. The televised trial became a national obsession, dragging on for months and eclipsing daytime soap operas and the Gulf War in the ratings (CBS News 2004).

The basic facts are these. Shortly before midnight on June 12, 1994, in Brentwood, a wealthy Los Angeles suburb, a dog, its paws splashed with blood, led neighbors to a woman's body. Her throat was slashed so deeply that the spinal column was visible. To the right, in a pool of blood, lay a man's body. He had been stabbed over and over. The woman was Nicole Brown Simpson. A beautiful blond, she was O. J.'s second ex-wife. The man was Ronald Goldman, a waiter at a fashionable restaurant. He had come to the courtyard of Nicole's condominium to return a pair of eyeglasses that Nicole's mother had left at the restaurant that evening.

The murderer, the district attorney's office soon decided, was O. J. Simpson. As they saw the case, the police had gone to Nicole's residence years earlier, when O. J. had kicked in the door, screamed obscenities, and beaten Nicole's Mercedes-Benz with a baseball bat (Buckleton 2005, 110). "On the day of the murders, [he] had attended a dance recital for their daughter. . . . They were barely civil to one another" (Sheindlin 1996, 144). The prosecution's theory was that O. J., "in a state of rage over Nicole's rejection," drove his car the short distance from his Rockingham estate to her condominium (ibid., 145). Carrying a knife and gloves, he lay in wait in the bushes. "He may have planned to kill Nicole, or he may have only intended to slash her tires—something he'd done before" (ibid.). In any event, when Goldman arrived, his "fury spilled over at seeing Nicole with another man." In the course of the killings, the knife slipped and cut O. J. on his hand. After the butchery, he returned to his car, "leaving a trail of bloody footprints" (as well as five drops of his own blood), drove back to his home (depositing traces of blood in his car), "climbed a fence in the back" (dropping a glove that was the mate to one found near the bodies), "and ran through the dark yard into his house, leaving drops of blood" in the pathway, foyer, and bathroom (ibid.). In his haste, he left a pair of socks with a bloodstain in his bedroom. DNA typing linked bloodstains leading from the bodies to O. J.; those in the car to Nicole, Goldman, and O. J.; those on the glove in O. J.'s estate to Goldman; those in the estate grounds and rooms to O. J.; and those on O. J.'s socks to O. J. and Nicole (Buckleton 2005, 115–116).

This was the prosecution's theory of the case. O. J. and his lawyers (whom journalists denominated "the Dream Team") had innocent explanations for almost all this and other evidence involving fibers, shoe patterns, and other matters. The police had handled the DNA evidence poorly, raising a possibility that they accidentally transferred some of the incriminating bloodstains from one place to another. Other stains were said to have been planted to create or nail down the case against O. J. The true killer's DNA never appeared in any of the five drops leading away from the bodies, the defense suggested, because the molecules had spontaneously degraded to the point of being undetectable (Levy 1996, 170; W. C. Thompson 1996, 832). The defense also attacked the quality-control procedures at some of the laboratories (but did not mention the results of the tests conducted by the laboratory it retained to retest certain samples). (For assessments of these theories, see Levy 1996, 170–185; Buckleton 2005; cf. Lynch et al. 2008.)

After being sequestered for nine months and enduring 133 days of televised testimony generating 50,000 pages of transcripts and 857 exhibits, the jury acquitted O. J. within five hours. It did so after listening to defense counsel call for an acquittal to "police the police" and to strike a blow against "genocidal racism." Among the public, reactions to this verdict of the predominantly African American jury ranged from outrage to jubilation and were clearly divided along racial lines.

Many authors have discussed whether the verdict was factually correct, and some of the dramatis personae in the trial have given their views in a bevy of books (scathingly reviewed in Posner 1996).[16] The literary contributions include accounts not only by the lawyers but also by "the world's greatest criminalist," Henry Lee, who advised the defense and testified that "something is wrong here" (Lee and Labriola 2001, 229). Three of the jurors wrote their own book, in which they complained that the attorneys and the prosecution's star DNA witness treated them like "a bunch of illiterate, ignorant people" (Cooley, Bess, and Rubin-Jackson 1996). The most bizarre book is the best-selling "hypothetical" confession, *If I Did It,* ghostwritten by O. J. twelve years after the murders (Simpson 2007).[17]

Rather than reexamine the evidence against O. J., the courtroom antics and tactics of the lawyers, or the judge and jury's performance, I want to focus on the aspects of the case that are more central to the history of the admissibility of DNA evidence.[18] The *Simpson* case made no new law, and it was nothing like the run-of-the-mill criminal case in which DNA evidence was admitted without much contest, but the litigation is significant for at least five reasons. First, it revealed the importance of collecting evidence from crime scenes in ways that preserve the integrity of the evidence. Second, it showed that DNA evidence, like ordinary fingerprints, can be used to frame innocent (or guilty) individuals. Third, it showed how the 1992 NRC report could be used to impeach prosecution experts or laboratories that had not implemented all the recommended reforms (Levy 1996, 169–170). Fourth, the televised presentations on forensic DNA technology enhanced the belief of the general public—and many trial and appellate judges—that the technology was ready for prime time. Finally, the efforts of the Dream Team to exclude the evidence reveal the kind of arguments that the most accomplished criminal defense lawyers then were making about admissibility. I shall focus on this last, less publicized part of the case.

A Motion to Exclude DNA Evidence

On October 5, 1994, three months before the trial began, Simpson filed a "Motion to Exclude DNA Evidence." Barry Scheck, Peter Neufeld, and William Thompson wrote a comprehensive, if somewhat discursive, memorandum of points and authorities in support of the motion. The three had joined the Dream Team to deal with the mass of DNA evidence that the state was accumulating. Ever since their partial victory in *Castro* (Chapter 4), Scheck and Neufeld had been the uncontested leaders of the defense bar with regard to DNA evidence. Thompson came to the case from a different background. With degrees in psychology and law, he was an associate professor in the Criminology Department at the University of California at Irvine. (Today, he is the head of the department.) Although relatively inexperienced in the courtroom, after writing an early and often-cited law review article on the emerging forensic technology (Thompson and Ford 1989), he frequently consulted on DNA cases. Indeed, his coauthor and colleague at Irvine, Simon Ford, traded university life for full-time testifying and consulting in DNA cases (Roberts 1992b, 735),[19] and today Thompson is a principal in a DNA-evidence consulting firm. Although it seems doubtful that Scheck and Neufeld were silent partners in the memorandum, it has been reported that "Bill was the brains behind the O. J. Simpson case" (Krane 2002) and that he "wrote the motion to exclude DNA evidence for the *Simpson* team" (Coleman and Swenson 1994, 103).[20]

Because the DNA tests had not all been completed and the defense would have access to some of the samples for its own testing to verify the accuracy of the state's results, to succeed in excluding all the DNA evidence, the defense would have to show that as of late 1994, laboratories could not perform RFLP-VNTR testing (or other tests) and assess the relative rarity of the resulting DNA profiles to the satisfaction of the legal system.[21] This would not be easy. The defense team would have to do more than rehash (as it did) the fact that in *People v. Castro*, Eric Lander "criticized [Lifecodes] for using bins that are narrower than their match criteria" (Defendant's Memorandum, *20) or that he had recommended using a confidence interval "to deal with sampling error" (ibid., *21). Although the memorandum implied that these were matters of substantial continuing debate, it was obvious that the forensic scientists could use bins of a suitable width and could calculate a confidence interval. Nevertheless, with regard to this issue and others raised in the memorandum, developments in science and law were subtly forced into a pattern to the defense's liking. Parts of the pattern were real enough, but the picture was much more complicated than the authors of the memorandum cared or dared to admit. To create a simple picture, parts of the scientific literature were read through a distorted lens, legal rules were described with little regard to the facts of the cases on which they were based (Kaye 1994a, 1995a), and parts of the law that did not fit the desired pattern were omitted or minimized.

The trio of DNA lawyers emphasized that whether "adjustments for sampling error are necessary and appropriate has been controversial." Given that confidence intervals have been used to estimate random sampling error since the 1930s, when Jerzy Neyman introduced them, the defense had to acknowledge that "statistical methods for incorporating sampling error are well developed" (Defendant's Memorandum, *21). Nevertheless, the lawyers insisted that

> [t]he scientific dispute over how to deal with sampling error is one facet of the continuing debate over forensic DNA statistics. It is one of the issues likely to be addressed by a new committee recently appointed by the National Academy of Sciences to make a second attempt to resolve the statistical controversy. The debate about sampling error, and how to deal with it, is therefore evidence that the National Academy's first effort to resolve the issue (the 1992 NRC Report) was unsuccessful—a conclusion confirmed by the Academy's current efforts to revisit the issue and try, once again, to find a resolution to the statistical controversy. (Ibid.)

But the motivation for empaneling a second committee had nothing to do with sampling error. It had everything to do with the whether the ceiling methods were necessary and appropriate. Could that issue provide richer fodder for the defense?

The 1992 NRC report, it will be recalled, took no position on whether population substructure was too severe to permit use of the basic product rule. Instead, it described the literature and chose "to assume for the sake of discussion that population substructure may exist" (NRC 1992, 80). Simpson's memorandum, however, maintained that the committee affirmatively "concluded that the concerns raised by Lewontin, Hartl and others were sufficiently serious that the statistical estimation methods developed by the forensic laboratories should not continue to be used." No doubt, this was the view of some members of the committee, but the committee as a whole never said this. The stated justifications in the report had to do with the perceived need to find a conservative method of estimating genotype frequencies that would be "applicable to any loci used for forensic typing" and "that is independent of the ethnic group of the subject" (ibid.). Significantly, these justifications are views about legal policy; they are not scientific propositions.

Thus Lander's fear had come true. The very act of commissioning a second report was being portrayed as evidence that even the exceedingly conservative interim ceiling rule was not ready for legal use. The defense DNA lawyers now were insisting that "it would make little sense to admit DNA statistical estimates before the new Committee has issued its report." Not only was this galling to members of the old committee, but such assertions led the new committee to discuss whether it should issue an interim statement clarifying the status of the ceiling principle in the scientific community—in essence, the stratagem that Lander had proposed in lieu of a new report. The idea was quickly quashed on the ground that the committee's task was not to intervene in pending cases. The judge trying the *Simpson* case would have to fend for himself.

The defense lawyers did not stop with the proposition that scientists had yet to agree on a single, best method for describing the import of a DNA match. They argued that the ceiling approach might not be conservative enough. This notion struck many scientists as incredible. As noted earlier, the typical criticism of the ceiling principle was just the opposite— that it was "absurdly conservative" (Morton 1992, 2560), a view shared by one of the two members of the NRC committee with expertise in population genetics (see Chapter 6). Still, publications by Joel Cohen, a letter by Hartl and Lewontin (1993), and an affidavit by Elizabeth Thompson in another case (Chapter 6) seemed to support the contention that prominent scientists believed that the interim ceiling method might not be sufficiently conservative. Thus Simpson's lawyers dramatically posed the rhetorical question, "Can anyone seriously argue that the ceiling principle is generally accepted in the scientific community when prominent geneticists and statisticians are comparing it to the 'flat-earth theory,' calling it 'pseudo-statistical,' saying it 'has no rational basis' and calling for it to be 'buried'?" (Defendant's Memorandum, *32).

The references to a flat-earth theory and burying the ceiling principle came from Morton (1994, 587). The use of them in the memorandum stands Morton on his head. Morton's letter flatly denied that there was any meaningful risk that the ceiling principle would not produce conservative estimates. It showed that the counterintuitive examples of underestimation in Cohen (1992) and Slimowitz and Cohen (1993) were "highly artificial" and would not occur "in practice" (ibid.). The very point of Morton's letter to the journal was to debunk Cohen's findings as a realistic concern. Yet the lawyers solemnly advised the court that Cohen's "showing has given impetus to a more general condemnation of the modified ceiling principle as scientifically indefensible and inadequate" (Defendant's Memorandum, *31)—as exemplified in Morton (1994). It was not Cohen's work that prompted Morton to condemn the ceiling methods as "50 years" behind the times and "a disservice" (Morton 1994, 587). Morton's point was that there was little scientific reason for them, and they failed to make use of modern population-genetics theory.

Despite the peculiar readings of particular articles and events, the defense motion had a good deal going for it. The supreme court of the state had not spoken, and the courts of appeal in California were all over the map on what the population-structure debate meant for the admissibility of DNA matches. Moreover, the standard for ascertaining general scientific acceptance in California (as elsewhere) was notoriously vague. Because of the importance of the general-acceptance standard in DNA cases, it is worth examining its history and status in California at the time the *Simpson* motion was filed.

General Acceptance of RFLP-VNTR Analysis

As noted in Chapter 1, the requirement of general scientific acceptance originated in the District of Columbia in *Frye v. United States*.[22] The court of appeals could cite no authority for the newly minted requirement that it used to affirm the exclusion of the systolic-blood-pressure test for deception, but "general acceptance" proved to be a catchy phrase. It certainly caught on in California, although it did not receive the imprimatur of the state's supreme court until 1966 or so. At that time, in *Huntingdon v. Crowley,* the defendant in a paternity action produced some evidence that the complainant was sexually active with a television producer and her hairdresser. The trial court ordered blood tests, and these excluded both men, but the exclusion of the hairdresser rested on the then-"experimental" Kell-Cellano antigens. The supreme court quoted from *Frye* and stated for the first time that *Frye* was the rule in California. Emphasizing testimony indicating a "lack of widespread acceptance and use of the Kell-Cellano tests," and pointing to the fact that the American Medical Association's Committee on Legal Medicine did not include the Kell-Cellano types in its recommendations for routine blood grouping in parentage testing, the supreme court affirmed the exclusion of that evidence.

In 1976, in *People v. Kelly,* the court "reaffirm[ed its] allegiance to the *Frye* decision." Terry Waskin was the target of a series of anonymous, threatening telephone calls. With Waskin's consent, police recorded two of these calls. An informant listened to these tapes and tentatively identified Robert Kelly as the caller. The police then taped Kelly's voice during a telephone call. They sent the tapes to Lieutenant Ernest Nash, the head of the Michigan State Police's voice-identification unit, for spectrographic analysis. Nash reported that the voices on the tapes matched, and Kelly was charged with extortion.

The prosecution asked the trial court for a pretrial hearing on the general acceptance of "voiceprints." After some vacillation, the court agreed. But prosecutors made a major blunder at the hearing. Apparently thinking that case law all but established the general acceptance of "voiceprints," they called only one witness—Lieutenant Nash. Nash testified that members of the scientific community involved in voiceprint analysis generally accepted the technique as extremely reliable, but he conceded that this "scientific community" consisted primarily of law-enforcement officers like himself. Kelly put on no expert testimony at all. He just insisted that the prosecution's proof fell short.

The California Supreme Court agreed with Kelly. First, it questioned "whether the testimony of a single witness alone is ever sufficient to represent, or attest to, the views of an entire scientific community regarding the reliability of a new technique." "Ideally," the court wrote, "resolution of the general acceptance issue would require consideration of the views of a typical cross-section of the scientific community, including representatives, if there are such, of those who oppose or question the new technique."

Second, the court was "troubled" by Lieutenant Nash's involvement in developing and promoting voiceprint analysis. Because "he has virtually built his career on the reliability of the technique," the court worried that "he may be too closely identified with the endorsement of voiceprint analysis to assess fairly and impartially the nature and extent of any opposing scientific views."

Finally, the court emphasized that Lieutenant Nash was "a technician and law enforcement officer, not a scientist." As a result, it feared that he might not give an "informed opinion on the view of the scientific community." The opinion suggested that "[t]his area may be one in which only another scientist, in regular communication with other colleagues in the field, is competent to express such an opinion."

Deeming the "essentially conservative" *Frye* standard to be the best test for the admissibility of scientific evidence and concluding that Nash's testimony and the opinions of other courts had not "demonstrated solid scientific approval," the supreme court held that the admission of the "voiceprints" was error and reversed Kelly's conviction. The court explicitly left open the possibility that "the future proponent of such evidence may well be able to demonstrate in a satisfactory manner that the voiceprint technique has achieved that required general acceptance in the scientific community."

However, the degree of acceptance that was required remained undefined. *Kelly* focused more on the nature of the proof than on the proposition to be proved. The court did not discuss the level of dissension within "a typical cross-section of the scientific community" that would demonstrate the lack of general acceptance. It did not need to. It had already decided that the testimony of a single, arguably biased technician was not likely to give a fair picture of scientific opinion.

Because *Kelly* fleshed out the bare bones of *Frye,* however minimally, California courts began referring to the general-acceptance standard as the "*Kelly-Frye*" test and struggled with its application to hypnotically induced memories, rape trauma syndrome, penile plethysmography, and other exotica. How would it be applied to DNA evidence? Unlike the prosecutor in *Kelly,* the prosecutors in *Simpson* could be expected to produce more than the self-interested testimony of a single technician. They could put one eminent scientist after another on the witness stand to attest to the validity of VNTR profiling. This the defense did not want to see. Simpson contended that the level of previously recorded dissent in the scientific community from the state's rosy view of DNA typing was so acute that the court should not even conduct an evidentiary hearing on whether there was general acceptance.

In making this argument, the defense memorandum assumed that "[t]he admissibility of DNA evidence must be determined under the standard of *People v. Kelly.*" This was less than certain. In the years since *Kelly,* more and more jurisdictions had been abandoning the *Frye* requirement of general acceptance in favor of inquiring more directly into the "reliability" of the scientific evidence. In these "relevancy-plus" jurisdictions, more than mere relevance was required, but no rigid rule of general acceptance was applied (Kaye, Bernstein, and Mnookin 2004, 192–194). The movement came to a crescendo in 1993, not long before the Brentwood murders. The lower federal courts then were divided over whether *Frye* was incorporated into the 1975 Federal Rules of Evidence. These rules do not explicitly distinguish between scientific and other forms of expert testimony, and they do not mention general acceptance. As originally enacted, Rule 702 majestically provided that "[i]f scientific, technical, or other specialized knowledge will assist the trier of fact to understand the evidence or to determine a fact in issue, a witness qualified as an expert by knowledge, skill, experience, training, or education, may testify thereto." Some courts construed the omission of any direct reference to "general acceptance" as evincing a legislative intent to overturn the well-established common-law requirement of this degree of acceptance.

Although the much more convincing view is that the rules left the viability of the general-acceptance standard open to further common-law development, the Supreme Court determined in *Daubert v. Merrell Dow Pharmaceuticals, Inc.,* that the federal rules "displaced" general acceptance as "the exclusive test for admitting expert scientific testimony" (509 U.S. at 589). As explained in Chapter 1, the Court read into the phrase "scientific . . . knowledge" in Rule 702 a requirement of a "body of known facts or . . . ideas inferred from such facts or accepted as truths on good grounds" in accordance with "the methods and procedures of science" (509 U.S. at 590). The Court then suggested that the requirement of scientifically "good grounds" might be satisfied by inquiring into such matters as the degree to which a theory has been tested empirically, the extent to which it has been "subjected to peer review and publication," the rate of errors associated with a particular technique, and the extent of acceptance in the scientific community (id. at 594). General acceptance, in other words, had been dethroned. No longer was it *the* test for scientific evidence. But it was not banished. It remained a major factor in the determination of whether a theory or method was good enough science to be admissible in court.

Although *Daubert* adopted the preexisting relevancy-plus approach already used in a large number of jurisdictions, the legal community saw it as a sea change. Some courts and commentators cheered it, others booed it, but everyone knew about it. Now there was a uniform, if malleable, rule for all federal courts. Lawyers and judges in practically every state began to debate whether *Daubert,* not *Frye,* was now (or should be) the standard under the rules of evidence for their state.

Simpson's DNA lawyers were keenly aware of the danger that *Daubert* posed for their argument that the discord among scientists was so extreme that a hearing to explore the disagreement was not needed. If a lack of general acceptance did not preclude admission—which is exactly what *Daubert* held—then the judge might have to hear from scientists to decide the underlying question of the scientific validity and reliability of DNA-typing methods and statistics. Also, the judge might side with the prosecution's experts, as the magistrate judge in *Yee* had done (chapter 5). That case was a bitter loss for Neufeld and Scheck. As both trial and appellate counsel in that case, they knew all too well that the court of appeals perceived little problem in deeming RFLP-VNTR profiling admissible under what it described as "the more liberal Rule 702 test adopted by the Supreme Court" in *Daubert* (*Bonds,* 12 F.3d at 557). "*Daubert,*" the court grandly (but incorrectly; see Kaye, Bernstein, and Mnookin 2004, 204) stated, "sets out a . . . more lenient test that favors the admission of any scientifically valid expert testimony" (12 F.3d at 565).

As if this were not bad enough, just a few months earlier, in *United States v. Chischilly,* the U.S. Court of Appeals for the Ninth Circuit, whose opinions determine federal law in California and other western states, had reached the same conclusion. Chischilly, "a Navajo Indian living in a remote rural part of a Navajo reservation," had been convicted of a rape and murder on the reservation in northern Arizona (30 F.3d at 1147). Like Simpson, he moved in advance of his trial to exclude evidence of a DNA match in the case. The district judge conducted a pretrial hearing that was something of a post-1992-NRC-report rerun of the many objections in *Jakobetz* and *Yee* (Chapter 5). After listening to eight scientists and admitting 152 exhibits, the court allowed Ranajit Chakraborty to testify at trial that "one in 2563 would be a 'conservative estimate' of the probability of a similar match between the DNA of a randomly selected American Indian and either the evidentiary sample or the defendant's DNA" (30 F.3d at 1149). In the spirit of the NRC's ceiling approach, Chakraborty obtained this number by looking "at the allele frequencies in several American Indian . . . tribal populations" and using the ones with largest frequency in a single multiplication (id. at 1158 n.29).

On appeal, Chischilly, like Simpson, argued that the laboratory procedures were not sufficiently exacting and that even this modest number was not reliable because of "the 'raging controversy' in the scientific community over DNA testing" (id. at 1152). The court regarded the population-structure objection as "perhaps his weightiest," being supported by "considerable scientific backing." The court of appeals did not challenge Chischilly's claim that the "scientific community is torn by controversy about the appropriateness of the FBI's databases, the effects of substructuring within their databases, and their statistical methodologies." It also did not deny that "[t]his controversy has divided the scientific community into opposing camps" with "prominent spokespeople" on both sides (id. at 1155). Instead, the court insisted that "these same statements take on the hue of adverse admissions under *Daubert*'s more liberal admissibility test: evidence of opposing academic camps arrayed in virtual scholarly equipoise amidst the scientific journals" (id. at 1155–1156). It seems perverse, however, to regard a prominent and persistent disagreement among distinguished and respected scientists as a reason to admit the evidence. Contrary to the suggestion in *Chischilly,* the lack of a scientific consensus always should count against admissibility under *Daubert.*

Nevertheless, *Daubert* and *Chischilly* spelled major trouble for Simpson's effort to exclude the evidence against him. Simpson's lawyers dealt with the problem by studiously avoiding it. There is not a word about *Chischilly* in the memorandum. As for *Daubert,* the memorandum relegates it to a footnote that blandly states that "*Frye* is no longer the rule in federal courts" and then blithely announces that "*Kelly* remains the law in California." This is true enough, but the memorandum neglects to mention that the continuing vitality of the general-acceptance rule was pending before the California Supreme Court. The case that raised the question was *People v. Leahy.* William Michael Leahy was stopped after a police officer observed him driving a car fifty-five miles per hour in a twenty-five-mile-per-hour zone. Leahy's face was flushed, his eyes were red and watery, his speech was slurred, his balance was unsteady, and he smelled of alcohol. The officer decided to give Leahy some field sobriety tests. Leahy passed the "internal clock" test and the "alphabet" test but flunked the "horizontal gaze nystagmus" (HGN) test. Nystagmus is an involuntary, jerking movement of the eyeball as it tracks an object moving to the side. It is generally thought that alcohol intoxication increases the frequency and amplitude of HGN and causes HGN to occur at a smaller angle from the forward direction. An intoxilyzer breath test revealed a .10% blood-alcohol level. Leahy was charged with drunken driving.

The trial court denied Leahy's motion to bar evidence of the HGN test on the curious theory that the general-acceptance rule "is inapplicable because the nature of this test isn't a specific test for the determination of alcohol; it is only a symptom that the officer is testifying to" (22 Cal. Rptr. 2d at 324). The California Court of Appeal rejected this attempt to evade the general-acceptance requirement and concluded that it was error to allow testimony about the link between HGN and intoxication "without a *Kelly-Frye* foundation, i.e., proof of general acceptance of HGN in the scientific community" (id. at 328). However, the appellate court cautioned that this result was only good with respect to "the law of California [as it] currently stands" (id.). It noted that the reasoning in *Daubert* undermined *Kelly* and suggested that "*Kelly* . . . might be ripe for reexamination" (id. at 324 n.4). The California Supreme Court accepted this invitation. It granted review and requested supplemental briefing on whether the *Kelly-Frye* standard should be modified following *Daubert*.

Whether the omission of *Leahy* in Simpson's memorandum was born of ignorance, chutzpah, or just a supreme confidence in the stability of California law proved to be a moot point. Just three weeks after the *Simpson* motion was filed, the California Supreme Court issued its opinion in *Leahy*. Concluding that the *Daubert* opinion offered no compelling reasons to abandon the *Frye* standard, the California Supreme Court agreed that the trial court improperly admitted police testimony regarding HGN without proof of general scientific acceptance. The state courts would march to the beat of a different drummer than the federal courts, and Simpson would not have to face the music of *Daubert*.

Although *Leahy* supported the defense's demand for viewing the DNA evidence through the lens of *Frye,* the question of what constitutes general acceptance remained. The courts in California, no less than those in other jurisdictions, had articulated inconsistent formulations of the meaning of "general acceptance." The *Simpson* memorandum chose one possible version: "The court need only conduct a 'fair overview' of the subject, sufficient to disclose whether 'scientists significant either in number or expertise publicly oppose [a technique] as unreliable'" (Defendant's Memorandum, *4). With this particular construction, the *Simpson* DNA team was in high gear. It pointed, for instance, to publications by "the world's leading population geneticist, Harvard professor Richard Lewontin" (ibid., *18) and the "distinguished academic scientist" Eric Lander (ibid., *7 n.9). Skipping over Lander's statement in May 1993 that "DNA fingerprinting evidence" presented in accordance with the ceiling method "is and should be legally admissible in all U.S. courts because it meets the test of 'general acceptance'" (Lander 1993a), and citing the opinions of one California appellate judge that were said to "establish conclusively that scientists 'significant in number and expertise' opposed the product rule as unreliable as recently as March, 1993," the lawyers insisted that "[t]he sole issue for the court is whether these critics have changed their minds in the past nineteen months. If the district attorney can offer no evidence of a change of scientific heart by the critics with respect to the product rule, then no hearing is necessary—the product rule must be rejected" (Defendant's Memorandum, *30).

But the meaning of "general acceptance" was not nearly as clear as the defense team thought. In dicta, the *Leahy* court addressed a concern articulated by the Appellate Committee of the California District Attorneys Association as amicus curiae, as well as the state's attorney general, that *Kelly* is unclear about the meaning of general acceptance by the scientific community. Simpson's DNA lawyers might have breathed a sigh of relief at *Leahy*'s retention of *Frye,* but this later portion of the opinion may not have been entirely to their liking.

In *Kelly,* the court had explained that it was necessary to look to "the views of an entire scientific community . . . including representatives, if there are such, of those who oppose or question the new technique." This does not begin to address the question of how wide or deep the opposition must be to preclude admission of the evidence. *Leahy*'s dicta do little to clarify the situation. The court began with language from a 1982 case, *People v. Shirley:* "[I]f a fair overview of the literature discloses that scientists significant either in number or expertise publicly oppose [the technique] as unreliable, the court may safely conclude there is no such consensus at the present time." In *Shirley,* the court had deemed hypnotically refreshed testimony inadmissible when "*major voices in the scientific community* oppose the use of hypnosis to restore the memory of potential witnesses" (*Leahy,* 882 P.2d at 336; emphasis in original). Score one for the defense position in *Simpson.* However, the *Leahy* court then retreated from this formulation. It cited *People v. Guerra,* a hypnosis case decided just two years after *Shirley* that called, not for "absolute unanimity of views in the scientific community," but rather for support from "a clear majority of the members of that community" (id.). This "clear-majority" standard, which was never mentioned in Simpson's memorandum, is not readily reconciled with the insistence that "the relative number of supporters and critics of the product rule is also irrelevant, so long as the critics continue to be 'significant either in number or expertise'" (Defendant's Memorandum, 30). The consensus of a clear majority would seem to be consistent with some dissent even from "major voices." Score one for the prosecution position that a "clear majority" would suffice. The defense claim that a hearing on general acceptance of RFLP-VNTR testing and statistics was unnecessary therefore stood on uncertain ground.

In the end, the trial court did not need to choose between the conflicting definitions of general acceptance. On January 4, 1995, Simpson's lawyers filed a notice that their client was withdrawing his objections to the admissibility of the DNA evidence. The notice reserved the right "to challenge the weight and reliability of the prosecution's scientific evidence at trial, including . . . the scientific reliability of DNA testing and statistical tech-

niques . . . and the performance of law enforcement authorities and forensic laboratories in their handling of the samples and evidence in this case
and in their slipshod execution of these techniques." A separate pretrial
hearing, the lawyers now stated, would be "extraordinarily time-consuming"
and "enormously expensive" and might expose jurors to "highly prejudicial
[press] reports about evidence." In a creative construction of due process
and speedy trial rights, they added that a hearing "lasting as long as eight
weeks" would violate "state and federal rights to a speedy trial" and "state
and federal constitutional rights to due process and to a fair trial before a
fair and impartial jury."

Unmentioned was the distinct possibility that the judge would deny the
motion. First, within a month or so of the filing of the motion, the California Court of Appeal cases that were said to "establish conclusively that
scientists 'significant in number and expertise' opposed the product rule as
unreliable" did not prove conclusive or even persuasive to another division
of the Court of Appeal. In *People v. Soto,* the court refused to follow these
cases, agreeing instead with the state that "the experts espousing application of the product rule reflected the 'support of the clear majority of the
members of the relevant scientific community' " (35 Cal. Rptr. 2d at 858).
Second, the defense team had put many of its eggs in one flimsy basket
when it described "[t]he sole issue for the court" to be "whether these critics have changed their minds in the past nineteen months" (Defendant's
Memorandum, *30). Although this remark was directed at the use of the
basic product rule, in *Soto* and again in *Simpson,* the state was able to
demonstrate various concessions or alterations in the writings of Hartl
and Lewontin at least about the interim ceiling product rule. Its memorandum declared that "[e]ven the staunchest critics must agree that the NRC's
ceiling principle, such as it is, is generally accepted" in the sense that it
"produces reliably conservative profile frequency estimates" (People's Reply Brief to Defendant's Motion to Exclude DNA Evidence, Nov. 2, 1994,
*34). Also resorting to some hyperbole, the prosecution's written response
added that the 1993 letter from Hartl and Lewontin—which the defense
cited as showing intractable disagreement over the ceiling principle—
actually "endorsed the NRC's 'ceiling' " (ibid., *32). The letter did not explicitly advocate the use of the interim ceiling rule, but it did state that
"[e]veryone agrees that it is conservative, and some believe that it is too
conservative. Whether or not it is excessively conservative is a matter that
can be resolved empirically by ethnic group studies. . . . Only additional
data will reveal the general robustness and degree of conservatism of the
interim ceiling principle" (Hartl and Lewontin 1993, 474). Then came the
headline that the defense never expected—"Two Chief Rivals in the Battle

over DNA Evidence Now Agree on Its Use" (Kolata 1994). Eric Lander, the "distinguished academic scientist" and "one of the earliest and most influential scientific commentators on DNA evidence" (Defendant's Memorandum, *6 n.9, *13), whose name appeared thirty-three times in the defense memorandum and not once in the prosecution's opposing memorandum, had joined with the FBI's pugnacious Bruce Budowle to put an end to the population-genetics debate. This turn of events was important not merely in California and in the *Simpson* case but in courtrooms from coast to coast.

Boodles and Louder

Despite the firestorm of protest over the need for and the logical foundations of the ceiling methods, Lander doggedly stood by the approach. As noted earlier, his 1993 letter to *Science* laid claim to its "general acceptance," and he cautioned NRC officials against empaneling a new committee for fear that it would undermine this legal victory. In 1994, reacting to the excesses of the *Simpson* case and the cases questioning the acceptance of the ceiling method, he pushed even harder. His newest effort took the form of a commentary in *Nature*. His coauthor, Bruce Budowle, was "one of the principal architects of the FBI's typing programme" (Lander and Budowle 1994, 735). Budowle had joined the FBI in 1983 after earning a Ph.D. in genetics from Virginia Polytechnic Institute and completing postdoctoral research in medical genetics at the University of Alabama at Birmingham. In the courtroom, where he frequently defended his work, and at scientific conferences, where he often presented it, he did not gladly suffer fools—or anyone else he disagreed with. Some of the players in the DNA drama might have described both Lander and Budowle as arrogant, but few would have suspected that they had a unified view on the status of DNA forensics. Paul Ferrara, the chairman of the laboratory accreditation board of the American Society of Crime Laboratory Directors and the director of the Virginia Division of Forensic Science, expressed amazement at learning that Lander and Budowle were writing together—and delight at what they had said (Kolata 1994).

For their manifesto, the duo chose the title "DNA Fingerprinting Dispute Laid to Rest." They announced their objectivity ("we can address the questions in an even-handed manner") and their ability to speak for the entire scientific community ("we represent the range of scientific debate") (Lander and Budowle 1994, 735). They laid out the FBI view of the early history of DNA typing: "Pioneered by biotech startup companies with good intentions but no track record in forensic science, DNA typing was marred by several early cases [especially *Castro*] involving poorly defined procedures and interpretation. . . . For its part, the U.S. Federal Bureau of Investigation (FBI) moved much more deliberately . . . and opted for conservative procedures" (ibid.).[23] Referring to the "extraordinary scrutiny" DNA profiling had received, the pair grandly announced that "it is time to declare the great DNA controversy over" (ibid.). They could "identify no remaining problem that should prevent the full use of DNA evidence in any court" (ibid.). After suggesting that the kinds of quality-control and assurance measures recommended in the 1992 NRC report seemed to be in place in "most forensic labs," they turned to "the subject most often said to remain problematical: population genetics" (ibid., 735–736).

In doing so, Lander placed a thick gloss on the 1992 committee's chapter on the subject.[24] He disclosed that "the committee was quite dubious that substructure had significant effects." Although "[t]he members agreed that the product rule was probably near the mark," they "were hard pressed to say how close." Consequently, they made "the worst-case assumption," "gave the benefit of every conceivable doubt to the defendant," and devised an "unabashedly conservative" and "lop-sided approach"— "the ceiling principle"—to "withstand attacks from the most stubborn and creative attorneys" (ibid., 736–737). The committee's high confidence in the basic product rule and its serious doubts about the significance of population structure were a revelation to many readers of its book-length report.

Even more surprising is Lander's recollection that the committee never meant for the ceiling method to supplant the basic product rule, but only for it to be a kind of add-on to the estimates that he and the other population structuralists had so fervently attacked. Now, for the first time, he wrote that "the ceiling principle was not intended to be exclusive. Expert witnesses were still free to provide their statistical best estimate . . . based on the product rule." If this was how the committee intended the ceiling product rule to be implemented—as an equal partner with the basic product rule—no one outside the committee knew it. How could they, when the report firmly recommended "that *estimates of population frequencies be based on existing data by applying conservative adjustments*" in the form of (1) the "readily understood" count in the population database and (2) the interim ceiling rule, which "represents a reasonable effort to capture the actual power of DNA typing" (NRC 1992, 91; emphasis in original)? Lander bemoaned what he called this "major misunderstanding of the report" (Lander and Budowle 1994, 737), although he himself had contributed to it in the writing both of and about the report.[25]

Predictably, the most adamant critics of the basic product rule were not willing to ratify the bilateral cessation of hostilities. Jerry Coyne, a former student of Lewontin and "a population geneticist at the University of Chicago who opposed DNA typing as court evidence," insisted that "[t]here's plenty of controversy in the field." Moreover, "[i]t just upsets me no end because of the way this was timed. It was carefully calculated to come out right before the DNA hearings for the Simpson case. The only motive I can see is to influence the judge" (Kolata 1994). Lewontin (1994) scoffed that the manifesto was "a piece of propaganda that completely distorts the current situation." Hartl (1994) worried that the article presaged the abandonment of both the interim and the ultimate ceiling methods, and he took the opportunity to attack Budowle's previous testimony and writing on the lack of any "forensic significance" of population structure.

Ironically, Hartl's letter also fits Lewontin's description of "propaganda." Budowle's point was that although population structure surely exists at some level, it has only a minor impact on basic-product-rule calculations. Statisticians describe such a situation as lacking "practical" or "substantive" significance. In medicine and psychology, the phrase "clinical significance" frequently is used. These terms remind us that "statistical significance" is not a measure of the size of an effect. With large data sets, even trivial differences between groups can be statistically significant—they are unlikely to arise when membership in the groups is independent of the variable being measured. For example, what would it prove if two populations differed in mean IQ scores by 0.001, or if a treatment for the common cold reduced the mean period of clinical symptoms by two minutes? The difference could be "statistically significant," but it would be meaningless for any practical purpose. Budowle coined the phrase "forensic significance" to make the same point in response to claims that "statistically significant differences among ethnic groups have been documented for markers used in DNA typing" (Hartl 1994, 398).

This is a point that should be addressed on its merits. Although Hartl had participated in one study that tried to do so (Krane et al. 1992), his letter to *Nature* resorted to claims that would make many statisticians cringe: "Statistical significance is an objective, unambiguous, universally accepted standard of scientific proof. When differences in . . . groups are statistically significant, it means that they are real—the hypothesis that . . . differences . . . are negligible cannot be supported" (Hartl 1994, 398). As the IQ and sniffles examples demonstrate, a difference can be real *and* negligible.[26]

Notwithstanding the last-ditch efforts of the remaining population structuralists to rebut Lander and Budowle's tendentious claims, the courts were impressed with the announcement of a consensus. For instance, the commentary helped sustain the conviction of Robert Johnson, a resident of an Arizona mining town. Early one July morning in 1991, in Sierra Vista, a man attacked and sexually molested the owner of a restaurant as she entered the restaurant to open it for the day. He then drove away in a small, red car. Two women in the town reported that Johnson, who drove a small, red car, was wearing clothing that morning that the victim had noticed on her attacker. DNA from Johnson was indistinguishable from DNA left by the victim's attacker. An analyst from the Arizona Department of Public Safety testified at trial that the probability of a random match, according to the interim ceiling product rule, was no greater than 1 in 312 million. The court of appeals in *State v. Johnson* (1995) found no error in the introduction of this estimate, and it affirmed Johnson's conviction.

The Arizona Supreme Court granted review of this and three other cases to consider admissibility of DNA evidence in the wake of its 1993 opinion in *State v. Bible* (Chapter 6). The justices clearly did not understand a great deal about Hardy-Weinberg equilibrium, linkage equilibrium, and random sampling (Kaye 1997a). For example, they thought that the 1992 report, of all things, "makes [it] clear" that the equilibrium assumption "is well-grounded and has been proved accurate for purposes of DNA profiles" (*Johnson*, 922 P.2d at 297). They also misread the report as requiring "samples drawn at random from designated populations" for estimating the allele frequencies to be used in computing the interim ceiling match probabilities. Then, confusing random mating in a population with random sampling from that population, they thought that the population samples were drawn at random because "there is linkage equilibrium and Hardy-Weinberg equilibrium" (id. at 298).

In contrast to its indigestion over the relevant science, the court had no trouble correctly recognizing that "two of the principal antagonists involved in the initial debate over forensic DNA typing" no longer saw any "scientific reason to doubt the accuracy of forensic DNA typing results, such as the modified ceiling method" (922 P.2d at 299, internal quotation marks omitted). Indeed, at the oral argument on the case, the court's chief justice asked about "these Boodles and Louder guys," and defense counsel, with unguarded candor, conceded that "we know which way the wind is blowing."

Likewise, in *State v. Copeland,* a Washington jury convicted William Copeland of murdering and raping a woman who had been beaten, strangled, and stabbed with a knife and a barbeque fork. The case against Copeland included RFLP tests conducted by the FBI, which used the basic product rule to estimate the probability of a random match to be between 1 in 2.8 million and 1 in 3 million for various databases. At a pretrial hearing on the general acceptance of the product rule, Copeland produced two experts, Laurence Mueller and Seymour Geisser, who questioned whether VNTR alleles and genotypes were in the proportions that would be expected in a randomly mating population. The trial court found these experts to be "uninformed" about the work of researchers who were much more active in studying the population genetics of VNTRs. The Washington Supreme Court agreed with the trial court that the defense testimony "was neither persuasive nor credible." It observed that "[a]lthough at one time a scientific dispute existed among qualified scientists, [it] was short-lived. . . . [T]he FBI has collected data from around the world, and one of the most vociferous opponents of use of the product rule has joined with an FBI scientist in declaring that the DNA wars are over" (922 P.2d at 1318).

In short, a fifth wave of cases was washing away the gains that the 1992 report and the rancorous scientific debate had created for defendants seeking to exclude DNA evidence on statistical grounds. Whatever the dissenting voices in the scientific community may have been saying, "Boodles and Louder" had drowned them out. Interim ceiling estimates, at least, were welcome in court.

The Second NRC Report in the Courts

The release of the second NRC report in May 1996 added power to the fifth wave of cases and guaranteed the admissibility not only of interim ceiling estimates but also of basic-product-rule estimates for general-population cases and of affinal estimates for subpopulations. As interpreted by Lander, the 1992 NRC report called on experts to present two or three numbers. They were encouraged to give the size of the population database plus a ceiling-product-rule figure or to provide these two figures plus the basic-product-rule number. The 1996 report, however, stated that the ceiling and counting approaches were unnecessarily conservative. It recommended basic-product-rule estimates for general-population cases and affinal estimates for subpopulation cases. The courts heard the basic message about excessive conservatism. They cited the report (with varying degrees of emphasis) in rejecting the kind of arguments against both basic- and ceiling-product-rule estimates that were made and then dropped in *Simpson*.

Two cases from the West illustrate how this part of the report was received. The Arizona Supreme Court case *State v. Johnson*, described earlier, relied on commentary and news reports in science journals, law review articles, other cases, and the new report to decide that the interim ceiling method was generally accepted for arriving at an upper bound on the random-match probability. It left open the question of the admissibility of a basic-product-rule estimate.[27] Cases like *People v. Soto* (1999) in California took the additional step of holding admissible in its own right a basic-product-rule estimate. In *Soto,* a seventy-eight-year-old widow was raped at knifepoint in broad daylight in her mobile home by a man with a stocking mask over his face. Her description of the rapist did not match her neighbor, Frank Soto, very well, but semen on the bedspread where she had been forced to submit did. A criminalist "testified at trial that there was a probability of only 1 in 189 million of finding the same DNA pattern [a four-locus VNTR genotype] in individuals selected at random from the population represented by the [county's] Hispanic database" (981 P.2d at 961).[28] Laurence Mueller testified for the defense that population structure rendered this computation "incorrect." A better estimate, in his judgment, was "1/250, since there were approximately 250 samples in the . . . laboratory's Hispanic database." William Shields, described by the California Supreme Court as "the State University of New York College of Environmental Science and Forestry professor who has done work in population and conservation genetics with small mammals" (id. at 972), agreed that this estimate was the "most reliable if you're going to put a number in at all" (id. at 969). Minnesota's Seymour Geisser concurred as well. A stroke prevented the elderly victim from testifying at trial. The jury acquitted Soto of forcible rape but found him guilty of attempted rape.

As indicated earlier, the California Court of Appeal upheld the admission of basic-product-rule estimates shortly before Simpson withdrew his motion to exclude such estimates in Los Angeles. The California Supreme Court eventually affirmed, observing that "the controversy over population substructuring and use of the unmodified product rule has dissipated" (id. at 975). It pointed to the FBI's "extensive, worldwide study of VNTR frequency data" collected in response to the 1992 report, Lander and Budowle's article, and, "of greatest significance," the 1996 NCR report's recommendation that in cases such as this one, "the calculation of a profile frequency should be made with the [basic] product rule" (id. at 975–976).

In this manner, in case after case, courts rebuffed the old arguments about population structure. The "propaganda" from "Boodles and Louder" espousing the basic product rule in conjunction with the ceiling product rule, followed by the second NRC report repudiating the ceiling formulas and describing the more sophisticated affinal method, ended the war over population genetics. But battles over DNA evidence did not end in 1996. A few soldiers were unwilling to accept an armistice even over the issue of population structure, and conflict flared on other fronts.

CHAPTER EIGHT

Moving Back to Errors
and Relatives

L ANDER AND BUDOWLE (1994, 738) declared that it was "now time to move on." As we have seen, not everyone agreed with them. Although Lewontin and Hartl stopped testifying for defendants (Aronson 2007, 187), they continued to write that the degree of population structure within the major racial or ethnic categories was substantial (e.g., Lewontin 1994a, 1997). Other experts continued to testify that population structure was a reason to exclude basic-product-rule estimates. For example, in 1997, Paul Gore was on trial in Snohomish County, Washington, for a series of rapes of young teenagers. At the pretrial *Frye* hearing, Laurence Mueller, Dan Krane, and Randall Libby maintained that there still was "heated debate in the scientific community as to the use of the product rule" (State v. Gore, 21 P.3d at 274). The Washington Supreme Court observed that their position was "at odds with the National Research Council" and attributed Mueller's and Libby's persistence to their "financial stake" in perpetuating the controversy, as well as to Mueller's "limited qualifications" (id. at 274 n.9).[1] By depicting the remaining witnesses on this issue as hired guns and their previous academic work as limited to "the DNA content of sewage" or of "frogs, sea urchins, sand dollars, zebra fish, and fruit flies," prosecutors had rendered witnesses like Mueller and his former colleague, Simon Ford, toothless (People v. Marlow, 41 Cal. Rptr. 2d at 18, 19, 25).[2]

By and large, then, the criticism of DNA evidence did move on, but it also moved back to earlier issues. Even if random-match probabilities could not be kept from the jury on the ground that because of population structure, they were not scientifically valid (as per *Daubert*) or generally accepted (as per *Frye*), they still might be excluded if a random-match probability was scientifically unacceptable for some other reason or if it was irrelevant or unfairly prejudicial. So defendants shifted from the once-promising attacks on the omission of population structure in the basic-product-rule estimates of random-match probability to two other concerns that had previously been raised about the random-match probability for unrelated individuals: (1) that this probability does not reflect the possibility that the laboratory made a mistake in ascertaining the DNA profiles being compared, and (2) that the random-match probability overlooks the possibility that a close relative (rather than an unrelated individual) might be the source of the crime-scene DNA. Furthermore, pointing to the 1992 NRC report, they pressed the argument that (3) without blind proficiency testing and other monitoring of forensic laboratories, the work of DNA laboratories was not reliable enough to be admissible in court. Finally, the litigation-centered debate moved on to a truly new point of contention—that of (4) the scientific validity and general acceptance of advancing methods for analyzing completely new forensic DNA polymorphisms. Some of the scientists who pressed the population-structure objection to random-match probabilities lent their acumen and prestige to these arguments, but here their testimony proved less influential. The courts almost never excluded DNA matches and the accompanying random-match probabilities on these grounds.

The *Simpson* case, with its emphasis on DNA evidence and the involvement of attorneys well versed in the nuances of DNA evidence, illustrates the transition to this set of issues. I therefore return to the motion to exclude the DNA evidence in that case. This chapter describes and evaluates the attack that Simpson and other defendants mounted on the admissibility of the random-match probability on the ground that it failed to supply the jury with sufficient information and hence was largely irrelevant or prejudicial. The next chapter explains the advancing technology, the newer loci it opened up to forensic science, and the courts' receptiveness to these technological enhancements in DNA identification technology.

Proficiency Testing and Error Rates

The limited nature of the random-match probability means that the statistic does not shed light on every issue that is relevant to deciding whether the samples that are being compared have a common source. Random-match probabilities and frequencies address a single question: How probable is it that two, correctly identified DNA genotypes would be the same if they originated from two unrelated individuals? By definition, they do not consider any uncertainty about the origins of the samples (the chain-of-custody issue), about the relatedness of the individuals who left or contributed the samples (the identical-alleles-by-descent issue), or about the determination of the genotypes themselves (the laboratory-error issue). The *Simpson* memorandum artfully turned the limited nature of random-match probabilities into a legal argument about relevance. The memorandum did not simply assert that small numbers are necessarily prejudicial. That claim had been widely rejected since the days of genetic markers (Chapter 2) and the dawn of DNA testing (Chapter 3).[3] So a more subtle series of arguments was advanced. The lawyers asserted that good statistical practice and the need to avoid misleading the jury demanded that the random-match probability be accompanied, modified, or replaced by the probability that the laboratory had erred in some way. Without such a supplementary or modified statistic, they said, the presentation of a random-match probability would not conform to generally accepted scientific practice, and it would be irrelevant and prejudicial. Then they maintained that no generally accepted procedure for estimating the error probability was available. Hence, they concluded, neither the random-match probability nor the bare fact of a DNA match could be admitted against Simpson.[4]

The motion to exclude was withdrawn (Chapter 7), but bits of it resurfaced in Defendant's Notice of Objections to Testimony Concerning DNA Evidence and Memorandum in Support Thereof [hereinafter cited as Defendant's Notice], written by Scheck and Thompson and filed on March 20, 1995, four and one-half months after the first, aborted effort. In the October memorandum, the lawyers had argued that there was no "general acceptance of the methods used to determine the false positive error rates of the laboratories for each test" (Defendant's Motion to Exclude DNA Evidence, Oct. 5, 1994, *1). In the March notice, they again argued that the error statistics were required on grounds of relevance and prejudice, and they sought a hearing to establish that "[a]ny statistical evidence proffered by the prosecution concerning the rate of laboratory error [was] derived from a method that is generally accepted in the scientific community" (Defendant's Notice, *10). "Such testimony," they again insisted, "is inadmissible under *Kelly* without a foundational showing that an accepted method exists for estimating the rate of laboratory error and that correct procedures were followed to generate the proffered statistics" (ibid.).

Parts of this argument were advanced in later cases and discussed in the second NRC report. The issue of error is crucial in devising a suitable system for generating and using the results of modern forensic technology (Koehler 2008, Aronson 2007; Murphy 2007; National Research Council Committee on Identifying the Needs of the Forensic Science Community 2009 [hereinafter cited as NRC 2009]). In mentioning "error rates," the Supreme Court's *Daubert* opinion underscored the issue for the legal world. The legal demand to incorporate error rates into the introduction of DNA evidence was closely connected to the demand for rigorous, external, blind proficiency tests of the DNA laboratories that were endorsed in the first NRC report and, to a lesser extent, in the second. The hope of the first committee was that such a testing program might lead to reasonable estimates of (or an upper bound on) the chance that a laboratory would declare a match when the sample actually did not match—a false-positive error. This would have been a welcome contrast to the testimony from some forensic analysts that false-positive errors are impossible.[5]

Historically, forensic scientists have been free from systematic, external proficiency testing. The results of the first national crime-laboratory testing program, sponsored by the federal Law Enforcement Assistance Administration in the mid-1970s, came as a shock to many crime laboratories. Of the 200 participating crime laboratories, 71% provided unacceptable results in a blood test, 51.4% made errors in matching paint samples, 35.5% erred in a soil examination, and 28.2% made mistakes in firearms identifications. The 1978 report concluded: "A wide range of proficiency levels among the nation's laboratories exists, with several evidence types posing serious difficulties for the laboratories" (Peterson, Fabricant, and Field 1978, 3; for summaries, see Giannelli 2007a, 72–74; Saks 1989). Some "efforts to improve conditions in the laboratories" ensued, "but these encounter[ed] institutional inertia against reform" (Peterson 1983, 645). After his initial encounter with the system, Eric Lander (1989a, 505) expressed amazement at this state of affairs, declaiming that "clinical laboratories must meet higher standards to be allowed to diagnose strep throat than forensic labs must meet to put a defendant on death row."

Although the pace of change is slow, the regulatory landscape is changing. The American Society of Crime Laboratory Directors–Laboratory Accreditation Board has been accrediting forensic laboratories since its formation in 1981, but until recently, accreditation always was voluntary (Giannelli 2007a, 75). New York and a few other states now require forensic DNA laboratories to be accredited (N.Y. Exec. Law § 995-b (McKinney 2006)), as does the federal Justice for All Act, enacted in 2004.[6] In authorizing "$755 million over five years to address the DNA backlog crisis in the nation's crime labs" and "more than $500 million in new grant programs to reduce other forensic science backlogs, train criminal justice and medical personnel in the use of DNA evidence, and promote the use of DNA technology to identify missing persons" (Leahy 2004), the Justice for All Act wields the power of the purse to promote mechanisms for quality control and assurance in forensic laboratories. In particular, it requires applicants for federal funds to certify that the laboratories use "generally accepted laboratory practices and procedures, established by accrediting organizations or appropriate certifying bodies" (42 U.S.C. § 3797k(2)). The standards of these accrediting bodies typically call for periodic open, external proficiency testing (Peterson et al. 2003, 24). Furthermore, the DNA Identification Act of 1994 explicitly requires proficiency testing for analysts in the FBI (42 U.S.C. §14133(a)(1)(A)), as well as those in laboratories participating in the national database (id. at § 14132(b)(2)) or receiving federal funding. Finally, the Justice for All Act requires applicants for federal funds to certify that "a government entity exists and an appropriate process is in place to conduct independent external investigations into allegations of serious negligence or misconduct" (id. at § 3797k(4)).

Of course, laws on the books and laws in action are often two different things. In some areas of forensic science, the rigor of the proficiency testing programs is dubious. Despite a disturbing history of scandals in forensic laboratories and medical examiners' offices (Chapter 13), compliance with the § 3797k(4) certification requirement has been problematic (Office of the Inspector General, U.S. Department of Justice, 2008). Thus even today, many observers regard the network of police and private forensic laboratories in America as a relic of an earlier era—insular and unscientific, unregulated and unexamined (e.g., Giannelli 1997, 2007b; Cooley 2004, 2007; Saks and Koehler 2005; Murphy 2007, 746–747; cf. NRC 2009, S-5).

The first NRC committee responded to this situation and the experiences in cases like *Castro* and *Yee* by announcing that "[n]o laboratory should let its results with a new DNA typing method be used in court, unless it has undergone . . . proficiency testing via blind trials" (NRC 1992, 55). In a "blind trial," the laboratory would not know that the case it was working on was not a real one but only a highly realistic mock-up. Concerned as well with the small random-match probabilities quoted in many cases, the committee added:

> Especially for a technology with high discriminatory power, such as DNA typing, laboratory error rates must be continually estimated in blind proficiency testing and must be disclosed to juries. For example, suppose the

chance of a match due to two persons' having the same pattern were 1 in 1,000,000, but the laboratory had made one error in 500 tests. The jury should be told both results; both facts are relevant to a jury's determination. (Ibid., 89)

Some scientists felt that merely presenting an error rate along with a population frequency was not enough. Lewontin and Hartl (1991, 1749) had contended that "probability estimates based on population data that are smaller than the false-positive rate should be disregarded." Lewontin (1994a, 260) reiterated that "[i]f the data themselves are unreliable, questions of probabilities of alternative suspects are irrelevant." Writing as an evidence law scholar and sociologist and not as a member of the committee, Lempert (1991, 325) also had argued that "jurors ordinarily should receive only the laboratory's false positive rate as an estimate of the likelihood that the evidence DNA did not come from the defendant." Criminal lawyers tried to transform these prescriptions into rules of evidence. Scheck (1994) maintained that *Daubert* itself somehow mandated that "in most cases, error rate should be the only probability offered about the likelihood that the defendant was not the source of DNA trace evidence" and predicted that "*Daubert* DNA litigation will inevitably focus on the problem of how best to assess laboratory error rate" (ibid., 1997).

The second NRC committee followed the first committee to the extent of urging that "[l]aboratories should participate regularly in proficiency tests, *and the results should be available for court proceedings*" (NRC 1996, 88, emphasis added)—something that the FBI had opposed (Aronson 2007, 115–116). However, the second NRC report stopped short of demanding blind proficiency testing as a precondition to admissibility. It strongly supported "[r]egular proficiency tests, both within a laboratory and by external examiners," as "one of the best ways of ensuring high standards" (NRC 1996, 88), but it was not prepared or equipped to make the policy judgment that the benefits of blind proficiency testing outweighed the costs of this particular procedure. The committee called for blind proficiency testing, but only "[t]o the extent that it is feasible" (ibid.).

Furthermore, the committee firmly opposed the idea of Hartl, Lewontin, Lempert, and Sheck of using only one figure to inform the jury about the meaning of a reported DNA match. As we have seen, the committee, like its predecessor, felt that both the population frequency and the chance of a false-positive typing error "are relevant to a jury's determination" (ibid., 89). However, it was less enamored of using proficiency test results to estimate false-positive error rates of DNA laboratories. The first committee proposed blind proficiency tests that would be rigorous and truly representative of actual casework, but it could be costly and difficult to stage a large number of these tests. They would require cooperation from police and prosecutors willing to pretend that the submitted samples were real and to answer possible questions about the case. On the one hand, if only a few tests of a laboratory were available, the observed error rate would be zero if there were no errors to date. On the other hand, the rate would be relatively large if even a single error was made, and this number would not apply directly to the case at bar if the laboratory had taken steps to prevent recurrence of the problem. Thus, although the second report firmly endorsed proficiency testing as part of a quality-control and assurance program (ibid., 76–78), calling it "essential" and "key," it placed much less emphasis on blind tests and far more weight on giving the defendant the opportunity to retest samples at another laboratory to avoid a false declaration of a match in a given case.

The second committee shied away from prescribing how the results of a proficiency testing program, blind or otherwise, should be used in court on the ground that there was no scientific reason that they must be part of the initial presentation of the evidence or that they even be admissible. As a legal commentator, I have suggested that "where external proficiency testing of the experts utilizing these techniques is feasible, such testing should be a prerequisite to admissibility; and . . . the jury should be informed of the results of the validity and proficiency tests" (*McCormick on Evidence* 2006, 1:834 n. 53). The committee, however, saw its role as more limited. It noted that whereas "[t]he 1992 NRC report stated that the probative value of such statistics, when balanced against their potential to mislead a jury, favored admissibility, . . . [because] the purpose of our report is to determine what aspects of the procedures used in connection with forensic DNA testing are scientifically valid, we attempt no such policy judgment" (NRC 1996, 185).[7]

The critics of the DNA laboratories' work, as well as several scholars, took the committee to task for not espousing their position that error rates ascertained from blind proficiency testing must be part of the prosecution's case. Lempert (1997) called it "a failure of common sense." Academic and defense experts Jonathan Koehler (1997) and William Thompson (1997) also were sharply critical. These three critics identified weaknesses in the committee's analysis of the usefulness of historical error rates and argued that jurors would be misinformed if an error-rate statistic was not somehow incorporated into the presentation. They concluded that an opportunity to improve the functioning of the criminal justice system had been lost.

Although the committee took no position on the evidentiary issue, its distaste for a single statistic that combined laboratory-error statistics and population frequencies was palpable, and its remarks on the difficulties in applying the historical results of proficiency tests in current cases reinforced the natural tendency of courts to treat the issue of laboratory error as distinct from the question of how rare a genotype is in a given population. The most recent opinion comes from the highest court for the District of Columbia. Accused of raping a thirteen-year-old girl in a southeast Washington parking lot, the defendant in *Roberts v. United States* (2007) sought to exclude an FBI report that the random-match probability "was no greater than 1 in 410,000 among four major population groups in the United States" (916 A.2d at 925). Following the formula in the *Simpson* memorandum, he argued that the statistic was inadmissible for want of general acceptance

> because the FBI "omit[s] from its statistical analysis any estimate of the chance of a false match caused by laboratory error," [and] scientists significant in number and expertise regard that method of analysis as unreliable and misleading. . . . In his view, to be admissible [the] estimate of the chance of a coincidental match had to factor in, or at least be accompanied by, an acknowledgment that the FBI's estimate of random match probability "typically overstates the significance of a DNA match by several orders of magnitude, as the chance of a false match by laboratory error typically ranges from 1 in 100 to 1 in 1,000." (Id. at 929)

The court denied that "industry-wide error rate data, including published results of double-blind proficiency testing, [were] necessary for an accurate and reliable statistical evaluation of a DNA match."[8] It did not question "a defendant's right to bring the frequency (or not) of laboratory error leading to false matches to the attention of a jury," but it relied on the "authoritative 1996 report" to reject the argument about general acceptance (id. at 930). Beyond that, the court emphasized that the

> jury was not denied "any consideration" of the risk of false positive matches occurring: [The FBI analyst] was cross-examined fully about the possibility and historical evidence of such errors, and nothing prevented appellant from presenting expert testimony—he did not—about the incidence of laboratory error industry-wide and how that could affect the random match estimates by the FBI or other laboratories. Finally, appellant could also have obtained independent testing of the DNA samples at issue to establish the fact or probability of a false match, but did not do so.

Thus the court concluded that "the *Frye* standards of admissibility did not require . . . testimony by the government about laboratory error rates not shown to have any relation to the FBI's testing practices in particular, or to the testing done in this case."

This result is both traditional and defensible.[9] In critiquing the legal arguments of Scheck (1994) and Koehler (1997), law professor Margaret Berger, who served on the second NRC committee (and who is widely known for her coauthorship of an evidence treatise written with Judge Jack Weinstein, a member of the first committee), emphasized that we do not ask the proponent of other forms of evidence, such as business records, to quantify and quote a probability of erroneously recorded information as a precondition to admitting the evidence. Neither is there a special rule that demands that the prosecution provide an estimate of the rate at which laboratories generally err in interpreting or processing samples that are subject to scientific tests (Berger 1997).

Nevertheless, a special rule about error-rate statistics that comes into play specifically when random-match probabilities are presented would not be outrageous. The fact that the risk of an error due to laboratory misconduct or imperfections could well be greater—indeed, many times greater—than the chance that a randomly selected individual has the incriminating genotype *is* important for the jury to know. But we should not delude ourselves into thinking that scientific practice or theory, operating through *Frye* or *Daubert,* requires this specialized rule. There is no disputed scientific technique or principle to be inspected under the general-acceptance or scientific-validity standards, and "'science' does not dictate the exclusion of DNA evidence unless a laboratory error rate is made part of the prosecution's case" (ibid., 1092). "Consequently," the second NRC committee "rightly declined to make recommendations about handling error in judicial proceedings as beyond its mandate to deal with scientific issues" (ibid.).

At bottom, the error-rate objection is that a scientifically valid computation of a population frequency or random-match probability should not be presented because it is not the figure that the jury really needs to know. This objection is one of relevance. Its resolution depends on ordinary logic and the psychology of human decision making, not on the opinions of geneticists or statisticians about what information should be kept from jurors. If jurors will ignore or undervalue the chance of error unless they are presented with a quantitative probability before cross-examination, then the argument has bite. But the claim is not intuitively obvious, and empirical research into this issue, some of which was conducted by members of the second NRC committee years later, does not establish its truth (Berger 1997; Schklar and Diamond 1999; cf. Kaye et al. 2007). Because knowledge of how rare correctly ascertained DNA genotypes are in the general population helps the jury assess the possibility that an unrelated defendant would just happen to share the crime-scene DNA profile, this statistic should be admissible. The defendant can present evidence about the risk of error, and a court can instruct the jury that the random-match probability reveals nothing about the chance that the laboratory is mistaken in its conclusion that the samples contain matching DNA. Certainly, it is important to understand that the random-match probability is incomplete—there is much more to think about—and that it is not the posterior "source probability" that the jury ultimately must assess. But it is not, for these reasons, inadmissible or unfairly prejudicial.

The Evil Twin and Other Relatives

The random-match probability and population frequency also are incomplete in that they do not focus on relatives. Product-rule calculations (whether basic, ceiling, or affinal) relate to the chance that an innocent person who is not a close relative of the actual source will have the misfortune of possessing the same genotype as that seen in the crime-scene sample. The probability that a close relative will have the genotype will be higher. Indeed, an identical twin of the true source is all but certain to have a matching genotype. With suspects being identified through searches of large databases of the DNA of convicted offenders, prosecutors are encountering the "evil-twin" defense more often (Willing 2004a, 2004b; Kulish 2009). And what about other relatives? Could the offspring of an errant husband unknown to the innocent defendant (who just happens to have all the same alleles) be the actual criminal? Take the case of Gary Curtis Faison, on trial in 1999 for the aggravated sexual assault of a woman in Dallas, Texas. Faison was adopted, and his adoptive mother testified that "she did not know whether his birth mother had twins . . . or if he had a brother" (Faison v. State, 59 S.W.3d at 235). "Robert C. Benjamin, a molecular biologist, testified on behalf of [Faison] that the odds were one in four that a brother could have the identical DNA" (id. at 237).[10] The state responded with evidence that Faison had committed other, similarly staged rapes, and the prosecutor assured the jury that "there is no identical twin running around in Dallas County" (id. at 243). The jury agreed, and Faison was sentenced to life imprisonment—his second such sentence for rapes prosecuted in separate trials.

The second NRC report (1996, 123) recommended that "[i]f the possible contributors of the evidence sample include relatives of the suspect, DNA profiles of those relatives should be obtained. If these profiles cannot

be obtained, the probability of finding the evidentiary profile in those relatives should be calculated." As with the laboratory-error rate, the committee saw this probability as a supplement to the random-match probability
in a large population. After all, the probability for a relative addresses the
hypothesis that such a person is the source. In contrast, the population
frequency addresses the hypothesis that an unrelated person is the source.
In a complete analysis, both hypotheses should be considered.

However, an ambiguity infects the committee's recommendation. On its
face, the proposal is to test all known relatives who are "possible contributors" if this is possible, and to calculate the probability of a match for any
such relatives who cannot be tested. However, testing and eliminating all
known relatives in every case consumes scarce police and laboratory resources. With a large number of loci, the probability of a match to an untested relative is so small that the effort arguably is not justified. Consistent
with this analysis, the report prefaces its recommendation for exhaustive
testing with the remark that "[i]n some instances, there is evidence that
one or more relatives of the suspect are possible perpetrators" (ibid.). The
obvious inference is that testing of relatives or calculation of relative probabilities is not necessary unless the police or the defendant can offer such
evidence.

This is the approach that most prosecutors have taken. They have no
desire to clutter a case with probabilities for events that, they might argue, have no basis in the evidence in their case. As with error probabilities, however, one must ask whether the prosecution should have the ob-
ligation to present, in its case in chief, probabilities for relatives (or a
revised random-match probability that incorporates them, as proposed
by Donnelly 1995). What if, as in *Faison,* there are no known relatives?
What if there are relatives, but no evidence points to any of them? Although the traditional approach leaves it to the defense to raise the possibility that the true culprit might be a relative, some critics of the 1996
NRC report questioned the committee's failure to prescribe that prosecution experts go beyond the typical random-match probability in all cases.
Lempert (1997, 461) maintained that as long as "the defendant [names]
any close relatives whom he thinks might have committed the crime, . . .
the state [must] replace its random match statistic with a statistic showing the likelihood that at least one named relative has DNA like the
defendant's."[11]

Lempert and David Balding, then a professor of applied statistics at the
University of Reading, also questioned the claim of the second committee
(NRC 1996, 113) that "[b]ecause one or a few relatives in a large popula-
tion will have only a very slight effect on match probability, the importance

of unknown relatives has been exaggerated." Their point was that the population of plausible suspects might not be so large (Lempert 1997), and "just one brother among the possible culprits can outweigh the effect of very many unrelated men for realistic values of the match probabilities" (Balding 1997, 474).

The issue of relatives is an extreme case of the subpopulation issue. Subpopulations are composed both of near relatives and of more distant relatives. The more general populations are composed of these groups plus very distant relatives. The committee drew on some of the work on population genetics and statistical inference regarding DNA matches published by Balding and his colleagues, but it did not accept his view on the need to "condition" on the defendant's genotype in explaining the evidential impact of a DNA match.[12] Although the committee outlined respectable procedures for estimating frequencies in large populations, in subpopulations, and in close relatives, it did not specifically address what might be called "micropopulations" dominated by close relatives.[13] Its report answers such questions as "How should we estimate a population frequency?" Arguments on whether there is a better statistic than frequencies and random-match probabilities would continue to be litigated.

The analysis of a legal rule forcing the prosecution to present a match probability for a close relative instead of a random-match probability parallels the analysis of the same proposal for error probabilities. The validity of the random-match probability as an indication that an unrelated individual is the source of the crime-scene DNA is not in question, and it is far from clear that juries will ignore the alternative hypothesis—in this case, a related individual being the source—if the defendant raises it. Likewise, the need for a rule requiring the prosecution to supplement the random-match probability with the relative probability is a departure from the traditional notion that it is up to the defendant to probe the limitations of a valid statistic about a relevant hypothesis.[14]

That said, the argument for insisting on relative probabilities in the prosecution's case in chief is stronger than the argument for demanding a supplemental error probability, if only because relative probabilities are subject to less dispute. False-positive probabilities, as we saw, are difficult to estimate, and the application of industry-wide figures to specific cases is itself open to a relevance objection. In contrast, the probability of two individuals in the same pedigree having identical genotypes is a well-studied question of genetics (NRC 1996, 113). Although a court-imposed rule may be too stringent, laboratories interested in scientific integrity and completeness ought to include relative probabilities along with random-match probabilities in written reports for the parties and the court. These reports also make it clear that the random-match probability pertains only to unrelated individuals (and they ought to caution against the ubiquitous transposition fallacy).

A postconviction challenge in *Brown v. Farwell* brings together the problems of untested relatives and transposition. According to a 2008 opinion of a panel of the U.S. Court of Appeals for the Ninth Circuit, a prisoner named Troy Brown was denied due process of law because of a "scientifically flawed DNA analysis" producing "false, but highly persuasive, evidence" (525 F.3d at 796). Yet the case involved no scientific flaw in the genotyping of the DNA, or at least none that was ever revealed. Brown had been convicted and sentenced to life imprisonment for the sexual assault of a nine-year-old girl in her bedroom in a trailer home in Carlin, Nevada. After various appeals and a state petition for postconviction relief, he filed a petition for habeas corpus relief in federal court on the ground that the evidence against him was insufficient to prove guilt beyond a reasonable doubt. The federal district court granted the petition. The U.S. Court of Appeals for the Ninth Circuit affirmed. According to the majority opinion by Judge Kim McLane Wardlaw, Nevada's "DNA expert Renee Romero of the Washoe County Sheriff's Office Crime Lab . . . provided critical testimony that was later proved to be inaccurate and misleading. . . . [A]bsent this faulty DNA testimony, there was not sufficient evidence to sustain Troy's conviction" (id. at 789).

The court identified two faults in Romero's testimony. One involved the chance that one of Brown's four brothers would have the same DNA profile. The other was the transposition fallacy noted in Chapter 2. Forty years after *People v. Collins,* the *Brown* court wrote:

> Here, Romero initially testified that Troy's DNA matched the DNA found in Jane's underwear, and that 1 in 3,000,000 people randomly selected from the population would also match the DNA found in Jane's underwear (random match probability). After the prosecutor pressed her to put this another way, Romero testified that there was a 99.99967 percent chance that the DNA found in Jane's underwear was from Troy's blood (source probability). This testimony was misleading, as it improperly conflated random match probability with source probability. In fact, the former testimony (1 in 3,000,000) is the probability of a match between an innocent person selected randomly from the population; this is not the same as the probability that Troy's DNA was the same as the DNA found in Jane's underwear, which would prove his guilt. Statistically, the probability of guilt given a DNA match is based on a complicated formula known as Bayes's Theorem, . . . and the 1 in 3,000,000 probability described by Romero is but one of the factors in this formula. Significantly, another factor is the strength of the non-DNA evidence. Here, Romero improperly conflated random match and source probability, an error

that is especially profound given the weakness of the remaining evidence against Troy. In sum, Romero's testimony that Troy was 99.99967 percent likely to be guilty was based on her scientifically flawed DNA analysis, which means that Troy was most probably convicted based on the jury's consideration of false, but highly persuasive, evidence. (Id. at 794)

This analysis is not entirely convincing because it misportrays Bayes's theorem—which is not particularly complicated here—and it fails to consider fully how the theorem operates. We need to consider two opposing hypotheses. Let H_P stand for the prosecution's proposition that the defendant is the source and let H_U be the defendant's hypothesis that some individual unrelated to the defendant is the source. The theorem states how the probability of these hypotheses changes as a result of the fact that Troy's DNA matches that of semen from the victim's clothing. We can abbreviate this fact as M. Suppose that before the DNA samples are tested, the odds of H_P (based on the other evidence in the case) are 1:1 (corresponding to a probability of 1/2). It is just as likely that Troy left the stain as it is that some unrelated person did. Bayes's theorem could hardly be simpler. It instructs us to update these prior odds by multiplying the prior odds of 1:1 by the likelihood ratio $LR = P(M$ given $H_P) / P(M$ given $H_U).$[15] If the DNA match really is 3,000,000 times more probable given that the defendant as opposed to an unrelated person is the source, then the updated or posterior odds are 3,000,000 × 1:1 = 3,000,000:1. In other words, the DNA match transforms the prior probability of 1/2 into a posterior probability of 3,000,000/3,000,001, or 99.999967%.

The transposition fallacy that produced the "false, but highly persuasive" expert testimony consists of equating (a) the probability that DNA would match if it had come from an unrelated individual, $P(M$ given $H_U)$, with (b) the probability that the DNA came from an unrelated individual given that it matched, the posterior probability $P(H_U$ given $M)$. Conceptually, these probabilities are quite different, although they often are approximately equal. *Brown* may well be such a case of approximate equality. Naive transposition yields the figure of 1/3,000,000 for $P(H_U$ given $M)$. The probability that Troy is the source is then $P(H_P$ given $M) = 1 - P(H_U$ given $M) = 1 - 1/3,000,000 = 2,999,999/3,000,000 = 99.999967\%$. If the court's description of the expert's testimony is correct, she made an arithmetic error in transposing, dropping one of the nines after the decimal point to testify that "there was a 99.99967 percent chance that the DNA found in Jane's underwear was from Troy's blood." Yet, contrary to the court's remarks, Bayes's theorem actually indicates that the statement may not be so far off if the only alternative worth considering is H_U, that an unrelated person is the source. Suppose that the other evidence was very weak, so weak that before considering the DNA match, it was 1,000 times more likely that an unrelated individual was the source (i.e., the prior odds were 1:1,000), Even then, if the DNA samples from the semen and the defendant have the profiles reported by the laboratory, the formula now shows that the odds that Troy is the source are $(1:1,000) \times [1/(1/3,000,000)]$, or 3,000:1. The corresponding probability is 99.96668%. This is smaller than the 99.99967% reported in the case, but the discrepancy hardly leaps out as a violation of due process.

One might argue that the error in the second decimal place or beyond rises to this level because the witness's description of the "chance that the DNA . . . was from Troy" invites a more serious error. It encourages the jury to think that this chance is 99.9+% even though the figure ignores the possibility that one of Troy's four brothers was the rapist, as well the other evidence in the case. Perhaps this is the basis of the majority's concern that the "complicated formula" requires other "factors" to be considered.

Once the argument is framed this way, however, it no longer is an argument about the sufficiency of the evidence. It is an argument about prejudice in the manner in which sufficient evidence is presented. But even this more tenable argument has to confront the fact that the jury was given a separate number for the chance that Troy Brown and any one of his brothers would match. Nevada's expert witness, Romero, testified that the chance of a match between two full siblings at any locus was 1/4. The DNA match in the case involved five independent loci, so the chance of a brother's having the same DNA profile then would be $(1/4)^5$, or 1/1,024. Somehow analyst Romero came up with a value of 1/6,500.[16]

All this provides ample material for cross-examination to bring out the fact that the expert's testimony could not be trusted because (1) the 99.99967% figure ignores the possibility that close relatives such as brothers would match; (2) the chance that a single brother picked at random would match was not 1/6,500; and (3) the chance that one or more of Troy's brothers would match Troy would be larger still. The last probability is approximately 1/512 for the two brothers living in Carlin and 1/256 if we toss in another two brothers living across the state line in Utah.[17]

Judge Diarmuid O'Scannlain grasped the relationship between these probabilities. His dissenting opinion maintained that transposition notwithstanding, "it was extremely unlikely that a random person committed the crime, and of the brothers, it was extremely unlikely that the specimen DNA would match not only Troy—as it did—but another brother. These probabilities put together still constitute overwhelming DNA evidence against Troy which the jury was entitled to consider" (525 F.3d at 798,

800).[18] Nevertheless, even Judge O'Scannlain's more perceptive discussion of the hypotheses about unrelated people and brothers does not necessarily dispose of the case.[19] The dissenting opinion means, as I indicated earlier, that the real issue is not the sufficiency of the evidence. The opinion shows that a rational jury that understood the evidence as Judge O'Scannlain did could have been persuaded beyond a reasonable doubt. The problem is that the evidence, as actually presented, was garbled by the failure of the prosecution, the expert, and the defense attorney to deal with a few simple probabilities intelligently. The jury might well have been confused about what the DNA actually proved and might have given it undue weight. Therefore, the system might have failed, as it too often does, because of the inadequacies of its participants.

At last, we arrive at the fundamental issue in *Brown.* Does this kind of system failure amount to a deprivation of due process, given that the defendant had the opportunity to correct the prosecution's overreaching? The majority thought that it could pretermit this issue. It wrote that "[b]ecause we affirm the district court's grant of Troy Brown's habeas petition on due process grounds, we need not reach his arguments regarding ineffective assistance of counsel" (id. at 798). But because the sufficiency theory of the majority is doubtful, the case leaves unresolved the basic questions: Did the prosecutor's (apparently) negligent presentation of scientific evidence in this case deprive the defendant of due process? Did the defense attorney's (apparent) failure to challenge the prosecution's presentation deprive Troy Brown of a constitutional right?

Brown and many other cases teach us that lawyers, judges, and DNA analysts need to understand that the random-match probability is not an index of guilt. It is not even an index of how likely it is that the defendant is the source. The probability bears on a single hypothesis—a coincidental match among unrelated individuals. But this limitation does not make the random-match probability irrelevant, because the number permits a jury to reject the hypothesis of coincidence and move on to other possible explanations for the reported match. These range from fabricated evidence or interpretative or performance errors at the laboratory to a close relative's being the source. These possibilities are not so esoteric as to be beyond a jury's comprehension. Counsel can direct the jury's attention to any of them during the examination of witnesses or in closing argument; or the court can instruct the jury on the meaning of the random-match probability or any other statistic. If, at the end of the day, the only reasonable explanation for the match is that the defendant is indeed the source, then the DNA evidence will have made a useful contribution to the proof of guilt. Properly presented and explained, the random-match probability can assist the jury in addressing the hypothesis of coincidence without distracting it from other hypotheses consistent with innocence. As a result, routinely excluding the statistic simply because it does not attend to the distinct probabilities of a false positive or of a match due to kinship would be a mistake.

CHAPTER NINE

Moving on to Short Tandem Repeat Loci

W<small>ITH THE EMASCULATION</small> of the population-structure objection, the defense bar, as well as the law-enforcement establishment, had to move on in another respect. The technology of DNA typing was changing. VNTRs have great power to discriminate among individuals, but the procedures for detecting RFLPs demand considerable time and technically challenging effort; they require that a great many copies of the relevant DNA be present in the crime-scene sample; and because gel electrophoresis cannot resolve the exact number of copies of the core sequence, they yield measurements that are inexact.

In 1983, a year before Jeffreys's "Eureka moment" with RFLPs (Chapter 3), another young biochemist, Kary Mullis, had an idea that would permit scientists to overcome all these drawbacks.[1] His idea—to mimic in the laboratory the natural process of copying strands of DNA—would advance research in molecular biology and genetics by leaps and bounds. Within ten short years, he received a Nobel Prize for his insight. In the subspecialty of forensic genetics, the polymerase chain reaction (PCR) procedure enabled new generations of forensic identification technology, rendering forensic VNTR typing a historical relic, important primarily for its past achievements in the administration of justice and for the judicial opinions and literature on the more enduring questions of laboratory quality and statistical inference of identity it precipitated.

To some courts, DNA was DNA, and PCR was just another arcane acronym (mercifully shorter than RFLP-VNTR). It was widely recognized that PCR-based methods, being more sensitive to contamination, required care in their execution, but this seemed to go to the implementation of a manifestly valid technique unanimously accepted by scientists for genetic research and medical diagnostics. No major judicial opinions expressing significant misgivings emerged to slow the transfer of the technology. The judicial vetting thus proceeded rapidly, if not always rigorously. To trace these developments, I begin with a description of PCR itself and then turn to the methods used to detect differences in the fragments of DNA generated by PCR, and the judicial opinions about them.

The Polymerase Chain Reaction

When a cell divides, its DNA must be duplicated for the daughter cells. The polymerase chain reaction (PRC) permits cells to accomplish the copying process. It also is the basis of a simple laboratory procedure for making not just two copies but exponentially large numbers of copies of short stretches of DNA (much shorter than most VNTRs). Here is a thumbnail sketch of PCR as it might be applied to the double-stranded DNA segments extracted and purified from a forensic sample. First, the purified DNA is separated into two strands by heating it to near the boiling point of water. This "denaturing" takes about a minute. Second, the single stands are cooled, and "primers" attach themselves to the points at which the copying will start and stop. (Primers are small, human-made pieces of DNA, usually between fifteen and thirty nucleotides long, of known sequences. If a locus of interest starts near the sequence ATCGAATCGGTAGCCATATG on one strand, a suitable primer would have the complementary sequence TAGCTTAGCCATCGGTATAC.) "Annealing" these primers takes about forty-five seconds. Finally, the soup containing the annealed DNA strands, the enzyme DNA polymerase, and lots of the four nucleotide building blocks (A, C, G, and T), is warmed to a comfortable working temperature for the polymerase to insert the complementary base pairs one at a time, building a matching second strand bound to the original "template." This "extension" step takes about two minutes. The result is two identical double-stranded DNA segments, one made from each strand of the original DNA. The three-step cycle is repeated, usually twenty to thirty-five times, in automated machines known as thermocyclers. Ideally, the first cycle results in two duplex DNA segments. The second cycle produces four; the third eight, and so on, until the number of copies of the original DNA is enormous (NRC 1996, 69–70). In practice, there is some inefficiency in the doubling process, but the yield from a thirty-cycle amplification is generally about 1 million to 10 million copies of the targeted sequence. In sum, the PCR magnifies short sequences of interest in a small number of DNA fragments into millions of exact copies.

Early PCR-Based Discrete-Allele Systems

PCR is not a typing method. It is just a preliminary step. But with the copious quantities of a DNA sequence that it produces, distinguishing among different alleles can be done quickly and (if all goes well) unambiguously. According to Edward Blake et al. (1992, 707), the first use of PCR in a criminal case occurred in a 1986 Pennsylvania case titled *Commonwealth v Pestinikis*. The test that Blake used could identify certain alleles at the *DQA* locus. *DQA* is a member of the group of highly variable genes responsible for recognizing foreign tissue (Chapter 1).

The different alleles of the *DQA* gene are nothing like VNTRs. Being versions of a gene, the *DQA* DNA alleles affect a biological trait (tissue types). The forensic VNTRs are not expressed as proteins and do not determine such traits. Moreover, instead of varying in their lengths, the *DQA* alleles differ in the precise order of the base pairs. For example, a G might be substituted for a T at a specific position in the gene (producing a single nucleotide polymorphism, or SNP).

The six or so different alleles commonly used in forensic work (in the 1990s) were distinguished by sequence-specific probes. Probes specific to individual alleles were placed in designated locations on a membrane. The amplified DNA then was poured onto the membrane, and a chemical reaction was used to produce a colored stain wherever the amplified *DQA* alleles had hybridized with the embedded probe. The positions of the "dot blots" on the membrane strip indicated which allele (for homozygotes) or which two alleles (for heterozygotes) were present (NRC 1996, 71–72).

By itself, the *DQA* system was not terribly powerful. Only six or seven alleles were recognized at this locus, but even these were enough to exclude a suspect quickly in many cases. On average, the *DQA* genotype of a given person is identical with that of about 7% of the population, so an innocent person could expect to be cleared 93% of the time without waiting months for the results of a VNTR test. In addition, the *DQA* system was used as part of a more detailed DNA profile. One popular system was the "polymarker" (PM) system. It analyzed the alleles of several other genes simultaneously. The chance that two randomly selected people would match at all these loci was about 1 in 4,000 (National Commission on the Future of DNA Evidence 2000, 45). Finally, a system that amplified a VNTR named D1S80 was used with a different, but still rapid system for identifying length polymorphisms. The PCR amplification works because the D1S80 VNTR is unusually short, which also permits the number of repeats of its sixteen-base-pair core sequence to be determined, thereby avoiding the statistical complications of match windows, binning, and ambiguous single bands (ibid., 151). Companies such as Roche Laboratories, Lifecodes, and Perkin-Elmer marketed these typing systems in the form of easy-to-use kits for ascertaining the discrete alleles. The FBI used them for small samples, such as saliva on cigarette butts and postage stamps, while continuing to use RFLP for large samples of blood or semen (Noble 1995).

In some cases, both VNTRs and the second-generation discrete-allele tests were used to really nail a defendant. When this happened in *Simpson,* the defense argued before the trial that the technique was not ready for prime time. First, it advanced the arguments I have already discussed in connection with VNTRs about population structure and the admissibility of a population frequency in the absence of a false-positive error probability. In an extreme reading of *People v. Collins* (Chapter 2), it also contended that the absence of "compelling proof" of the statistical independence of the new discrete-allele systems, both among themselves and with respect to the VNTR alleles, required their total exclusion from evidence on relevance grounds. This putative requirement of "compelling proof" of independence, however, is nowhere to be found in *Collins* itself. *Collins* stands for a more limited proposition: that when there are strong reasons to believe that events are not statistically independent—remember the beard and the mustache that Malcolm Collins sported—prosecutors may not blithely assume the opposite. On the other hand, when there are reasons to believe that the events of interest are at least approximately independent, but the basis for such judgments is debatable, *Collins* does not seem to dictate any particular result. Likewise, a prosecutor who does not attempt to multiply well-founded probabilities of individual events to arrive at the probability of a joint event remains outside the forbidden zone of *Collins.*

Second, referring to reservations or cautions about the use of PCR on crime-scene samples expressed by Kary Mullis, Richard Roberts, Richard Lewontin, and the 1992 NRC committee, the memorandum stressed the technical challenge of amplifying the desired DNA rather than that of a contaminant. Specifically, it contended that "[t]he methods used in this case for the collection, preservation, handling, and processing of crime scene samples for forensic PCR-based DNA testing—the DQ alpha, D1S80, and polymarker techniques—are not generally accepted as reliable among molecular biologists" (Defendant's Memorandum of Points and Authorities in Support of Defendant's Motion to Exclude DNA Evidence, Oct. 5, 1994, 36).

Mullis, as the inventor of PCR, might seem like a powerful witness for the defense. The Dream Team's Johnny Cochran told the jurors that Mullis "is the man who invented PCR. . . . He is the man who received the Nobel Peace Prize [*sic*] for this invention. And he will come in here and tell you about this evidence, how sensitive it is, and how these police departments are not trained in the collection [and] use of it, that this is by all accounts, twenty-first century cyberspace technology that is used by these police departments with covered wagon technology" (Lynch et al. 2008, 101 n.25). Yet the defense did not deliver its promised witness, probably because Mullis was (and is) something of a bête noire among many scientists. In his Nobel Prize acceptance speech, he told the world of the "tragedy" of his girlfriend's leaving him "for reasons that seemed to have everything to do with me but which I couldn't fathom" (Mullis 1997), and he was said to be more devoted to seduction than science (Yoffe 1994). He was on the defense list of witnesses for the trial itself. In anticipation of the event, the lawyers sparred over the prosecution's right to inquire on cross-examination into Mullis's eccentric views and behaviors. Neither side acquitted itself very well in that exchange. The prosecution implied that it might portray Mullis as a mind ravaged by LSD, while the defense characterized Mullis's heterodox views on the cause of AIDS as comparable with Albert Einstein's sometimes naive political positions.

In any event, the essential idea about police and laboratory "reliability" was stated well by Lewontin in response to a review article by Kathryn Roeder (1994) in *Statistical Science*. Roeder had been part of the Yale group that maintained that were no significant unresolved biostatistical issues in VNTR testing (Chapter 6). Rather than respond to her detailed analysis, Lewontin (1994a, 259) suggested that Roeder had missed "the real problems," such as the emergence of PCR-based tests. He then expostulated:

> The problem with the PCR technique is that because of its chain nature, contaminant molecules in the original sample may also be amplified and, since the original crime scene sample contained so few molecules, contaminants may overwhelm the original in the amplification. In addition, small differences in DNA sequence can have very large effects on the relative amplification of the components in an original mixture.
>
> Now consider the actual practice in a forensic DNA laboratory. A technician is handling two samples. One is the very large DNA sample from the suspect's blood, the other is the minuscule DNA sample from the crime scene,

which is then amplified by PCR. The situation is ideal for PCR contamination, with the result that the suspect's DNA will not really be compared with that from the crime scene, but with his or her own DNA that has just been replicated in the PCR reaction. The result will be a perfect match.

Speaking from his academic pulpit, Lewontin (1994a, 259–260) added:

All of us who use the PCR technique regularly are acutely conscious of the contamination problem, and the best laboratories have suffered occasionally from it. The perspiration and "oils" on fingertips have provided enough DNA contamination in PCR experiments to give completely artefactual results. Only careful replication catches these errors, and some errors have not been caught until much later when another laboratory found conflicting results. In the forensic context, where the liberty and even life of the suspect is in question, it is essential that courts be assured that laboratories are taking careful precautions against these contamination errors, not to speak of grosser errors of recording etc. Representatives of commercial laboratories that have previously been found to provide erroneous results have told interviewers that they have "cleaned up their act." Perhaps they have, but we cannot know without independent checks, and anyway what about the people convicted before they "cleaned up their act"?

In a forensic context, where the liberty and even life of a suspect is at stake, there must be frequent, independent and unannounced inspections and tests of DNA laboratories, on the model of the inspections carried out by radiation safety officers and the Department of Energy of laboratories using radioactive materials. . . . If the data themselves are unreliable, questions of probabilities of alternative suspects are irrelevant.

Neufeld, Scheck, and Thompson quoted all this, but the implications for admissibility were not clear. Lewontin did not deny that PCR-based methods, even in the forensic context, *could* produce powerful evidence. Like the 1992 committee and many other thoughtful observers, he wanted a system with far more institutional safeguards—and for good reason. Under the general-acceptance standard as traditionally understood, however, the *Frye* question for the trial court was whether PCR, as performed in a poorly self-regulated system, was part of a procedure in analytical chemistry capable of producing accurate results. If it was, then the prosecution could argue that there was acceptance of the requisite scientific foundation for this first generation of PCR-based tests, and that the risk of a false result due to contamination should remain a matter for the jury to consider.[2] Indeed, as Simpson's trial unfolded, the defense argued that the Los Angeles Police Department's crime lab was "a cesspool of contamination" (Margolick 1995).

Lewontin also wanted an end to the debate about probabilities. He concluded with a call for "idiotyping":

If, like real fingerprints, profiles were unique to individuals, no probability statements would be required. Such methods are now under development and are in the testing stage. They are based on differences in actual DNA sequence among the "repeats" in the VNTR loci, so that each person can be recognized by a unique signature. If our interest is, indeed, to correctly identify the perpetrators of violent crimes, then it is unclear why we continue to argue about probability calculations and statistical artefacts in place of carrying out the necessary research to create a real "DNA fingerprinting." (Lewontin 1994a, 261–262)

But using the fine structure of VNTRs to increase the probability of a unique DNA genotype (identical twins excepted) would not have the advantages of PCR-based systems in terms of efficiency, economy, and sample size. It would be a second generation of PCR-based tests using far more loci that would lead to claims of unique identification (Suro 1997).[3]

The courts, almost without exception, approved of the admission of the early PCR-based systems despite defense arguments that the procedures lacked scientific validity or general acceptance. Sometimes they did so in disturbingly shallow opinions. For instance, the Texas Court of Appeals in *Clarke v. State* held "DQ alpha DNA testing" admissible under the "relevancy standard which we find is consistent with *Frye*." The court did not explain how the two standards coexisted, and it did not discuss the state of the science of *DQA* testing, except to note that "[t]he State presented expert testimony from an academic biologist, a forensic serologist, and a professor of medical genetics" (813 S.W.2d 655). This is the same court that decided *Kelly v. State,* the case with the defense testimony that the radioactivity in radioactive probes was not well understood (Chapter 4). As in *Kelly,* the defense in *Clarke* apparently presented no effective evidence to counter the state's experts. The state's highest court for criminal appeals affirmed both cases. Its cursory opinion simply announced that "DNA evidence is admissible when relevant" and cited previous Texas cases (including *Kelly*) (839 S.W.2d at 93). Only one of those cases involved a PCR-based method.

Another way to circumvent *Frye* or *Daubert* objections to new techniques is legislative. Statutes providing for the admission of police radar measurements and blood-alcohol, breath-alcohol, and parentage tests are widespread, and several states have comparable statutes designed to expedite the admission of DNA evidence in criminal cases. A Tennessee law provides that "the results of DNA analysis . . . are admissible in evidence without antecedent expert testimony that DNA analysis provides a trustworthy and reliable method of identifying characteristics in an individual's genetic material" (Tenn. Code Annot. § 24-7-118(b)(1)). The point of the law is to avoid drawn-out pretrial hearings on the admissibility of a technology that the legislature has determined is suitable for courtroom use. But can the legislature be said to have rationally determined whether methods of DNA analysis that were unknown or unused when the law was adopted have sufficient scientific validity? Although the Tennessee law was adopted in 1991, when the controversy was about RFLP-VNTR typing, the Tennessee Supreme Court, starting with *State v. Begley* in 1997, applied it to a variety of PCR-based techniques that were not part of the early forensic armamentarium.

Sweeping all forms of DNA evidence under the same evidentiary rug is ill advised. That the system of gel electrophoresis of RFLP-VNTRs has been validated does not reveal much about the dot-blot system for characterizing the alleles of vitamin-D binding gene (part of the "polymarker" system). That adequate databases for estimating VNTR allele frequencies were in place was no assurance that *DQA* allele frequencies also were known. Therefore, the 1996 NRC report suggested that "courts in each jurisdiction must decide whether this new mode of [PCR-based] DNA typing satisfies the applicable test for admitting scientific evidence, regardless of whether RFLP-based evidence has been admitted" (NRC 1996, 177).

Although this much might seem obvious, not all lawyers agreed with it. In 1998, Attorney General Janet Reno, a former Dade County, Florida, prosecutor, was impressed that DNA testing was exonerating a growing number of men, many of them on death row, of committing murders and rapes. The original evidence in such cases sometimes included eyewitness identification, other kinds of trace evidence, and even confessions. As noted in Chapter 4, in 1992, Scheck and Neufeld had founded a new kind of student law clinic, the Innocence Project, at Cardozo Law School to use DNA to rectify false convictions. They encountered formidable obstacles. Many law-enforcement agencies did not collect DNA then or did not preserve samples. Legal procedures for discovering and analyzing old evidence and for obtaining judicial or other review of closed cases were limited. Prosecutors and courts were not always cooperative (Garrett 2008b; for a description of the current situation, see Garrett 2008a). Scheck (2006, 598) reminisces:

> We would have law students calling on the phone, and they would call a clerk in Virginia or North Carolina or wherever, and they would say, "Would you please go look for evidence in an old case that is twenty years old so that we can do DNA testing and prove that somebody is innocent and perhaps demonstrate that there was misconduct or mistake and upset everybody in the jurisdiction? And by the way, we are calling from Yeshiva University, Cardozo Law School, New York City. And the lawyers—yes, it is true that the lawyers

that we are working for were involved in the defense of O. J. Simpson. All right? Could you give us the stuff?

As the country's chief law-enforcement officer, Reno established a National Commission on the Future of DNA Evidence to offer guidelines on postconviction review, to assess current legal and ethical problems and needs with DNA evidence, and to consider the implications of further technological advances. The commission's Legal Working Group included Rockne Harmon, a California prosecutor sometimes accused of strong-arm tactics in his advocacy of DNA typing, and Barry Scheck, Harmon's adversary in the *Simpson* case. Another law professor, Edward Imwinkel-ried, and I prepared a survey of the legal issues and the relevant law for the committee to approve and transmit to the full commission. Scheck opposed adopting the report. I believe that he regarded it as insufficiently supportive of some positions taken by defense attorneys. For the opposite reason, Harmon opposed a statement that as new forensic DNA technologies arose, they would have to be evaluated anew under *Frye, Daubert,* or the like.[4]

The vast majority of courts, however, agreed with the Supreme Court of Indiana's statement in *Harrison v. State* that "'DNA test results' are not magic words which once uttered cause the doors of admissibility to open." James Harrison was on trial for stabbing a woman to death and then burning her two infants to death by setting the house ablaze. There was semen in the woman's mouth, but Cellmark was unable to detect any RFLPs except those of the deceased. A "new" (but unspecified) type of PCR-based test picked up enough DNA to match Harrison. He moved for a pretrial hearing on the scientific soundness of the new procedure. The state did not oppose the hearing, but in the judge's opinion, a hearing was neither required nor recommended under "the current case and the statutory law of Indiana" (644 N.E.2d at 1251). He had some basis for this ruling because Indiana had the same prosecution-friendly law as Tennessee that made "the results of forensic DNA analysis . . . admissible . . . without antecedent expert testimony." The state supreme court, however, held that the denial of the hearing was error (albeit harmless). The legislative shortcut was of no moment: "Rules of procedure, including rules of evidence, established by this court prevail over any statute" (id. at 1251 n.14).

Courts like this one were unwilling to admit new forms of DNA evidence solely on the coattails of an earlier and substantially different technology. Even so, the final outcome was the same. Despite an occasional dissent, one court after another applied anew the normal tests for scientific evidence, only to hold PCR-based techniques sufficiently reliable to establish matches between samples. The Supreme Court of Montana, for example, deemed *DQA* test results admissible despite the concern about contamination expressed in the 1992 NRC report because "the experts handling the piece of brain tissue were aware of the possibility of contamination, and took appropriate steps to avoid and detect contamination" (State v. Moore, 885 P.2d at 474–475).

More PCR-Based Systems: Short Tandem Repeats and Alphabet Soup

The genes of the *DQA* and "polymarker" systems and the diminutive D1S80 VNTR were all well and good, but they lacked the individualizing power provided by four- or five-locus VNTR matches. A better system would have the discrimination of VNTRs (it would narrow the pool of suspects to a small number of individuals even in the biggest populations) and the advantages of PCR (it would be sensitive, economical, and fast). Large VNTRs were out because they were too big to amplify. Genes that could be amplified did not come in enough common varieties because they are subject to natural selection. (Most mutations would be dysfunctional, so those new alleles would be weeded out of the population by the greater reproductive success of individuals with the fitter alleles.)

Another class of repetitive DNA provides the best of both worlds. Discovered in 1989 (Litt and Luty 1989; Weber and May 1989), they are known as "short tandem repeats" (STRs) or "microsatellites."[5] A colleague of mine calls them "God's gift to forensics." An example of an STR is the one known as D16 or D16S539, which is located on chromosome number 16. The shortest allele has five repeats of the four-letter sequence GATA, so its full sequence is GATAGATAGATAGATAGATA, or [GATA]$_5$ for short. I will just call it allele 5. The longest allele is [GATA]$_{15}$ (Butler and Reeder 2008).[6] Other STRs are liberally sprinkled along all the human chromosomes. Being far shorter than VNTRs, these STRs can be amplified, and differences in length on the order of a single repeat unit or even a single base pair can be detected.

Like VNTRs, STRs do not seem to do much of anything except litter the genetic landscape (Kaye 2007b). Because they are not subject to the evolutionary pressure of natural selection, their varied mutations can radiate through the population. Hence STRs have a fairly large and well-distributed number of alleles. Thus in a sample of Argentinians, the most common version of D16S539 was allele 11, which was seen 27% of the time. Among two samples of Italians (Cerri et al. 2004; Lancia et al. 2006), the percentages of this allele were 27 and 35. For a sample of Ecuadorian Kichwas, it was 26 (González-Andrade and Sánchez 2004, 728). These percentages are larger than VNTR allele frequencies, meaning that per locus, STRs are less discriminating than their big brothers. Nevertheless, it is possible to analyze several STRs at a time (a trick called "multiplexing"). The extra loci make up for the lower level of discrimination per locus. Thus the chance of two randomly selected individuals having the same STR alleles at thirteen loci has been estimated at about 1 in 10^{15}, making them more discriminating than the normal five or even six VNTR loci (National Commission on the Future of DNA Evidence 2000, 41).

In the most commonly used analytical method for detecting STRs, the STR fragments in the sample are amplified by using primers with fluorescent tags. The tags become part of the amplified STRs. These fragments are separated according to their length by electrophoresis. Instead of using large gels, Southern blotting, and x-ray film, though, laboratories can take advantage of automated "genetic analyzer" machinery—a by-product of the technology developed for the Human Genome Product (Butler et al. 2004; Jobling and Gill 2004).[7] In these machines, a long, narrow tube is filled with an entangled polymer or comparable sieving medium, and an electric field is applied to pull DNA fragments placed at one end of the tube through the medium. This capillary electrophoresis procedure is faster and uses smaller samples than gel electrophoresis. A laser beam is focused on a spot near the end of the tube, causing the tagged fragments to fluoresce as they pass under the light. The intensity of the fluorescence is recorded on a graph (an electropherogram), which shows a peak as an STR flashes by. On the basis of the time it takes the tagged DNA fragments to pass through the capillary, a computer program determines which alleles are present at each locus. Figure 9.1 is a sketch of how alleles 5 and 8 might appear in an electropherogram.

Medical and human geneticists were interested in STRs as markers in family studies to locate the genes that are associated with inherited diseases, and articles on their application to human identity testing began to appear in the early 1990s (Edwards et al. 1991, 1992). Developmental research to pick suitable loci moved into high gear in England, Europe, and Canada. The United Kingdom's Forensic Science Service applied a four-locus testing system in 1994. Then it introduced the "second-generation multiplex" (SGM) for simultaneously typing six loci in 1996. These tests soon were used to build England's National DNA Database. First called NDAD and later designated NDNAD, the system allows a computer to check the STR types of millions of known or suspected criminals against

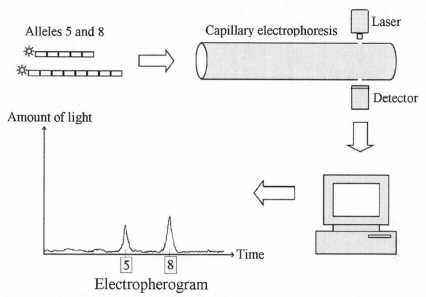

Figure 9.1. Sketch of an electropherogram for two D16 alleles. One allele has five repeats of the sequence GATA; the other has eight. Each GATA repeat is depicted as a small rectangle. Although only one copy of each allele (with a fluorescent molecule or "tag" attached) is shown here, PCR generates a great many copies from the DNA sample with these alleles at the D16 locus. These copies are drawn through the capillary tube, and the tags glow as the STR fragments move through the laser beam. An electronic camera measures the colored light from the tags. Finally, a computer processes the signal from the camera to produce the electropherogram.

thousands of crime-scene samples. A six-locus STR profile can be represented as a string of twelve digits; each digit indicates the number of repeat units in the one or two alleles at each locus. These discrete, numerical DNA profiles are immensely easier to compare mechanically than the complex patterns of fingerprints.

In the United States, the FBI began coordinating a major project to evaluate protocols and kits from Promega Corporation and Applied Biosystems and to establish reference databases for allele frequencies. The FBI settled on thirteen "core loci" to use for the U.S. national DNA database system. All these loci—and more—now can be typed in a single reaction (J. M. Butler 2006). The result, as we will see in Chapter 11, is an identification technology that approaches uniqueness—the idiotyping that Lewontin wanted. Figure 9.2 displays 203 possible alleles at fifteen STR loci that can typed in one multiplex PCR reaction.[8] The electropherogram of an individual would show only one or two of these peaks at each locus.

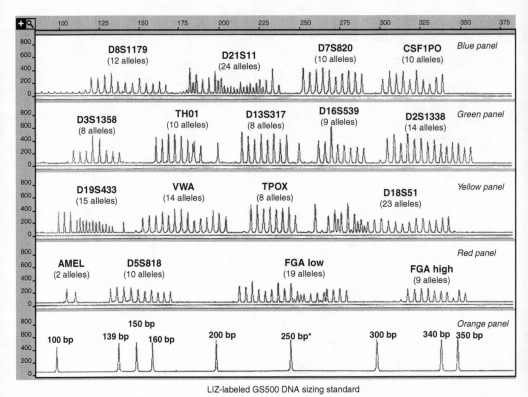

Figure 9.2. Alleles of fifteen STR loci and the amelogenin sex-typing test from the AmpFISTR Identifiler kit. The bottom panel is a "sizing standard"—a set of peaks from DNA sequences of known lengths (in base pairs). The numbers in the vertical axis in each panel are relative fluorescence units (RFUs) that indicate the amount of light emitted after the laser beam strikes the fluorescent tag on an STR fragment. This figure was published in J. M. Butler (2005, 128), copyright Elsevier (2005).

Initially, the courts were asked to exclude all STR results from evidence. The grounds were familiar: STR profiling uses PCR, which demands careful sample handling and controls; moreover, the reference databases on STR allele frequencies are small and not geographically extensive enough to establish the validity and accuracy of the random-match probability estimates. The courts were not receptive to these claims. To them, "STR" was just another ingredient being stirred into the alphabet soup of DNA, RFLP, VNTR, *DQA,* PM, and PCR that had been or were being held admissible now that the "DNA wars" had been declared over. The 1996 NRC committee had expressed optimism about the emerging technique, and the DNA forensic science community had mastered the art of publishing. Indeed, the *Journal of Forensic Sciences* continues to publish tedious (but reassuring) reports on allele frequencies "for the record"—meaning for the courts.

Given these developments, the courts were easily convinced that enough research and development had been undertaken for STRs to be admissible. In *State v. Gregory* (2006), for example, the Supreme Court of Washington reviewed a pair of appeals brought by Allen Eugene Gregory. In 2000, Gregory was convicted of a 1998 rape. In 2001, he was convicted of a 1996 murder committed in the course of a rape and robbery. The evidence included the Washington State Patrol's RFLP-VNTR typing for which the "random match [probability] was 1 in 235 million" and a "private lab['s STR] testing" for which "the chance of a random match in the African American population was 1 in 190 billion" (147 P.3d at 1229). Reviewing the case law and referring to "[h]undreds of scientific articles [that] have been published regarding the use of STR technology" and extensive validation in "inter-laboratory comparisons conducted throughout the world," the court held that the trial judge was within his discretion in denying Gregory a pretrial hearing on general acceptance. Gregory had supported his request for a hearing with statements from University of Washington neurogeneticist Randall Libby, a burr in the side of prosecutors,[9] who denied that "STR testing, use of the profiler plus testing kit, or capillary electrophoresis, is . . . generally accepted," and from Laurence Mueller, who "raised questions about linkage equilibrium in STR databases." The court wrote Libby off as "a defense expert whose conclusions this court has questioned before" because of his "personal financial interest in having the courts hold that there is significant disagreement in the scientific community." Mueller fell into the same category, and his ideas about population genetics were passé. Hence "questions concerning linkage equilibrium in STR databases would be more properly discussed by experts at trial whose testimony has been evaluated under the [usual relevance] standard" (id. at 1240–1241).

In sum, the courts were losing their patience with challenges to forensic DNA technology writ large. But several special problems cannot be brushed aside. They continue to provoke more substantial dissent or dispute among experts.

CHAPTER TEN

Transcending Race and
Unscrambling Mixed Stains

O NE MIGHT THINK that with an end to the controversy over population
structure and a loss of traction for arguments about error and kinship probabilities, the statistics of DNA matches were no longer a barrier
to their admission as evidence in criminal cases. But there were more obstacles to overcome, and qualms about DNA statistics continue to surface.
This chapter takes up two such concerns. First, it returns to the use of ill-
defined, socially constructed racial categories in estimating random-match
probabilities. This time, the objection is not that the census-type classifications mask genetically varied subgroups that mate preferentially among
themselves (population structure), but that the broad classifications themselves are biologically problematic. Second, this chapter discusses the vex-
ing problem of interpreting the mixtures of DNA sometimes found at
crime scenes. To solve this problem, the concept of the likelihood ratio as
a measure of probative value (Chapter 2) will prove useful. Some symbols
and numbers also will come in handy, but these are for clarity and con-
creteness, not computation. As with the population-structure issue, a few
products and sums will go a long way toward analyzing the basic issues.

Deracializing DNA Statistics

Even after the California Supreme Court laid the population-structure
objection to rest in *People v. Soto,* a line of cases in that state continued to

reject DNA statistics on distinct relevance grounds (Kaye 2008). The Court of Appeal for the Fifth District of California initiated this misadventure in *People v. Pizarro*. This case is a holdover from the early days of the population-genetics debate. In 1990, Michael Pizarro was sentenced to spend the remainder of his life in prison for raping and suffocating his thirteen-year-old half sister, Amber, in 1989. Vaginal swabs revealed semen that matched Pizarro's VNTRs at three loci. An FBI analyst testified that "[t]he likelihood of finding another unrelated Hispanic individual with a similar profile as Mr. Pizarro is one in approximately 250,000" (3 Cal. Rptr. 3d at 97–98).

Pizarro appealed, contending that the DNA evidence was inadmissible because the prosecution had failed to demonstrate that the DNA test procedure and the within-race multiplications were generally accepted in the scientific community. The court of appeal remanded for a hearing on these matters. By the time of the rehearing in 1998, the California Supreme Court had resolved these questions in favor of admissibility. Deprived of his original arguments, Pizarro raised "several new issues concerning the reliability and relevance of the DNA evidence presented to the jury" eight years earlier. "The trial court again ruled that the evidence was admissible and reentered the judgment" (id. at 29).

Pizarro appealed once more. In 2002, the appellate court reversed the conviction because of a variety of perceived defects in the analysis and presentation of the random-match probability. After publishing this opinion, the court decided to rehear the second appeal a second time "to ensure that the complex issues in this case were thoroughly examined and briefed by both parties" (id.). In its final opinion in 2003, the appellate court determined that the trial court erred in two respects. The conviction of Michael Pizarro was finally, firmly reversed.

In part, this reversal came about because the prosecution presented DNA genotype frequencies for only two groups—Hispanic and Caucasian—when there was no proof (other than that pointing to Pizarro) that the perpetrator of the crime was Hispanic or Caucasian (or half of each, as Pizarro was). Elsewhere I have criticized the court's peculiar theory that "in the absence of sufficient evidence of the perpetrator's ethnicity, any particular ethnic frequency is irrelevant" (3 Cal. Rptr. 3d at 104). Because seeing how much random-match probabilities vary within major population groups conveys useful information about the distinctiveness of the DNA profile, the usual practice of giving several statistics is both logically and legally appropriate. When a range of statistics is introduced to show how robust a result is, it makes no sense to demand evidence that every one of the varied assumptions is true. The point of the exercise is to show that whatever possible assumption one makes, the outcome is essentially the same (Kaye 2004). Nevertheless, the court's conclusion that the testimony of the FBI analyst, Dwight Adams, prejudiced Pizarro is justifiable. If the perpetrator could have come from any of several groups, looking only to the defendant's group for a random-match probability could lead the jury to jump to the conclusion that the perpetrator must have come from that same group. The manner of presenting the DNA statistics therefore could be regarded as unfairly prejudicial.[1]

In any event, there was something more to *Pizarro* and its progeny than a mistake in logic. Not far beneath the surface was an aversion to the notion of race itself. In *Pizarro,* Presiding Justice James Ardaiz referred to the need for "cautious evaluation . . . because of the ambiguous nature of artificially defined ethnicities and the uncertainties connected to use of an ethnic database" (3 Cal. Rptr. 3d at 104). Race, he correctly suggested, is an ambiguous category, and the criteria for the racial classifications in the reference databases are not clear cut.

Other judges have expressed similar or more extreme discomfort. In *Wilson v. State,* Jackie Barron Wilson had been convicted of capital murder for abducting, raping, strangling, and suffocating a five-year-old girl and then driving over her body in Grand Prairie, Texas. Wilson applied for postconviction DNA testing, but the trial court denied the petition. In 2006, the Texas Court of Criminal Appeals, the state's highest court for criminal cases, affirmed on the ground that there was ample evidence linking him to the abduction[2] and that inasmuch as Wilson did not deny that he was involved,[3] more testing "would not exonerate appellant because it would show nothing more than there was another party to the crime, at best" (185 S.W. 3d at 485). Judge Cheryl Johnson wrote an opinion concurring in the result because of perceived weaknesses in DNA profiling when used as the sole evidence of guilt. Apparently, she believed that a DNA match to someone other than Wilson would not be conclusive proof of that individual's guilt, and therefore Wilson should not be entitled to postconviction DNA testing.[4]

Judge Johnson, who holds degrees in chemistry and crystallography, questioned the statistics of DNA matches. She opined that "the evidence that inserts numbers into the legal equation is new and marginally understood."[5] She then posed a "basic question"—"what makes one 'Hispanic'?" (id. at 489). This is a good question, but the judge slipped into the fallacy, described in Chapter 6, of thinking that it is important to define the ancestry of the defendant in a case rather than to identify the population of plausible suspects:

Appellant's surname is Wilson, a name not ordinarily thought to be Hispanic. May we assume that appellant's father was not Hispanic? Part Hispanic? Part

African-American? Part Western European? Eastern European? Asian? Were the differing probabilities of a non-Hispanic gene pool taken into account in calculating probabilities? How are probabilities for racial groups calculated in general? How do we calculate reasonably accurate probabilities for people like that famous self-described "Cablinasian," Tiger Woods? We do not know. (Id. at 489, footnote omitted)

But we know exactly what to do about defendants of recently mixed race. The issue in a criminal-stain case is whether the match between a suspect and the stain results from the fact that the suspect left the stain. With no disrespect intended to the phenomenal golfer Tiger Woods, let us imagine that he is the suspect in a criminal case. DNA recovered from the crime-scene matches his. The prosecution's hypothesis (H_p) is that Woods is indeed the source. As in the *Brown* case (Chapter 8), another hypothesis (H_U) is that an unrelated individual is. To evaluate the probative value of the match in choosing between these hypotheses, we form the likelihood ratio. If the laboratory has identified the genotypes correctly, the conditional probability in the numerator is one. Regardless of how mixed up Woods's racial heritage is (and we all have mixed ancestry to some degree), his genotype will be in the stain.[6] The conditional probability in the denominator also does not depend on the intricacies of the suspect's ancestry. If the crime-scene stain came from an unrelated person, then we want to know how rare the genotype is in various populations (the random-match probability). Stratifying by self-declared race gives a rough idea of how greatly the frequencies of forensic DNA genotypes vary among the major population groups in a mixed population.

Admittedly, the census-type categories used in constructing reference databases are socially, not biologically, derived. Ancestry and geography are not race (Marks 2002). But forensic DNA analysts do not need—and do not seek—to force individuals into some mythical racial taxonomy. By examining appropriate polymorphisms, anthropologists and forensic scientists can make inferences about the perceived race or ethnicity of the source of a DNA sample (Shriver et al. 1997; Frudakis et al. 2003; Bamshad et al. 2004; Phillips et al. 2007; cf. Mountain and Risch 2004). These inferences are probabilistic, of course, but self-declared racial classifications correspond surprisingly well to a large set of STRs (Tang, Quertermous, and Rodriguez 2005). Despite the imprecision and errors in whose alleles get counted in the reference population databases, the division of people into these databases is not entirely arbitrary (Walsh et al. 2003).

That said, there is something appealing about getting away from the fuzziness of these cultural constructs. Why not eliminate the reference to races or to ethnic or geographic subgroups? In *Pizarro* and again in the 2005 case *People v. Prince*, Judge Ardaiz suggested that perhaps one could dispense with the race-based statistics by presenting "only the most conservative frequency, without mention of ethnicity; or . . . the frequency in the general, nonethnic population" (*Prince,* 36 Cal. Rptr. 3d at 324). These remarks are reminiscent of the 1992 NRC committee's desire to avoid considering the relevant population by using a procedure—the ceiling method—that is supposed to apply (as an upper bound) to all ethnic and racial groups.

When a sufficient number of loci can be typed, the only thing lost by the proposal to use a single number is the information on variability that comes from hearing how much the profile frequencies differ for three or four races. It is doubtful that the jury will find this variation of much importance, at least when the members of the set of frequencies are in the millionths or beyond. Consequently, the proposal to pick the largest frequency in any of the major races generally is reasonable.

The second approach floated in *Prince,* to use a racially blended database, is also feasible. In effect, one would pool the reference databases for the various races and then use the allele frequencies in the pooled database to obtain a profile frequency estimate. The population supposedly represented by the pooled reference database is not randomly mating, but the affinal model (Chapter 7) can handle this complication. As the technical working group of the National Commission on the Future of DNA Evidence (2000, 5), headed by James Crow, put it, "[T]he necessity for group classification could be avoided by using an overall U.S. database and an appropriately increased value of [the population-structure parameter] θ. . . . A θ value of 0.03 would usually be appropriate."

Finally, the need to refer to specific groups could be avoided by computing the probability that the defendant would match given that the true source is a close genetic relative. Even when the suspect population is composed entirely of unrelated individuals, this number can provide an upper bound on the frequency in that population. Thus Crow's group recommended using such a computation for a sibling when "there is uncertainty about the population substructure, as with isolated tribes or communities, or possible unsuspected relatives" (ibid., 4). The formulas for this "sib method" are well known, and "[s]ince no other relatives are as close as sibs, the match probability for sibs provides a rough upper limit for the actual match probability" (ibid.; compare Belin, Gjertson, and Hu 1997). In addition, the sib method is almost free of assumptions about population structure and linkage equilibrium. (The formula is dominated by the factor $1/4$, which depends only on Mendelian rules (NRC 1996, 113).) Although using the probability of a match in a full sibling as opposed to an unrelated person is very conservative, it stands on a sounder scientific footing than the various ceiling methods of an earlier era.

Thus there are scientifically defensible ways to avoid referring to racial population frequencies or probabilities and to convey a rough sense of the power of the evidence in a general-population case. The rules of evidence should allow these "deracialized" statistics to be used, but neither evidence doctrine nor scientific knowledge dictates that this be done.

Unscrambling Mixtures

The dubious reasoning about relevance in *Pizarro* was not confined to the simple question of how frequent a particular DNA type is in the general population. A complication in the case was that the vaginal swab contained DNA from more than one individual. This is typical of rape cases, where DNA from the victim's epithelial cells is present together with DNA from spermatozoa. Often, a chemical procedure can remove the female DNA first, leaving the male fraction for analysis.[7] In *Pizarro*, however, this differential extraction was not successful. At two of the VNTR loci, more than two bands were present. Because one individual can have no more than two alleles per locus—one from each parent—it was clear that the laboratory was dealing with an intractable mixture.

Two Missing Alleles

Interestingly, at the third locus (D2S44, abbreviated by the court as D2), only two alleles could be detected. What was missing? The state's theory was that the perpetrator happened to have the same single-locus genotype as the victim. Then the male and the female DNA would show only those two bands (which I will call A_1 and A_2). As half siblings, Pizarro and the victim would be expected to have more than the normal number of alleles in common. Sure enough, they had the same genotype at this particular locus. So the FBI just used the (small) frequency of such heterozygotes in the population in computing the multilocus random-match probability.

This was a mistake. It could have been avoided if the likelihood theory of relevance had been used. The point of the statistical analysis is to express the value of the evidence in distinguishing between the prosecution's theory that Pizarro's sperm was in the sample and the defense theory that someone else was the source of that DNA. Under the prosecution's theory, the chance that the mixture would be type A_1A_2 was 1: A_1 and A_2 were the only alleles that Pizarro and his half sister possessed. Under the defense theory, which the FBI analyst should have been evaluating, one needs to consider all the men whose DNA would produce the A_1A_2 type when combined with the deceased girl's A_1A_2 type. Certainly, the unknown perpetrator might have been just like Pizarro and his half sister at that locus, as the analyst had assumed. But he also might have been homozygous (either A_1A_1 or A_2A_2). Let us say, purely for the sake of illustration, that each allele occurs in 10% of the population. Then the chance of randomly drawing a man with both alleles would be $2p_1p_2 = 0.02$. That is like the FBI's figure. In addition, we need the chance of randomly drawing a homozygote of either type A_1 or A_2. That is just $p_1^2 + p_2^2 = 0.02$. Everyone else—the remaining 96% of the population—would be excluded by testing at the D2 locus. Framed as a likelihood ratio, the match at the locus is thus $1/0.04 = 25$ times more likely to arise if Pizarro is the source than if a randomly selected, unrelated man is.

Yet Pizarro was able to find an expert to testify that the match at this locus should be tossed out because it could not be interpreted. According to biologist William Shields, "If . . . the perpetrator is homozygous and the defendant is heterozygous, the defendant is actually excluded as a potential perpetrator (i.e., he is exonerated). There is no way to know what the mixture means. . . . [W]hen there is even one shared band between the victim and the defendant, the autorad should be excluded entirely" (3 Cal. Rptr. 3d at 72). The other two defense experts in the case, geneticist Laurence Mueller and Northwestern University statistician Sandy Zabell, did not subscribe to Shields's peculiar take on the mixture. Needless to say, neither did the prosecution's experts, Ranajit Chakraborty, Patrick Michael Conneally, and George Sensabaugh.

The court of appeal nevertheless embraced Shields's recommendation that "autorads with mixtures such as this be entirely excluded from the statistical calculations, in part because two of the three possible perpetrator profiles would actually exclude defendant as a suspect" (id. at 77). At first, the court correctly recognized that "if the perpetrator's D2 genotype was discerned solely by reliance on defendant's D2 genotype, the perpetrator's genotype was discerned by an improper procedure" (id. at 78). The failure to consider all the relevant alternative hypotheses was another example of an FBI analyst not getting the statistics right. However, the court then "digress[ed] to discuss the suggestion . . . that because the perpetrator's genotype cannot be discerned, all possible genotypes should have been accounted for" (id.). Building on Shields's idea, Justice Ardaiz proffered the following analogy about hair color:

When evidence is lacking on a certain fact such that the fact cannot be established, the situation does not justify consideration of all possible alternatives to that fact. Only the one fact is relevant. If [an] eyewitness is uncertain about the perpetrator's hair color, but can narrow the color down to black, brown, or blond, should all three possibilities be taken into account? The logic supporting an affirmative answer states: all possible perpetrators have black, brown, or blond hair; the defendant has black hair; therefore, the defendant is a possible perpetrator. Although initially appealing, this logic ignores the fact that the perpetrator has only one hair color and thus only that one hair color is relevant to his profile; more importantly, it ignores the fact that if the perpetrator actually has brown or blond hair, the defendant simply is not the perpetrator. The correct logic requires a choice of these three possible syllogisms: (1) all possible perpetrators have black hair; the defendant has black hair; therefore, the defendant is a possible perpetrator; (2) all possible perpetrators have brown hair; the defendant has black hair; therefore, the defendant is not the perpetrator; (3) all possible perpetrators have blond hair; the defendant has black hair; therefore, the defendant is not the perpetrator. (Ibid.)

The court continued:

It would defy the principles of evidence to allow the eyewitness to testify that the perpetrator has black, brown, or blond hair when there is no way of establishing which one hair color the perpetrator actually possesses. This testimony is neither relevant nor probative, but it is potentially damning because it draws the defendant into the pool of possible perpetrators when in reality it more likely excludes him—two of the three possibilities exonerate him. (Ibid.)

Applying this logic to the DNA results, the court reasoned:

Similarly, only the perpetrator's one D2 genotype was relevant to his genetic profile. If the prosecution could not establish which genotype the perpetrator possessed at that locus, there was no relevant evidence to admit from that locus. But, as in the analogy, the most compelling reason for demanding proof of the perpetrator's genotype and for refusing to admit evidence of all three possible genotypes was that the other two possible genotypes were more than irrelevant—they potentially proved defendant's innocence. Thus, the evidence that was admitted to incriminate defendant actually had a greater chance of exonerating him. If the perpetrator was not heterozygous (i.e., if he was either homozygous for the top band or homozygous for the bottom band), defendant did not match the perpetrator and he was excluded as a possible perpetrator. Only if the perpetrator was heterozygous did defendant match and become a possible perpetrator. (Id. at 78–79)

These dicta are extremely problematic. All manner of evidence in criminal cases narrows the class of suspects but still admits of heterogeneous subclasses. That membership in some of these subsets can be inconsistent with guilt does not render the evidence irrelevant or prejudicial. Consider the court's own illustration. If an eyewitness's testimony can eliminate everyone with red, white, orange, and purple hair (10% of the population, let us say), leaving only the 90% drab browns, blacks, and blonds, then the witness's observation advances the inquiry. It makes it more probable that the defendant, who has black hair, is the criminal. An accurate witness's account will occur 100% of the time when the suspect is the culprit, but only 90% of the time when someone else is. The resulting likelihood ratio is $1/.9 = 1.11$. This ratio tells us that the hair color is weak evidence—it is close to the value of 1 that would obtain if the report were totally irrelevant. But it has a slight tendency to prove that the suspect is the culprit. Therefore, it is relevant and should not be ignored. Because there is no particular reason for the jury to give the testimony about hair color undue weight, it should not be excluded.

In contrast, the court's suggestion that evidence that a criminal has either black, brown, or blond hair exonerates a suspect with black hair is absurd on its face. It rests on the premise that because two of three logically enumerable hypotheses imply innocence, a suspect is probably innocent. This assumes that every state of nature is equally probable—an assumption that is clearly implausible. The alternative hypotheses listed by the *Pizarro* court—that "the perpetrator was . . . either homozygous for the top band or homozygous for the bottom band" (id. at 79)—are relatively improbable. Most people are heterozygous, and (without further information on the genotype of the perpetrator) it is more likely that the unknown perpetrator in this case was too. DNA mixtures can be complicated to interpret, but this does not make those interpretations inadmissible.

As for Pizarro himself, the state conducted new DNA tests. One STR locus after another in the DNA from a vaginal swab showed alleles that were not the victim's. All of these alleles were part of Pizarro's genotype. With this additional evidence, the jury in Michael Pizarro's third trial returned another verdict finding him guilty of the rape and murder of his half sister on a June night nineteen years earlier.

Mixing It Up in *People v. Simpson*

Mixtures are not limited to sexual-assault cases. In the *Simpson* case, various bloodstains were mixtures, as would be expected if the blood of the murderer and a victim (or that of the two victims) had commingled. But a stain that is consistent with the presence of DNA from Nicole Brown and O. J. Simpson could be a mixture of blood from individuals with DNA profiles that are different from those of Brown and Simpson. That is, the profile of each individual could be different, but because no one can tell which individual contributed which DNA alleles, the composite pattern could be the same. A set of four alleles $A_1, A_2, A_3,$ and A_4 at a locus could originate from pairs of people with genotypes A_1A_2 and A_3A_4 or from pairs with A_1A_3 and A_2A_4 or with A_1A_4 and A_2A_3. Mixtures from all such people will show the same four alleles A_1 through A_4. The trial court ruled that the DNA evidence for the mixed stains could not be admitted without quantifying the probability or frequency of finding such patterns in mixed samples in which Simpson was not a contributor.

This quantification might be accomplished in two ways. Because the interpretation of mixtures is becoming an increasingly significant issue with the advent of exquisitely sensitive methods of detecting stray DNA, it is important to understand the logical foundations of mixture analysis, and the *Simpson* case is a good starting point.[8] The defense wanted to constrain the prosecution to a probability or frequency of inclusion. Suppose that each of the four alleles A_1 through A_4 occurred in 10% of the population. Everyone with the following genotypes would be included as a possible contributor: $A_1A_2, A_1A_3, A_1A_4, A_2A_3, A_2A_4,$ and A_3A_4 (the heterozygous types), and $A_1A_1, A_2A_2, A_3A_3,$ and A_4A_4 (the homozygous types). Each of the heterozygous types occurs with frequency $2(0.1)(0.1) = 0.02$, and each of the homozygous types occurs with frequency $(0.1)^2 = 0.01$. The proportion of the population that would be included as possible contributors is therefore $6(0.02) + 4(0.01) = .16$. This leaves 84% of the population excluded. Applying this type of reasoning to one of the mixed stains in the *Simpson* case, defense lawyer William Thompson advised the court that "45.4 percent of Caucasians would match . . . , 59.2 percent of African Americans, [and] 48.8 percent of Hispanics . . . would be included" (Transcript of Examination of Robin Cotton, May 15, 1995, *5).

Such numbers made it seem that the mixed stains had little value in linking Simpson to the killings. A second approach produced more impressive numbers. Instead of asking what the probability is that a randomly selected individual would be excluded as a possible contributor to the mixed stain, one can pose a more specific series of questions about subclasses of these individuals. Suppose that one of the contributors to the stain on the steering wheel of Simpson's car was Simpson himself (with genotype A_1A_2) and the other was not Nicole Brown (with genotype A_3A_4). What is the probability that a randomly selected person would have contributed the look-alike alleles A_3 and A_4? The answer (for our illustrative numbers) is $2p_3p_4 = 0.02$. Or, suppose that the blood came from neither Nicole nor O. J. What is the chance that two randomly selected individuals would produce an $A_1A_2A_3A_4$ mixture? This probability is $(2p_1p_2)(2p_3p_4) + (2p_1p_3)$ $(2p_2p_4) + (2p_1p_4)(2p_2p_3) + (2p_2p_3)(2p_1p_4) + (2p_2p_4)(2p_1p_3) + (2p_3p_4)$

$(2p_1p_2) = 24p_1p_2p_3p_4 = 0.0024$. There are many other possibilities to consider. Maybe there were three contributors to the stain, but they happened to have the same collection of the A_1 through A_4 alleles, or maybe they had some other alleles as well, but these failed to show up in the laboratory test. The probability depends on the facts on which one chooses to "condition" the computation.

The advantage of this condition-by-condition approach is precisely that it lets us consider the value of a match under the different conditions. Simpson had testified to having cut his hand recently. A juror might well ask, "OK, if part of the stain on the steering wheel came from Simpson and the other part came from someone else driving the car at some other point, what is the chance that this unknown driver happened to be type A_3A_4?" How many times more likely is it that the evidence would be the way it is under the different possible scenarios that are consistent with Simpson's guilt as opposed to his innocence? Thus the conditional approach leads to a series of likelihood ratios.

The prosecution turned to Bruce Weir to satisfy the court's demand for quantification. A native of New Zealand then teaching in North Carolina (he is now chairman of the University of Washington's Department of Biostatistics), Weir is an accomplished statistical geneticist who had done extensive analyses of VNTR and other data from the commercial DNA testing labs, the FBI, and other law-enforcement agencies around the world. He was at the forefront of the defenders of independence—the kind of person Lewontin feared would end up on a second NRC committee (Chapter 7). Weir was precise and meticulous in his work and in his speech. Lander and Budowle (1994, 737) had recently referred to him as "a vigilant commentator." His preliminary report proposed to take the conditional approach and to provide likelihood ratios because, as he later put it, the exclusion probabilities favored by the defense "often rob the items of any probative value" (Weir 1999, 29).

The defense would have none of this. Its March 20 Notice of Objections contended that likelihood ratios are not scientifically accepted in interpreting DNA mixtures and that no jury could understand them. Only the simple, aggregate inclusion probability was scientifically valid or generally accepted. The only authority that Scheck and Thompson's memorandum cited for their scientific claim was a recent law review article by Thompson. On May 9, however, Neufeld unveiled something better. He called the court's attention to a sentence in the 1992 NRC report that would dominate Judge Lance Ito's thinking, at least for several weeks. The report asserted—without explanation, analysis, or any references to the relevant statistical literature—that "[i]f a suspect's pattern is found within the mixed pattern, the appropriate frequency to assign such a 'match' is the sum of the frequencies of all genotypes that are contained within (i.e., that are a subset of) the mixed pattern" (NRC 1992, 59). Neufeld assured the court that this "is the way that any experienced scientist would do it" (Transcript of Examination of Robin Cotton, May 9, 1995, *5), and his cross-examination got Cellmark's Robin Cotton—the prosecution's star DNA expert—to agree that she was not aware of any writing on any other approach besides the NRC committee's idea of "aggregating genotypes for mixed stains" (ibid., May 10, 1995, *10). Judge Ito finally ruled, fuzzily, that he would require "the prosecution to present some type of statistical analysis with regards to the mixtures" (ibid., *16).

The prosecution seemed prepared to give in to the defense's demand for the NRC-blessed statistic. On May 15, San Diego deputy district attorney (now judge) George "Woody" Clarke, who had been recruited to help with the DNA evidence in the case, advised the court that "there is no intent on our part to introduce likelihood ratios" (ibid., May 15, 1995, *3) Then, late on June 15, just several days before Weir was to testify, Clarke telephoned Simpson's lawyers. The prosecution had given Weir the OK to talk about likelihood ratios.

The defense DNA lawyers were livid. Back on May 15, Thompson had argued to Judge Ito that "we know of no precedent for the admission of likelihood ratios in a criminal matter" (ibid.). On June 13, Neufeld reiterated that "there was no precedent for using likelihood ratios in a criminal case" (Transcript, Examination of Lakshmanan Sathyavagiswaran, June 13, 1995, *1). Judge Ito seemed convinced. On June 16, he told Thompson that he had located "only three likelihood ratio cases, reported cases nationwide" and that these did not count because they involved "likelihood ratios in paternity cases, which is obviously not the situation we are dealing with here." Thompson could not agree more. "Mixtures, et cetera, et cetera," the judge murmured. Thompson was elated: "If I understand you correctly, you are saying you view this as an unprecedented matter. So do we" (Transcript, Examination of Richard Rubin, June 16, 1995, *2). Judge Ito jumped aboard: "It is from beyond novel to being unique. . . . And I made a ruling . . . that I think was pretty clear." A happy Thompson concurred: "It sounds like your thinking coincides with ours, that the prosecution should simply be precluded from presenting statistics in the likelihood ratio form." But Judge Ito backed away: "Well, I think at the very minimum, if they insist on proceeding with that, I think we need [an admissibility] hearing out of the presence of the jury" (ibid., *3).

The stage was set for Weir to defend his preference for likelihood ratios outside the presence of the jury. Weir's scientific position was impeccable.

Likelihood functions and ratios are a fundamental and well-established concept in statistics (e.g., Cramér 1946). Using them in analyzing mixtures had been discussed in the forensic science literature (Evett et al. 1991).[9] Weir testified that the NRC's recommendation was either ambiguous (and consistent with the conditional-probability approach) or wrong-headed, but that in any event he placed little stock in the committee's treatment of statistical issues (Transcript, Examination of Bruce Weir, June 22, 1995). The second NRC committee, charged with examining the statistical issues more carefully, took the same view of the probability-of-inclusion calculation. "That calculation is hard to justify, because it does not make use of some of the information available, namely, the genotype of the suspect. The correct procedure was described by Evett et al. (1991)" (NRC 1996, 130; for more extended criticism of the probability-of-inclusion statistic, see Clayton and Buckleton 2005, 219–223; Buckleton and Curran 2008).

Thompson's cross-examination of Weir was tedious. Lawyer and statistician quarreled over Thompson's choice of words.[10] An exasperated Judge Ito finally declared: "Mr. Thompson, if it is of any benefit to you, I understand the two assumptions that underlie Dr. Weir's testimony and the calculations that he makes and the fact that you will get different results depending on what the assumptions are. I understand that. I understood that an hour ago" (Transcript, Examination of Bruce Weir, June 22, 1995, *40). Eventually, Thompson called his own witness, William Shields, to testify, among other things, that the NRC's recommendation was clear, that Shields agreed with it, and that Weir was not abiding by it. Although Shields's research at the State University of New York at Syracuse's College of Environmental Science and Forestry was half "in the area of behavioral ecology, primarily with birds" and "half . . . about . . . the evolution of population structure and how population structure influences evolution" (Transcript, Examination of Bruce Weir and William Shields, June 22, 1995, 2), Thompson asked the biologist to opine on whether Weir's assumptions "are easy assumptions for people to understand" and whether it "would it be easy to explain what's wrong with Dr. Weir's assumptions" (ibid., 9).

Judge Ito grasped the essentials of the debate. He agreed with the defense that Weir was not following the NRC recommendation, but then he decided to allow this deviation. Out of the entire "185-page report," Judge Ito pointed out, the committee devoted "all of one sentence to the analysis of mixtures." He ruled in favor of the prosecution:

Having listened to the testimony of Dr. Weir and Dr. Shields, I agree that the manner proposed by the prosecution to present the mixture evidence is appropriate as long as it is made clear that it is based upon the two basic assumptions and that there's no reference to the known samples, that it be presented merely that Dr. Weir has the certain qualifications, that he's analyzed the data with regards to these mixtures based upon the evaluation of the databases, the ranges of frequencies are from a to z, period, thank you very much.

"That should take half an hour," he concluded (ibid., 30). Then he asked Weir to produce calculations on the assumptions that there were two, three, or even four contributors to mixed stains.[11]

Weir never had the chance to discuss likelihood ratios with the jury. To avoid the defense argument that they were prejudicial and to accommodate the judge's reluctance to be the first to admit them in a criminal case involving DNA mixtures, the prosecution limited its questioning to the conditional frequencies that form the denominators of these ratios. For example, Weir testified that if two randomly selected people of various races contributed to a stain, the probability of the alleles found on the steering wheel would range from 1/60 to 1/11,000; that if three people did so, the range would extend from 1/9 to 1/3,500; and that for four contributors, the range was from 1/1 to 1/3,000 (Transcript, June 23, 1995).

From the likelihood perspective, these numbers were unfair to the defense. They do not include the conditional probability for the prosecution's hypothesis. The probability computed under the prosecution's theory belongs in the numerator of the likelihood ratio, and the probability computed under an alternative hypothesis belongs in the denominator. We are asking how many times more likely it is to see the evidence under the hypothesis related to guilt than under the one related to innocence. In the case of a stain of type A_1A_2 from a single person, the probability that the stain will contain the defendant's genotype under the prosecution's hypothesis that the matching defendant is the source is simply 1. The probability of a random match is $2p_1p_2$. Hence it is $1/2p_1p_2$ times more likely to have the incriminating genotypes when the defendant is the source than when an unrelated individual is. This is the likelihood ratio that grades the probative value of the matching genotypes. For the conveniently uniform allele frequencies of 10% postulated earlier, its value is $LR = 1/2(0.1)(0.1) = 50$. The matching genotypes are 50 times more probable if the defendant is the source than if he is not.

The situation is, perforce, more complicated with mixtures, but the basic logic is identical. Under the prosecution's theory that Nicole's DNA was transferred to the steering wheel, if there were exactly two contributors, the other part of the stain could have come from Simpson, but it also could have come from an unknown individual. The chance of a mixed stain from Nicole (with type A_1A_2) and a random person matching the mixture A_1 through A_4 is just $2p_3p_4$. The chance of the mixed stain for two unknown individuals is $24p_1p_2p_3p_4$. The likelihood ratio applicable to these two hypotheses is therefore $2p_3p_4/24p_1p_2p_3p_4 = 1/12p_1p_2 = 8.33$. The point is that with various hypotheses involving mixtures, the numerator of the likelihood ratio will be less than 1, which diminishes the ratio. Attending solely to the conditional frequencies in the denominator, as Weir was forced to do, therefore gives an inflated impression of the value of the match in a mixed sample. In the *Simpson* case, the defense's insistence on frequencies rather than likelihoods for the mixed stains may not have redounded to its benefit.

Could the trial judge have allowed the more complete presentation? Certainly. Likelihoods have a well-established statistical pedigree. The derivation of the applicable expressions was published, and in any event, it was just a matter of algebra that need not be subjected to *Frye* or *Daubert* tests (Kaye, Bernstein, and Mnookin 2004). What about the lack of precedent? Well, there is a first time for everything, but likelihood ratios were no strangers to the courtroom. They were (and are) used every day in civil cases of disputed paternity to indicate that the genotypes of the mother, the child, and the alleged father were much more probable if the alleged father was the biological father than if an unrelated man was the father. There is no basis for thinking that likelihood ratios are sound statistics in civil cases but not criminal ones or that they are useful to a jury in evaluating allegations of genetic relatedness but not genetic identity. Although Judge Ito had found only three previous criminal cases in which likelihood ratios had been admitted, appellate courts had approved of the use of the ratio for genetic relatedness or identity (either alone or as part of a more dubious Bayesian analysis) in criminal cases in Indiana, Iowa, Kansas, Maryland, Missouri, Montana, and North Carolina. Today, federal district courts, as well as appellate courts in Arizona, the District of Columbia, Massachusetts, Montana, New York, and Washington, have held that likelihood ratios are an admissible method of analyzing mixed stains.

Simpson had two other arguments against them, though. One was a version of the untenable no-foundation argument in *Pizarro*. Thompson argued that "the . . . ratios . . . are lacking in foundation because the assumptions that are predicate to the conditional probabilities don't necessarily jibe with the facts of this case and in some ways are completely inconsistent with the prosecution's theory of the case" (Transcript, May 15, 1995, *3). Shields testified along the same line, stating that conditioning on the number of possible contributors to a stain or which individuals might have contributed to it "makes assumptions that go beyond the genetic evidence in my opinion. The genetic evidence of a mixture only allows one to say that one or more individuals contributed to that. It doesn't tell you who. If one assumes that one knows who produced a mixture, one is no longer doing a test of a hypothesis" (Transcript, Examination of Bruce Weir and William Shields, June 22, 1995, *6).

This criticism makes the same mistake that the *Pizarro* ruling did. The prosecution's theory of the case affects the numerator. The alternative hypotheses in the denominator are intended not to "jibe with" that theory but to contradict it. By analyzing their implications, the statistical analyst can assist the jury in deciding between competing hypotheses. In other words, the alternative hypotheses about the number and nature of contributors are not factual assumptions that are part of a party's chain of proof. Those would require a foundation. These hypotheses are premises adopted solely to show how strong that proof is. Being postulated for the sake of argument, they do not require genetic or any other proof. The likelihood ratio is not pulling any hypothesis up by its own bootstraps. It is unbuckling the bootstraps to see what fits.

Simpson's final argument was that jurors would be misled by likelihoods. For example, they might think that a likelihood ratio of 100 means that the prosecution's hypothesis is 100 times more likely to be true than the defense's hypothesis. All it really means is that if the prosecution's hypothesis is true, then the observed genotypes are 100 times more likely to be in the sample than otherwise. This is easily the strongest argument. Misstatements of this sort can be found in judicial opinions (Kaye, Bernstein, and Mnookin 2004). They are a variant of the transposition fallacy. As with the transposition of individual conditional probabilities, however, the tendency of experts and lawyers to transpose the probabilities in a likelihood ratio calls for care in the presentation of the evidence rather than its outright exclusion.

This is not to say that all testimony about mixtures is admissible or that the best methods for handling them are in place. Deciding whether a mixture is present and systematically coping with the inherent ambiguity in the number and nature of contributors to a mixed stain remains a challenge. Peter Gill of the United Kingdom's Forensic Science Service has been quoted as advising forensic scientists, "Don't do mixture interpretations unless you have to" (J. M. Butler 2005, 166). The *Simpson* prosecutors did not follow this precept, and Weir did the best he could with the mixtures he was given. As a statistical geneticist, he basically listed many possible combinations and computed relative frequencies or probabilities for these combinations. This is a comfortable way for statisticians to proceed. But laboratory workers might want to use additional information to help disentangle mixtures. Looking at an autorad with four bands at a single locus, for instance, they might be tempted to arrange them into pairs if two were faint and two were dark. The darker bands could have come from the major contributor to the sample, because more DNA fragments should pick up more of the radioactively tagged probe and thereby produce greater exposure on the x-ray film. Likewise, in an electropherogram of STRs, the DNA from a major contributor should produce bigger peaks than the DNA from a minor contributor.

The 1992 NRC committee warned against trying to tease much out of the intensities of bands on an autorad: "Interpretations based on quantity can be particularly problematic—e.g., if one saw two alleles of strong intensity and two weak intensity, it would be improper to assign the first pair to one contributor and the second pair to a second contributor, unless it had been firmly established that the system was quantitatively faithful under the conditions used" (NRC 1992, 65).

Autorads have been supplanted by new technologies, however, and the case for using intensity information (the heights or areas of the peaks in an electropherogram), as well as the presence and frequencies of STR alleles, is stronger (e.g., Clayton and Buckleton 2005; Cowell, Lauritzen, and Mortera 2007). Expert computer systems have been devised for facilitating the probabilistic analysis, and algorithms have been developed to use clues about the relative quantities of DNA from different sources (e.g., Cowell, Lauritzen, and Mortera 2007).

As currently conducted, however, most mixture analyses involving partial or ambiguous profiles entail considerable subjectivity. Subjective judgments about ambiguous output are not necessarily bad, and there are general guidelines and protocols for complex mixture interpretation (Gill et al. 2006).[12] But when different experts might interpret the same data differently, procedures such as blinding the examiner to an expected or desired result become important. Analysts and fact finders need to understand that likelihood ratios or inclusion probabilities for a particular profile give an incomplete and overstated impression of probative value when other profiles might reasonably be said to be consistent with the laboratory data (W. C. Thompson 2009).

As scientists push PCR to detect STRs from only a few cells (Buckleton and Gill 2005), the number of cases with mixed profiles and artifacts—and the room for debate about the interpretation of these results—will only increase (Gilder et al. 2008; McCartney 2008). Procedures for typing as little as a single cell's worth of nuclear DNA have been shown to work, to some extent, with trace or contact DNA left on the surface of an object such as the steering wheel of a car. The most obvious strategy for typing such small samples is to increase the number of amplification cycles. The danger is that chance effects might result in one allele being amplified much more than another. Alleles then could drop out, small peaks from unusual alleles at other loci might "drop in," and a bit of extraneous DNA could contribute to the profile. Such "low copy number" STR profiles have been admitted in courts in a few countries, but given the lack of standardized methods and the effort required to obtain useful results, it does not appear that courts in the United States have encountered this form of DNA evidence.

Y Chromosome STRs

In sexual-assault cases, additional information about the contributors to mixtures can be obtained with Y chromosome STRs (Y-STRs). High-school biology teaches us that females have two of the same kind of sex chromosome (XX), while males have two distinct sex chromosomes (XY). In human beings, the Y chromosome carries a gene, *SRY,* which stands for "sex-determining region Y." This gene codes for a protein that results in the development of the male's sperm-producing factory, the testis. The gene is a master switch. Unless it is thrown, the fetus will develop into a female.

The intricacies of the developmental process are not fully understood (Polanco and Koopman 2007), but what is important in forensic applications is the simple fact that a father passes his Y chromosome to his sons. Consequently, all the genes, STRs, and other loci on the Y chromosome are inherited paternally, as one package or "haplotype."[13] All men in the same paternal lineage should have the same Y-STRs (subject to occasional mutations).

The population genetics of Y-STRs therefore is much simpler than that of the "autosomal" STRs—the ones on the other chromosomes—that we have considered so far. Statistical independence, Hardy-Weinberg equilibrium, and linkage equilibrium are not applicable. A set of ten Y-STRs, for instance, will be inherited together, as part of the whole chromosome that is transmitted intact from one generation to the next. In this regard, they are like a single allele, and in estimating the frequency of a particular ten-locus Y-STR haplotype, no multiplication of allele frequencies is involved. To establish that other paternal lineages do not, by chance, have the same ten-locus haplotype, one needs to collect and analyze population samples to verify that the haplotype is unique to that family line or that such lines are at least unusual. If a ten-locus haplotype is not rare, perhaps a twenty-locus one will do. Thus much work has been done to develop a large panel of Y-STRs and to examine their frequencies in various populations (J. M.

Butler 2006). As with STRs on other chromosomes, commercial kits are available to amplify many of these markers in a single reaction.

This patrilineal inheritance of Y-STRs has made them a powerful tool in genetic studies of human migrations. The South African Lemba, for instance, describe their ancestors as Jews who sailed from Judea about 2,500 years ago. Like many Jewish populations, about 10% of the members of this tribe possess a particular haplotype that is rare among other groups (Thomas et al. 2000). Y-STRs are used in producing personal genetic histories as well (Shriver and Kittles 2004). Indeed, a genetic study helped convince many historians that Thomas Jefferson was the father of some descendants of the slave Sally Hemings (Foster, Jobling, and Taylor 1998; Abbey 1999; Wade 1999). Other researchers claim that some 8% of all the men now living within the boundaries of the empire ruled by Genghis Khan 1,000 years ago are his patrilineal descendants (Zerjal et al. 2003). This would mean that throughout the world, 1 man in 200 was fathered by the male children of the Khan and his brothers. Similar stories have been told about the Y chromosomes of Europeans (Sykes 2004).

In crime detection, the patrilineal inheritance of the Y chromosome creates an opportunity to use Y-STRs as a clue to the name of the man whose DNA is discovered in the investigation of a sexual assault or other crime. Fathers do not just give their children a random half of their genes. In many cultures, they also give them their family name.[14] Such surnames are a relatively modern invention (introduced about 700 years ago in England, when feudal record keepers needed to distinguish tenant farmers with the same first name). To the extent that men in the same paternal line originally adopted the same surname and that men in other lineages chose different ones, these surnames (or the modern, mutated spellings of them) will be correlated with Y haplotypes (Sykes and Irven 2000). Could the police exploit this correlation to generate a list of probable names from the DNA sample? Turi King, a postdoctoral researcher at the University of Leicester, and her colleagues (2006) tested the idea by recruiting a random sample from the electoral rolls of 150 men with different surnames born in the United Kingdom, taking care to avoid known or likely relatives. They typed the men using a set of seventeen Y-STRs. Then they sampled and typed another random set of 150 men with the same names from the electoral rolls. The first sample provided a table of haplotypes and names. King guessed the names of the men in the second sample by assigning each of them the name of the man in the first sample with the closest haplotype. The ratio of correct haplotype-driven guesses was almost one in five (28/150). For the 80 men with the least common names (but still common enough to cover more than 40% of the U.K. population), the ratio of correct guesses was one in three. The Leicester group noted that "DNA-based surname prediction is in principle applicable to any society having diverse patrilineal surnames of reasonable time-depth" (ibid., 387). These geneticists and others have advised the British government to develop and deploy a surname Y-STR database on a trial basis (Sharp 2007).

Surname prediction is a concept that has yet to be implemented in forensic science, but Y haplotypes have shown their mettle in resolving mixtures. In rape cases, any Y-STRs that can be detected will pertain to men only. Because each man has only one Y-STR allele per locus—again, we are dealing with only one Y chromosome per man—the number of Y-STRs at the various loci should indicate the number of male contributors to the sample (or at least the number of distinct male lineages in the mixture). Knowledge of this number simplifies the likelihood-ratio approach to evaluating the match at the autosomal loci because other possibilities can be eliminated. Furthermore, a match between a defendant and the Y-STRs in a sexual-assault sample is additional evidence in and of itself that the defendant is a contributor. If the haplotype is distinctive of a particular family line—as shown by an adequate database—it suggests that the male contributor is a member of that line.

At this juncture, hardly any reported opinions have examined the admissibility of Y-STR haplotyping. In *Curtis v. State*, a Texas court held a pretrial hearing on general acceptance, and the state's intermediate appellate court affirmed its decision to admit the evidence. Gloria King was strangled to death in her bed in 1995. In 1997, suspicion focused on her nephew, Clifton Earl Curtis. He denied any sexual intercourse with King, but DNA from autopsy swabs matched his DNA. In 2001, Curtis was convicted of capital murder, but the conviction was reversed because of an erroneous jury instruction. On remand, Curtis was tried and convicted of aggravated sexual assault. A ten-locus Y-STR test performed by Cellmark was part of the evidence against him.

He appealed again. He had evidently challenged the test under *Frye,* and the trial court had held a pretrial hearing, but the hearing had been somewhat perfunctory. The state had Cellmark's forensic laboratory director, William Watson, who "holds a Bachelor of Science degree in microbiology and a Master of Science degree in biology," testify that "YSTR employs the same technology" as the more established autosomal STRs (205 S.W. 3d at 661). Watson further testified that Y-STR typing "underwent validation studies prior to use in forensics," that "those studies have been published in peer-review journals," and that "there is a general consensus in the scientific community that YSTR testing is reliable, accurate, and valid" (id.).[15]

Like the defendants in the early Texas cases on VNTRs, Curtis did little to counter Cellmark's claims. He called no experts of his own but did establish on cross-examination that different laboratories used somewhat different sets of Y-STRs. Unsurprisingly, "the trial court found that the YSTR methodology had been validated 'internally and externally' and subjected to peer review, that it was generally accepted in the scientific community, and that the YSTR evidence was reliable and relevant" (id.). On appeal, Curtis contended that this was error because "there was no evidence that YSTR test results have been admitted into evidence in a criminal trial anywhere in the United States and . . . the YSTR test is not standardized because different labs use different markers for the test" (id. at 660). The court of appeals noted that the trial court acted well within its discretion.

Courts are likely to continue in this pattern. The typing technology for Y-STRs does not differ from that for autosomal STRs. The population genetics are different, but no less understood. Unless the defense bar can recruit experts to uncover serious problems in the validation studies or the Y-STR databases (many of which have been developed for other kinds of genetic research), courts will treat Y-STRs as just another form of STR. Indeed, it might well take experts with the prestige of a Lewontin and the persuasiveness of a Lander to reverse the momentum.

CHAPTER ELEVEN

Outside the Nucleus: Mitochondrial DNA

TWO BILLION years ago, a catastrophe struck Earth's atmosphere. As blue-green algae (the technical term is cyanobacteria) multiplied in the warm seas, they took in carbon dioxide and sunlight to produce carbohydrates. As a by-product of this photosynthesis, they spewed oxygen into the atmosphere. Within the space of a mere 200 million years, the partial pressure of this gas went from 1% to more than 15% of current levels. Oxygen tends to form free radicals, which are atoms or groups of atoms with one or more unpaired electrons. The unpaired electrons make these radicals highly reactive and thereby able to damage biologically important molecules, including lipids, proteins, and nucleic acids. In sum, toxic oxygen was being pumped into the atmosphere (Abele 2002).

But one bacterium's poison is another's treasure. Some bacteria already were adapted to local environments rich in oxygen and could use oxygen in reactions that produced ten times as much energy as their anaerobic cousins. Bacteria are basically bags of chemicals, including free-floating circular loops of DNA and a longer strand compressed like a ball of string. The DNA molecules are the bacterial genome; they code for the proteins that matter to bacteria. Some of these bacteria (probably purple nonsulfur ones) were engulfed by another type of cell. The host cell provided the new resident with organic molecules that it needed, while the smaller cell used oxygen molecules in making energy-rich molecules for the host, keeping the toxic oxygen within its own membrane. Over time, some genes from the bacterium migrated into the nuclear DNA of the larger cell, and others that no longer were necessary in the stable environment provided by the host were lost. Today, the modified bacterium is an "organelle" (the microscopic analog of an organ within a single cell) known as a "mitochondrion." In this way, increasingly abundant oxygen has allowed complex organisms with large energy demands to burn an energy-rich fuel.

As a rough outline, this theory of "endosymbiosis" is widely accepted as the explanation of how mitochondria developed (Kurland and Andersson 2000). Each mitochondrion is a miniature power plant where fuel derived from the food we eat is combined with dissolved oxygen to produce the high-energy chemical adenosine triphosphate (ATP). "Molecules of ATP pour out of the mitochondria to other parts of the cell that need energy; when they reach their destination, they are discharged like a battery, then sent back to the mitochondria for a boost" (Sykes 2004, 287). Human cells are peppered with hundreds to thousands of these little infernos, floating between the nucleus and the cell membrane, each encased in its own membrane and each containing two to ten copies of its own bacteria-like genome. Cells that require lots of energy, like brain and muscle cells, have more mitochondria than other cells. Although the nuclear genome has over 3 billion base pairs, the mitochondrial genome is a puny, circular molecule only 16,569 base pairs long. This minichromosome contains the thirty-seven genes that code for the structure and synthesis of the mitochondrion's oxygen-capturing enzymes. As in bacteria—and unlike the DNA in the nucleus of the cell—the genes are arrayed one after another, with small intervening sequences and few introns within the genes. However, two hypervariable regions of more than 300 base pairs compose what is called the "control region" or "D-loop." Mutations (SNPs) can accumulate in these noncoding sections without altering how the mitochondrion functions. As a result, the particular sequence in the control region varies substantially within and across populations (Stoneking 2000; Melton et al. 2001). Sequencing certain coding regions provides additional discrimination (Nilsson et al. 2008).

The multiplicity and small size of mitochondrial DNA (mtDNA) give it a special place in forensic DNA testing. First, because a cell has only one nucleus, for every copy of chromosomal DNA, there will be hundreds or thousands of copies of mtDNA. This "high copy number" means that it is possible to detect mtDNA in samples containing too little nuclear DNA (nDNA) for conventional typing. Second, being a relatively short sequence to start with, mtDNA is more stable than nDNA. Thus mtDNA is the genetic system of choice in cases where tissue samples are very old, very small, or badly degraded by heat and humidity.

One other feature of mtDNA has immense forensic significance. Mitochondria are inherited from mother to child. Fathers contribute only nuclear DNA to the next generation. The explanation goes like this:

The mammoth egg cell is endowed with perhaps 100,000 mitochondria. The tiny sperm's mitochondria (perhaps 50–100) are concentrated at the base of the tail, where they provide energy to power the flagellum. Very few male mitochondria are believed to enter the egg at the moment of conception, and those that do can easily be lost by "dilution" when the egg cytoplasm is partitioned during mitosis. Moreover, there is evidence that any remaining male mitochondria are actively eliminated from the egg. It appears that ubiquitin, expressed on the mitochondria surface, marks the male mitochondria as "foreign" and targets them for destruction. (Cold Spring Harbor Laboratory n.d.)

Thus the only mitochondria present in the newly fertilized cell originate from the mother. The fertilized egg divides again and again, forming an embryo, a fetus, a baby, and eventually an adult. All the mitochondria in all the cells are copies of the original from the mother's egg. From one generation to the next, only women pass their mitochondria to their offspring. Consequently, siblings, maternal half siblings, and others related through the maternal lineage possess the same mtDNA sequence (subject to infrequent mutations).

On the one hand, this is a drawback in some forensic applications because it means that mtDNA sequences are not unique identifiers. (Nevertheless, as we shall see, they still can be valuable in associating an individual with a hair or other material that has too little nDNA to type.) On the other hand, the maternal pattern of inheritance also makes them particularly useful for establishing membership in a family. Just as the paternally inherited Y-STRs of Chapter 10 are used to trace paternal lineages, the mtDNA types, or "mitotypes," can establish whether an individual is part of a family line going back through generations upon generations of mothers. This chapter travels from the jungles of Colombia to the capital of the Urals and to the hideouts of Mafia dons to provide examples of such forensic family-association studies. The next chapter turns to efforts to associate suspected criminals with human hairs found on victims or at or near crime scenes. It discusses the relationship between mtDNA typing and the traditional comparison of hairs under the microscope, and the arguments that have been made—unsuccessfully thus far—against the admissibility of mitotyping.

The Ordeal of Clara and Emmanuel Rojas

An example of the use of mtDNA to establish a family association comes from the current guerrilla war in Colombia. By 2008, the Revolutionary Armed Forces of Colombia (FARC) was holding more than 750 hostages. Among the captives in the jungle were Ingrid Betancourt, a French-Colombian presidential candidate, and Clara Rojas, her running mate.

They had been prisoners for six years, enduring aerial raids, long treks through the forest, and time spent in chains for trying to escape (James 2008). In a difficult delivery in 2004, Clara Rojas gave birth to a child, whom she christened Emmanuel, a gift from God. When the boy was eight months old and suffering from malaria, a broken arm, severe malnutrition, anemia, a high fever, diarrhea, and leishmaniasis (a serious parasitic skin disease common in the jungle), her captors took him away, promising to obtain medical care.

They left the suffering child with José Crisanto Gómez, a thirty-seven-year-old peasant, in the rebel-controlled town of La Paz, saying that they would return in several days. It was not uncommon for the guerrillas to leave their children in the care of other families, and Gómez did not know who the mother might be. More than a month passed. Seeing the boy's health worsening and one of his own sons ailing, Gómez took the two boys to a public clinic in a nearby town for treatment. He pretended that "Juan David Gómez" was the son of a niece who had died. The clinic immediately transferred the child to the provincial hospital. A day later, the boy was declared a ward of the government's Colombian Institute for Family Welfare. Soon, he improved enough to be moved to Bogotá for further treatment. While recuperating from surgery on his arm, "Juan David" was placed with a foster family. When Gómez sought to recover the boy, the Family Welfare officials ruled that "Juan David" had been abandoned (Gonzalez 2008).

The matter mushroomed into an international incident. Not realizing that the boy was out of their control, the rebels had pledged to release Emmanuel, his mother, and former Congresswoman Consuelo González in a deal brokered and publicized by Venezuela's flamboyant president, Hugo Chávez. FARC pressed Gómez for the child's return. Still unaware of the boy's true identity, Gómez thought that the child was his "life insurance." He feared that if he told FARC that Emmanuel was with Family Welfare, FARC would kill him. With nowhere to turn, Gómez went to the police. The government put the family in its witness protection program (ibid.).

By that time, Chávez had gathered a slew of international observers to supervise the hostages' release and sent several aircraft to the city of Villavicencio, on the edge of Colombia's jungle, to await the return of the three hostages (ibid.). To Chávez's embarrassment, FARC reneged, claiming that government attacks had made the transfer impossible. Colombian president Alvaro Uribe denied this. He responded that FARC canceled the promised release because it was unable to retrieve the child (Brodzinsky and Carroll 2008). Investigators followed the trail to the three-and-one-half-year-old foster child. To verify that he was actually Clara Rojas's son, they analyzed mtDNA from the boy and Clara's relatives—her brother, Ivan Rojas, and their mother, Clara González de Rojas. Given the maternal inheritance of mitochondria, all three—and, of course, Clara Rojas—would be expected to have the same sequence of base pairs in their mtDNA if the boy was Clara's child. (We soon shall see how such sequencing can be accomplished.) Colombia's attorney general announced that there was a "total" match (Forero 2008).

When Clara Rojas finally was released, she was reunited with her son (Allen-Mills 2008). Former presidential candidate Ingrid Betancourt remained in captivity, seriously ill with leishmaniasis and in deteriorating condition (Guillen and Gonzalez 2008). In July 2008, government troops rescued her and fourteen other hostages in the jungle of south central Colombia (Romero 2008).

The Ghost of Corleone

The "maternality" of mitotyping also helped track down Italy's most wanted man.[1] Bernardo Provenzano, also known as "the tractor," rose through the ranks of the Mafia by mowing people down. He was a member of the Corleone clan, which took control of the Sicilian Mafia after a power struggle in the early 1980s that left hundreds dead. Provenzano was the right-hand man of Salvatore "Toto" Riina, the Corleone boss and Sicilian godfather until his arrest in 1993. Then Provenzano took over.

Giovanni Falcone, the prosecutor who was most responsible for Riina's arrest, and another anti-Mafia judge were killed in a bombing in 1992. Provenzano was convicted in absentia and sentenced to life imprisonment for his involvement in the murders. Italy's national prosecutor of organized crime, Pietro Grasso, vowed to find the killers. But Provenzano had another nickname, "the Ghost" or "the Phantom of Corleone." He moved from one safe house to another, communicating through tiny typed notes, *pizzini,* passed back and forth through a labyrinth of secret relays. All the police had to go on was a photograph taken in 1959 and a computer-generated "identikit" of what he might look like now. Provenzano's lawyer even insisted that he was dead.

But Grasso never gave up. In 2005, he asked Guiseppe Novelli, head of the medical genetics laboratory at Tor Vergata University in Rome, to help identify Provenzano. "I was taken aback at first," said Novelli. "I mean, how can you identify someone who you have no information about?" (Butler 2007, 811). Grasso had one crucial lead, however. A Mafia informant, or *pentito,* said that Provenzano had gone to Marseille in 2002 under a false name to be treated for a prostate tumor. French police raided the clinic. In the files was a case history for a man who had never been admitted to the hospital. The description of one "Gaspare Troia" matched Provenzano.

Novelli was given hospital tissue samples for "Troia." Of course, he had no DNA from the elusive Ghost to compare it to, but the Ghost had a brother. Brothers have the same father, which means that they have the same Y chromosomes and hence the same Y-STRs (Chapter 10). Having the same mother, they also have the same mitochondrial chromosome and hence the same mtDNA sequences. If Novelli could compare these two haplotypes in "Troia's" DNA to Provenzano's brother's DNA, he would know whether "Troia" was Provenzano. All he needed was some DNA from the brother. There are many ways police can acquire DNA from individuals of interest. They can do it directly, with a judicial order. They also can do it surreptitiously. They might follow the individual into a restaurant and retrieve the coffee mug he drank from, wipe up his saliva if he spits on the sidewalk, or pick up a discarded cigarette butt (Imwinkelried and Kaye 2001). In this case, Novelli acquired blood samples from a Palermo hospital where Provenzano's brother had been admitted for surgery.

The results "showed that they shared the same mother and father—they were brothers," said Novelli. Grasso was ecstatic. "Novelli's group did a terrific job. You cannot even imagine my enthusiasm when I had the proof of the complete match between the profiles." Knowing that Provenzano was alive, the investigators focused their attention on the ring of people who had provided the logistics of Provenzano's trip and hideout in Marseille. By tracking a packet of laundry sent by his wife, police captured Provenzano in an isolated farmhouse just a mile from his birthplace in Corleone. For his part, Grasso was overjoyed "to have fulfilled the promise I made on the grave of Falcone, that for the rest of my life I would pursue a sole objective: capturing all those responsible for his death" (Butler 2007, 811).

The Last Tsar

The cases of Clara Rojas and Bernardo Provenzano involved living relatives and fresh samples of DNA. The same ability of mtDNA to trace maternal ancestors and progeny, combined with the durability of this tiny genome, makes it invaluable in identifying human remains as well. From the tsunami-ravaged coast of Thailand to the rubble of the World Trade Center, from Bosnia to Haiti, from Vietnam to Kuwait, mtDNA analysis has been used to identify the remains of the victims of natural disasters, accidents, and human conflict and cruelty (Holland et al. 1993; Owens,

Harvey-Blankenship, and King 2002; Biesecker et al. 2005). In fact, mtDNA dating from between 29,000 and 100,000 years ago has been extracted from the bones of Neanderthals and compared with the mtDNA of living people. The degree of similarity is the same across continents, undermining the theory of interbreeding between ancient Europeans and Neanderthals proposed by some anthropologists (Krings et al. 1997, 1999, 2000; Ovchinnikov et al. 2000; Serre et al. 2004).

A much-discussed example of the analysis of human remains is a small-scale but intensive study of mtDNA from a grave in Siberia. In the chaotic spring of 1918, Tsar Nicholas II and his family were transported as prisoners from St. Petersburg to the town of Ekaterinburg, beyond the Ural Mountains. According to most histories (e.g., Radzinsky 1992; Massie 1995), on July 17, with an anti-Bolshevik force approaching, the Bolsheviks marched the tsar, Tsarina Alexandra, their children, their doctor, a maid, and two waiters to their deaths in a cellar. Precisely what happened next is a little murky. According to one account, the bullet-ridden bodies of the family were soaked in acid and deposited in an abandoned mine. But when rumors of the location spread, the bodies were retrieved and loaded onto a truck. The truck broke down or became mired in the mud, and the bodies were left in a hastily dug grave. A simpler version of the story holds that the bodies never reached the mine shaft because the truck developed a mechanical fault earlier, and sulfuric acid was thrown into the open grave (Gill et al. 1994, 130). In still another variation, the bodies of Tsarevich Alexei and one of his sisters were burned and left in a pit elsewhere (Attewill 2007).

So things remained until the summer of 1991. Following the discovery in 1989 of a communal grave site twenty miles from Ekaterinburg, Russian Federation officials unearthed a large number of bones. After painstakingly organizing and fitting the fragments together, they concluded that they had a total of nine skeletons. The facial areas of the skulls were destroyed, and some of them showed bullet wounds and bayonet marks. Several of the deceased had been aristocrats (judging by the platinum and gold dental work). Russian medical examiners concluded that the bodies in the grave included those of the tsar, the tsarina, and three daughters. The bodies of Tsarevich Alexei and one of the daughters were missing (Gill et al. 1994, 130).

The British Study

At the request of Russian authorities, the British Forensic Science Service (FSS) applied DNA-typing methods to verify these findings. Working with

Pavel Ivanov, a molecular biologist at the Engelhardt Institute of Molecular Biology in Moscow, Peter Gill and his colleagues quickly established that four of the bodies were male and five were female. They made this determination by amplifying part of a gene that codes for a protein found in developing tooth enamel. This amelogenin gene occurs on both the X and Y chromosomes. As with STRs, capillary electrophoresis can separate distinct alleles on each chromosome. Because females are XX, only a single peak is observed when their nDNA is tested, while DNA from males, who possess both X and Y chromosomes, generally exhibits two peaks (Haas-Rochholz and Weiler 1997; Santos, Pandya, and Tyler-Smith 1998). From the small amount of DNA extracted from the bones, five STRs also were amplified. The STR patterns were indicative of a family of two parents and three children, along with four unrelated individuals.

Finally, DNA sequencing was used to compare the mtDNA in the skeletons with that of known relatives of the tsar and tsarina. A common sequencing procedure employs synthetic, single nucleotides with a fluorescent tag of a different color depending on whether the nucleotide is an A, T, G, or C. The specially constructed nucleotides are like a brightly painted engine that comes in four colors at the front of a train. The normal nucleotides that follow are like the boxcars of the train. (This train has no caboose.) The synthetic nucleotides—the engines—attach to a boxcar (normal nucleotide), but no boxcars will attach in front of them.

To use the special nucleotides to determine the order of the base pairs, the DNA fragment to be sequenced is prepared as a single strand. Consider the sequence AACAGCT. In a solution with the right enzymes, other chemicals, and lots of normal nucleotides, a complementary strand of increasing length will form. "Reading" in reverse order, it will start with an A to match the T at the far end of the target, then become a GA that complements the CT, and then grow to become a CGA, a TCGA, a GTCGA, a TGTCGA, and finally a TTGTCGA. Because most of the nucleotides in the solution are boxcars, most of the complementary fragments that assemble themselves into a little train become a normal TTGTCGA strand—they will not include the brightly colored engines. At any step in this process of elongation, however, a complementary synthetic nucleotide—an engine of the right color—could be incorporated. For example, a synthetic T (put in bold to show that it is an engine) could attach to the normal CGA to form **T**CGA. These strands will stay as they are because the engine, in this case the **T**, couples only to boxcars behind it. When all the synthetic nucleotides are used up in forming strands, there will be a collection of the following sequences: **A**, G**A**, CG**A**, **T**CGA, G**T**CGA, TG**T**CGA, and T**T**GTCGA. These strands can be separated by length, and an electropherogram will show a colored peak at the position of each engine. The sequence of colors reveals the sequence of nucleotides in the target strand (Sanger, Nicklen, and Coulson 1977).

To compare sequences of the mtDNA in the skeletons with those of known family members, the FSS turned to none other than His Royal Highness Prince Philip, the Duke of Edinburgh and the husband of Queen Elizabeth II—and a grandnephew by maternal descent of Tsarina Alexandra. Sure enough, his mtDNA sequences in the hypervariable regions matched the sequences in the bones of the tsarina and her children. As for the tsar, Nicholas II's maternal grandmother was Denmark's Queen Louise of Hesse-Cassel. Two of her living descendants provided DNA samples for comparison. These samples had virtually the same mtDNA sequence as the one detected in the bones thought be the last tsar's. However, the putative Nicholas II sample differed at a single base pair.

What could account for this discrepancy? The tsar traced his heritage to Louise of Hesse-Cassel, as did the donors of the other two samples. Could there have been a secret adoption or a mix-up at birth in his line or theirs? That hardly seemed possible, especially when one considers that the mitotypes of unrelated individuals generally differ at five or more base pairs, not just one. Could there have been a different family that just happened to have a mother and three daughters whose mtDNA was identical to the royal line, as well as a father whose mtDNA was only one base pair away? That too seemed like a remote possibility. Could the laboratory have made errors in ascertaining the sequences that just happened to produce all the matches and the near match? The scientists reexamined the electropherogram. Beneath the blue peak indicating a C for the putative tsar was a small red blip indicating a T (Sykes 2001, 69).

The forensic scientists decided to clone the anomalous DNA. This kind of cloning involves inserting a bit of DNA—here, an mtDNA sequence—into the DNA in bacteria. If this foreign DNA is successfully incorporated, the bacteria do not know about it, and every time they divide, they copy the inserted sequence along with the rest of their genome. Bacteria are prolific. On a culture dish with plenty of nutrients, a new generation appears about every twenty minutes. Given enough food, over a billion such bacteria can grow in ten hours. Now suppose that a mixture of DNA is injected into the bacteria. Part of the mixture is the sequence found in the known descendants of Louise of Hesse-Cassel, and the rest is that sequence with one T changed to a C. Some of the bacteria will incorporate one sequence; some will get the other version. When the sequences are cloned and extracted, there will be many copies of each version. In contrast, if the inserted DNA is not a mixture—if all of it differs from the original, royal sequence—only the nonroyal sequence will be found among the millions of clones.

When the FSS cloned the puzzling sample of mtDNA from the putative tsar, there were two clear mitotypes. Most of the mtDNA from the bone had the sequence with the C, but it also had some with the T instead. Evidently, the variant was amplified as well. The more prevalent sequence showed up as the blue peak, and the less prevalent sequence yielded the much smaller red peak.

But how could Nicholas II have had such a mixture of mtDNA in his body? In this case, the mechanism was a mutation in the germ line. Mutations in DNA sequences can arise if the original sequence in a mitochondrial chromosome is altered by exposure to "mutagens" such as radiation or certain chemicals and the damage is not repaired. In addition, when DNA is copied, mistakes can occur. Most of these changes never reach the egg cells that are the source of the mitochondria for the next generation. Through a mechanism whose details are still being elucidated, a bottleneck restricts the flow of the new sequences between the cells that produce the eggs and the ova (Cree et al. 2008; Khrapko 2008). When these founding germ cells go through a series of divisions to give rise to immature egg cells, they appear to take only a small number of mitochondria with them. A new mutation, being a small fraction of all the mitochondrial sequences, generally does not get through, but if one does, it will make up a larger proportion of the mitochondrial DNA in the new cells. When these cells divide again, there is a greater chance for the new mutation to be further enriched. Given the limited number of divisions between the founding cells and the egg cells, a new mutation rarely displaces entirely the original sequence in a single generation. Rather, the fertilized cell will have a mixture of two mitochondrial sequences—the old one, which is the same as the mother's, and the new one, which began as a mutation somewhere in the mother's germ-line cells. After several more generations, the new mutation may have replaced the old, or it may have "slipped back into obscurity and disappeared" (Sykes 2001, 158). In the interim period, the line of children going back to the egg cell with the partly established mutation may show both sequences—heteroplasmy—when tested.

Because the cloned mtDNA from the remains included the original sequence of the tsar's grandmother, Gill and his colleagues concluded that the skeleton really was Nicholas II's. The second, almost identical sequence simply showed that Nicholas II was in the early stage of this process of drift. On average, two unrelated Caucasians in the United Kingdom would differ at eight or nine base pairs. A difference at only one point would occur in only about 1 in 100 unrelated people (Gill et al. 1994,

133). "Sex testing and mtDNA sequencing have established that the remains were almost certainly those of the Russian royal family" (ibid. 134).

The American Study

Although the British team considered the case closed, the discrepancy in the one base pair was described by others as a "mismatch." The Russian government postponed issuing an official report. It called on Ivanov to analyze the skeletal remains of the tsar's brother, Georgij Romanov, the Grand Duke of Russia, to see whether the mtDNA sequences were the same as the putative tsar's. So the documented remains of the grand duke, who had died in 1899, were removed from a cathedral in St. Petersburg. This time, Ivanov worked with the U.S. Armed Forces DNA Identification Laboratory. The AFDIL, to use its military acronym, had acquired considerable expertise in identifying the remains of soldiers from as far back as the Civil War (Fisher et al. 1993; Holland et al. 1993). The lab amplified mtDNA from bones in the grand duke's leg. The sequencing confirmed the FSS's deductions. The sample from Georgij Romanov showed the same heteroplasmy at the same point in the sequence. Louise of Hesse-Cassel or her daughter had passed on two strains of mtDNA to the duke. The tsar would be expected to share this particular heteroplasmy. An unrelated individual would be extremely unlikely to carry both sequences. Indeed, the investigators reported a likelihood ratio of 380,000 for the match. In other words, this degree of matching was 380,000 times more likely to occur if the Ekaterinburg bones really were those of the tsar than if they were from an unrelated individual. Combining this result with the match between Prince Philip and the putative tsarina, Ivanov et al. (1996, 419) reported that the mtDNA matches were about 100 million "times more likely if the remains are those of the Romanovs than if they were an unrelated family." As far as they could see, there could be "no reasonable scientific objection to accepting the authenticity of the remains" (ibid.). In 1998, the government agreed. A special panel affirmed that the bones were those of the Romanovs, along with their doctor and three servants.

The California Dissent

The Russian Orthodox Church, a number of Russian expatriates, and some historians were not persuaded. A law student in Sacramento also had doubts. After reading a history of the Romanovs, Daryl Litwin "just kept finding contradiction and discrepancy from point to point" (Landhuis 2004). Feeling "kind of befuddled," Litwin approached a Russian history expert at the Hoover Institution, a conservative think tank at Stanford University, who agreed that the Romanov verdict was worth reexamining. Litwin then went to see Alec Knight, "a senior scientist in the Stanford lab of anthropological sciences Assistant Professor Joanna Mountain" (ibid.). Knight was skeptical that the remains were in good-enough shape to allow the extensive mtDNA sequencing that the FSS had accomplished. Like the O. J. Simpson defense team, he sensed contamination. In his view, DNA from skeletal remains that had spent over seventy years in a shallow, earthen grave would have been too degraded to have sequences longer than 250 bases. "Based on what we know now, those bones were contaminated," Knight declared. As he saw it, the successful amplification of the long sequence from all nine skeletons was "certain evidence" that the bone samples were tarnished with fresh, less degraded DNA—perhaps from an individual who handled the samples (ibid.).

Fueling Knight's suspicions was a blood-soaked handkerchief that Tsar Nicholas II had used to treat a head wound suffered after he was struck by a would-be assassin in Japan in 1891. Ivanov examined the handkerchief but refused to disclose experimental details and claimed that the handkerchief DNA was too degraded for a meaningful analysis (ibid.).

Lacking that source of the tsar's DNA, Knight sought some other sample that might prove him right. In a wooden box at the New York home of Bishop Anthony Grabbe sat a finger of Grand Duchess Elisabeth Feodorovna, Alexandra's older sister. Her bones were well traveled. The finger came to rest in New York following the opening in 1982 of Elisabeth's coffin in Jerusalem. The Russian Expert Commission Abroad, a group that challenges the assertion that the bones are royal, paid for Knight to travel to New York. "They just bought me the plane ticket and got me the sample. They had no control over the work," said Knight (ibid.). Knight's sequencing confirmed the commission's position—the sequence did not match the reported sequence of the tsarina. To Knight's group, this meant that "it is probable that the Ekaterinburg remains were misidentified" (Stone 2004).

"That's nonsense," fumed Ivanov. He maintained that Knight's study was the one that suffered from contamination (ibid.). As a member of the British team explained, "We were able to conclude that the remains were those of the Romanovs because they match the DNA of known living maternal relatives of the tsar and tsarina, including Prince Philip, all of which were analysed after the results were generated from the bones" (Highfield 2004). Thus the only contamination that could explain the matching mtDNA from the maternal lines of both the tsar and the tsarina would be the planting of DNA from some maternal relative of each of them. Who would have been able to do that?

Even more devastating for Knight's theory is the fact that "[t]he DNA result generated from the shrivelled finger is different to that of Prince Philip and therefore could not have come from the Grand Duchess Elisabeth or any other maternal relative" (ibid.). Tsarina Alexandra's mtDNA should match both her mother's (of which no sample exists), her sister's (in the finger), and a pristine sample from her living grandnephew, Prince Philip. The mtDNA from the woman's bones in Ekaterinburg did. Several mitotypes were derived from the eighty-six-year-old finger that traveled from Jerusalem to New York to California.[2] That one of these mitotypes differs from the Ekaterinburg one is not much of a refutation of the interlocking web of DNA findings from the British and the American teams. "The most logical explanation of the results by Knight et al. is that the shriveled finger was not from Elisabeth or that the DNA sequence they recovered was the result of contamination" (Gill and Hagelburg 2004). Acknowledging that contamination of the finger was a possibility, Knight still insisted that he had "overwhelming evidence to reject the conclusion of the identity of the remains as those of the Russian royal family" (Stone 2004; see also Knight et al. 2004).

Two More Bodies

The final gruesome pieces of the puzzle seem to have been locked in place by a Russian builder's discovery, in August 2007, of more bones in a scorched area near the first Ekaterinburg site. Anthropologists determined that they were the remains of a boy and a young woman roughly the ages of the missing son, Alexei, and daughter, Maria (Attewill 2007). "Citing preliminary forensic and DNA tests," Vladimir Gromov, the deputy forensic chief scientist, announced that "it is possible to conclude with a large degree of certainty that parts of the skeleton . . . belong to Tsarevich Alexei and his sister, Grand Duchess Maria Nikolayevna Romanova" (BBC News 2007). After additional testing in Russia, the United States, and Austria, the official Russian verdict was announced. The two burial sites, only 70 yards apart, contained the remains of the entire family (Weaver 2008).

CHAPTER TWELVE

Stray Hairs

THE MITOCHONDRIAL DNA TYPING we looked at in Colombia, Italy, and Russia established the identity or family relationship of an individual via comparisons with known relatives when the person was not available for testing. In the case of "Juan David Gómez" in Bogotá, Clara Rojas could not be tested, so the Colombian government used her brother's and mother's mtDNA as substitutes for Clara's. In the case of Bernardo Provenzano, the Italian government used Bernardo's brother's mtDNA to ensure that it was on the trail of the Ghost himself. In the case of the Ekaterinburg grave site, it was necessary to test the living relatives of the long-deceased tsar and tsarina to identify the bones whose origins were in question.

Most criminal investigations do not require such surrogate DNA testing. Family relationships are not the question of interest, and it is not necessary to test other family members. Instead, mtDNA can link a suspect to a crime more directly. If cells are deposited at a revealing location, the mtDNA in them can be sequenced and compared with the sequence in a sample taken directly from the suspect. We shed cells all the time—from our skin, scalp, and elsewhere. The first cases to approve of mtDNA testing in court involved a few, telltale strands of hair.

Ware's Hair

The FBI began mitochondrial casework in June 1994, and the use of mtDNA analysis as evidence in a criminal case soon followed. The first case to admit such evidence appears to have been *State v. Ware* (Curriden 1996). The body of four-year-old Lindsey Green was found in a utility room of her mother's home in Chattanooga, Tennessee, lying at the feet of Paul Ware, who was drunk and unconscious. When the little girl was rushed to the hospital, she was dead—blue from lack of oxygen—and "her vagina was torn, her rectum was torn." The emergency-room physician called it "the most horrible thing I'd ever seen" (1999 WL 233592, at *4). Ware was indicted for felony murder and rape.

Part of the evidence against him was a hair found in the pharynx and other hairs lying on the bed in which Lindsey had been sleeping. Since the nineteenth century, criminalists have been associating trace hairs with suspects by comparisons of both gross and microscopic features. These comparisons are not individualizing, population frequencies for hair characteristics are not known, and the reliability of the examinations is open to question (Smith and Goodman 1996). This state of affairs has prompted substantial attacks, especially under the scientific-validity standard of *Daubert,* on the use of the physical comparisons to infer identity. The challenges have generally failed, however, partly because there clearly is useful information in the features of the fibers that the analysts examine.

In *Ware,* Special Agent Chris Hopkins of the FBI Hair and Fibers Unit testified at trial that the hair in the girl's throat was a "red Caucasian pubic hair" that had been "naturally shed" (1999 WL 233592, at *7). He stated that this hair and the ones from the bed were "consistent with originating from the defendant" (id. at *8). He further explained that microscopic hair comparison is "not a means of personal positive identification" (id.).

DNA can provide greater specificity and objectivity. When hairs recovered from crime scenes contain enough DNA, these molecules can forge a stronger link to the suspect or exclude him or her. Hair itself has two separate structures: the follicle, a stockinglike structure in the skin, and the protruding shaft that we see. Cells at the bottom of the follicle divide every twenty-three to seventy-two hours, faster than any other cells in the body. Dead cells get crammed together in the hair shaft. These cells are mostly protein (keratin), and the shaft also contains a pigment (melanin) and traces of metals. The only living part of the hair is the cells in a bulb at the bottom of the follicle.

Thus hairs torn from the scalp may have enough cells to permit nuclear DNA testing, and even a single plucked hair should have cells from the bulb. Standard STR testing then might be possible. More commonly, though, dead hair shafts are simply shed, as in the *Ware* case. Frequently, there is enough mtDNA encased in these strands to sequence. In fact, a complete analysis of the mitochondrial genome was successfully undertaken with a 36,000-year-old woolly mammoth dug out of the permafrost in 1804–1806 and stored in a Russian museum for 200 years at room temperature (Gilbert et al. 2007).

The mtDNA sequences in human hair shafts can be an invaluable supplement to the information gleaned from peering through a comparison microscope. To study the possibilities, two FBI researchers took human hairs submitted to the FBI laboratory for analysis between 1996 and 2000. Of 170 hair examinations, there were 80 microscopic associations. Of these 80 pairs, 9 turned out to be hairs from different individuals as shown by the mitotyping. In a sense, these were false positives (although they were not due to any fault of the microanalysts). Only 6 hairs did not provide sufficient mtDNA, and only 3 yielded inconclusive results. The mitotyping confirmed 18 out of the 19 exclusions made on the basis of microscopy. (The other one produced an inconclusive mtDNA result.) The researchers therefore recommended "that both microscopic and mtDNA analysis be used for analyzing hair evidence" (Houck and Budowle 2002, 4).[1]

This procedure was followed in *Ware*. Not only did Agent Hopkins testify to the morphological similarities in the hairs from the bed and the one in the victim's throat, but another analyst, Mark Wilson, testified to matching DNA sequences between those hairs and a saliva sample taken from Paul Ware. In interpreting this result, Agent Wilson steered clear of probabilities, stating instead "that '[t]he average number of differences between any two Caucasian individuals is approximately six,' . . . that the sample hairs were consistent with having originated with [Ware], . . . [and] that the tests could not show that the sample hairs belonged to [Ware] to the exclusion of all others" (1999 WL 233592, at *13). He added "that the sequence had not before been observed in the FBI's database of 742 individuals" and concluded:

> All I'm saying is we have a database of a certain size, and this particular sequence has not been observed before. I am not saying that it's a particular frequency, one over this or that, because it cannot be expressed that way because the database is not large enough at the present, in its present form, present size to be able to assign a frequency, you know, like one percent or whatever. This . . . event would have to be observed many more times in order to assign it a frequency, so what we do is state a fact. (Id. at *14)

The defense opposed the introduction of such testimony. Citing *Daubert,* it produced an affidavit from "Dr. William M. Shields, a geneticist," who had moved from the State University of New York at Syracuse to be "a scholar in residence at the University of Virginia law school," but it chose not to call him as a witness at the trial (id.). Although Shields apparently did not dispute the results of the FBI's sequencing, his affidavit expressed his "opinion that mitochondrial DNA typing as proposed by the Federal Bureau of Investigation, is not yet sufficiently reliable to be scientifically reliable. The major problem is that critical pieces of the validation process have yet to be done or have been done with insufficient sample sizes to be statistically reliable" (id.).

The court did not agree, the evidence was admitted, and Ware was convicted. He asked the court for a new trial, arguing that the mitotyping should have been excluded. At this point, Shields made an appearance. He delved into issues of law and psychology, as he had done in the *Simpson* case (Chapter 10), explaining that "until there's a database that allows you to look at how frequently these kinds of matches are going to occur, to talk about identity is certainly misleading" (id.).

In its unpublished 1999 opinion affirming the conviction, the Tennessee Court of Criminal Appeals was impressed with Agent Wilson's assurances that "mtDNA is 'widely used' to 'identify the remains of servicemen that have been killed in Vietnam or Korea'" (id. at *17). But Shields's argument that a jury needs a frequency derived from a large database also had an effect. Without such a statistic, the court thought that "it is somewhat questionable whether the DNA testimony presented was such as to substantially assist the jury to determine a fact in issue" (id.). Nevertheless, the court concluded "that the trial judge did not abuse his discretion in admitting the results of the mitochondrial DNA tests into evidence" (id.).

There is a certain incongruity about worrying that "[a]bsent a frequency rate or some similar interpretation of the test results, the [mtDNA] testimony does not provide a strong basis for scientific conclusion by a layperson" (id.) while not evincing the same concern about the even less quantitative testimony on the morphology of the hairs. It may well be that "[t]he only result that an individual untrained in the analysis of DNA could reach after hearing Wilson testify is that the common DNA sequence shared by the hairs tested and the Defendant's saliva had never before been noted in the 742 individuals that comprised the FBI's then-current database" (id.). But this is more informative than Agent Hopkins's testimony that the hairs were "consistent with originating from the defendant" and that the match is "not a means of personal positive identification" (id. at *8).

Whatever limitations and infirmities the FBI's mtDNA database might have,[2] the jury had more specific information about the power of the scientific evidence of the mtDNA match than it did about the power of the morphological match. Also, both analysts wisely avoided making the more extravagant claims, based largely on intuition and impression, that one sometimes hears about the significance of other kinds of matching traits.

Saving Simpson's Scalp

Microscopic hair analysis and mtDNA typing, in a strange way, entered the O. J. Simpson case. Early in the case, the state secured an order from the Los Angeles municipal court compelling Simpson to submit to the removal of hairs from his head. These were to be compared under a microscope to "black curly hairs . . . determined to be of African-American origin" found in a blue knit cap apparently discarded or dropped near the bodies on the night of the murders (People's Motion for Order Directing Defendant Orenthal Simpson to Supply Hair Sample, June 27, 1994, *1). Such "nontestimonial orders" are routinely issued to force suspects to provide hairs, blood samples, voice exemplars, and the like.

Defense attorneys Gerald Uelmen and Robert Shapiro responded by demanding that the state limit the sampling to a single hair. Prosecutor Marcia Clark was flabbergasted. Hairs vary both within a region of the scalp and from region to region. Comparing just one hair to those few that were in the cap would be pointless. But Judge Kathleen Kennedy-Powell, who had been a prosecutor on the sexual-assault unit before joining the bench, apparently knew very little about the process. Clark informed her that about 100 hairs might be necessary, but the exact number would have to be decided by a criminalist. "Any scientist," Clark insisted, "no matter how inexperienced, is aware of that fact" (Transcript of Preliminary Hearing, June 30, 1994a, *1). Shapiro countered that he knew a scientist who disagreed—Henry C. Lee, "the greatest forensic scientist in the world" (Lee and Labriola 2001, dust cover). Shapiro told the judge that "Dr. Henry Lee, our chief criminalist, . . . tells us one to three hairs are sufficient" (Transcript of Preliminary Hearing, June 30, 1994a, *2). Shapiro wanted a hearing. "I've been practicing for a long time," he reminded the court, and "I have never, ever heard of such a request" (ibid.).

The reaction was surreal. Judge Kennedy-Powell ordered "no more than 10 hairs at this point. If the prosecution at some point feels that they are in need of more hairs than the 10 that I'm going to allow at this point, then we will have that hearing with experts, if necessary" (ibid.).

Clark could not believe her ears, or if she could, she would not accept their message. She addressed the court: "Your Honor, I don't think that counsel has conferred with Mr. Lee concerning the nature of the tests involved. . . . I'm not talking about anything unusual here, and what Mr. Lee is talking about—." The judge interrupted, but Clark plowed on, "excuse me, is P.C.R. work. We're talking about comparison for microscopic as well. We're not just talking about P.C.R. only" (ibid.). What Clark was thinking was that Lee must have informed Shapiro that mtDNA sequencing, which presupposes PCR amplification to generate enough mtDNA for the sequencing reactions, could be accomplished with a few well-chosen hairs and that Shapiro was twisting this around to prevent the more conventional, comparative microscopy from occurring.

Judge Kennedy-Powell did not seem to appreciate the distinction. She told Clark that "[t]he bottom line is that if, in fact, you do need more than that, you are going to have to produce an expert to justify that to the court. And if indeed that justification is made, I will sign the appropriate order." Clark took what she could get: "We will present that testimony today," she promised. "Today?" the surprised judge asked. "Yes," said Clark (ibid.). Shapiro would get the hearing he requested.

That afternoon, Michele Kestler, a criminalist for the Los Angeles Police Department, took the witness stand. Judge Kennedy-Powell asked her, "[T]o perform a microscopic analysis and comparison of hairs, how many head hairs do you need to retrieve . . . ?" Kestler was well prepared: "I have a couple of references here, and the references all indicate 30 to 100 hairs" (Transcript of Preliminary Hearing, Mar. 30, 1994a, *5). She blandly continued: "[O]ne of my references is from a book, *Forensic Science: An Introduction to Criminalistics,* by Peter De Forest, Robert Gaensslen and Henry C. Lee" (ibid.).

The judge perked up: "I'm sorry, is that the same Henry Lee that we had testimony concerning?" "Yes, it is," said Kestler. "It's the only Henry Lee I know" (ibid.). For good measure, Clark had Kestler discuss an FBI publication that recommended "that at least 20 hairs from each of 5 different areas of the scalp . . . be obtained by both pulling and combing" (Transcript of Preliminary Hearing, Municipal Court, June 30, 1994b, *6). Clark was after those 100 hairs, and the judge more or less agreed, ruling that "the prosecution be entitled to obtain at least 40 but no more than 100 hairs from the various areas, the five areas as denoted in the scalp of Mr. Simpson" (ibid., *11). Marcia Clark had had her revenge on Shapiro.

As to what came of all this hairsplitting, no mention of mtDNA results on the hairs appears in the trial transcript. At trial, Clark and her team relied on more conventional DNA testing of bloodstains (Chapter 7), and it had the special agent in charge of the FBI's Hairs and Fibers Unit testify only about the microscopic comparisons of the hairs. In his opinion, Simpson "had very distinctive hairs" that were "consistent with" hairs in the blue knit cap (Transcript of Examination of Douglas Deedrick, July 6, 1995, 1995 WL 429478, at *25).

A Connecticut Yankee in Court

In *State v. Pappas,* the Supreme Court of Connecticut produced the first high-court decision on the admissibility of mitotyping. On an October morning in 1994, a man "wearing a navy blue hooded sweatshirt, baggy navy blue pants, and turquoise gloves entered a branch of Citizens Bank on Ocean Avenue in New London" (776 A.2d at 1095–1096). The hood was drawn up "over his head, and his face was covered by a piece of cloth" (id. at 1096). "This is a robbery," he shouted, and "climbed over the teller counter, removed approximately $5,530 in cash from the teller drawers, and stuffed the money into his pants and sweatshirt" (id.). The robber's disguise was less than perfect. At one point, he lifted the cloth, and Linda Schwartz, a customer at the bank, saw "a prominent nose, a very sallow complexion and a few scars or shaving nicks" (id.).

As soon as the masked man fled, heading in the direction of some railroad tracks, the branch manager called the FBI. New London police rushed to the area and saw a man "holding his midsection" and running along the railroad tracks. They could not catch him, but they found a pair of "tan to teal-colored" gloves, a blue hooded sweatshirt, and $2,231 in cash (including "bait money" from the bank) strewn along the tracks. The FBI gave Schwartz an array of photographs to study, and she identified Stephen Pappas as the robber. The FBI also picked two hairs off the sweatshirt. The police then obtained a search warrant for Pappas's hair and pulled some from his head. The FBI wanted combed hairs, however, so the New London police went back with a comb and a second warrant. This kinder, gentler search produced the desired hairs. The FBI extracted and analyzed the mitochondrial DNA from the hair shafts. Pappas was charged with robbery, larceny, and being a persistent felony offender.

The *Daubert* Hearing

Before trial, Pappas moved to exclude all evidence regarding mtDNA analysis. At a pretrial *Daubert* hearing, FBI special agent Mark Wilson explained the genetics and genomics of mtDNA and the procedures for extracting, amplifying, and sequencing the hypervariable regions. He spoke about contamination, which, he assured the court, could produce only false negatives. He stated that "a match . . . means that the questioned sample cannot be excluded as deriving from the same maternal lineage from which the donor's sample is derived." He added "that the examiner cannot positively establish identity on the basis of mtDNA because all those having a common maternal lineage, absent mutation, share the same mtDNA" (id. at 1104).

William Shields, the biologist who had questioned the adequacy of the FBI's validity studies in *Ware,* appeared in person at this hearing. He reiterated that the FBI had "not performed validation studies concerning the extent of contamination by the DNA of others resulting from their handling of mtDNA" (id. at 1105). In addition, he raised the issue of heteroplasmy—the presence of more than one mtDNA sequence in the same individual (Chapter 11). Wilson had acknowledged that "heteroplasmy is observed in approximately 5 to 10 percent of cases," but he saw it as a problem only for exclusions and emphasized that "there was no evidence of heteroplasmy in the present case" (id.). Shields thought that "point heteroplasmy, a difference at one base pair in a sequence from samples of the same individual, occurs in between 10 to 20 percent of all people, and may occur in hair samples in 100 percent of the population" (id.). In his view, the existence of heteroplasmy impugned the FBI protocol for declaring a match (id.).

Both Wilson and Shields were correct about one thing—heteroplasmy is not uncommon (Lo et al. 2005). Everyone is probably heteroplasmic in that somewhere, in some tissue, some mitochondria carry some DNA sequence that varies from the predominant sequence in most of their mitochondria. Even a single mitochondrion could contain two or more DNA molecules with slightly different sequences (Holland and Parsons 1999; Melton 2004). When DNA, whether it is from a nucleus or a mitochondrion, is copied, mistakes can occur. Most are corrected by biochemical error-checking mechanisms, but a few sneak through. These are mutations. Changes also can arise if the original sequence is altered by exposure to mutagens such as radiation or certain chemicals. The mutation rate is remarkably low for nDNA—roughly 1 nucleotide base in 1,000 million will mutate during cell division (Sykes 2001, 55). The mitochondria are less fastidious. They allow at least ten times as many mutations (per nucleotide base pair) to slip by (ibid.; Melton 2004; Pakendorf and Stoneking 2005, 168). In addition, their DNA lacks the protective proteins that are part of the more complex chromosomes in the nucleus, and they are exposed to the mutagenic radicals of oxygen. There are mechanisms for repairing this DNA damage, but they are imperfect (Krishnan et al. 2008).

If mutations occur in the sex cells—eggs or sperm—they can be carried forward to the next generation. Even for mtDNA, the mutation rates per cell are so small that the limited number of germ-line-producing cells rarely mutate. Hence mtDNA is passed faithfully from mothers to children for many generations. The precise mutation rate in the inheritance of the mtDNA is difficult to pin down (Gibbons 1998), but studies indicate that a detectable, new sequence appears in the range of once in 33 generations to once in 300 generations (Ivanov et al. 1996, 419; Parsons et al. 1997, 363). Therefore, a long line of maternal ancestors and descendants should have the same DNA. That is why even the small difference between Tsar Nicholas II and his grandmother discussed in the last chapter came as a surprise. Gill and his colleagues had caught a germ-line mutation in the process of becoming established in the offspring of Queen Louise of Hesse-Cassel (Sykes 2001).

Mutations that occur in the other body cells—the somatic cells that do not produce sex cells—are much more prevalent than such germ-line ones because there are many somatic cells that can mutate. Think of a person as a colony of cells—tens of trillions of them (not to mention something like twenty times that number of microbes happily sharing the space). As an individual ages, point mutations are probable in at least some of the incredible number of mitochondria that are reproducing for replacement or growth. When a cell containing the mutated DNA divides, it can give rise to a line of cells with mitochondria that contain only the new sequence. Or a new line might consist of cells, each of which is a mixed bag of mitochondria with the mutated sequence and mitochondria with the original sequence. Either way, the body will have a mixed population of mitotypes, but the original sequence will predominate.[3] The new cell lines might be confined to a few tissues, such as a skeletal muscle or some of the rapidly dividing hair cells.

Because a DNA sample from hair or anywhere else usually is dominated by copies of the original sequence, only that one sequence ordinarily will be amplified. Thus heteroplasmy may well exist in everyone, but most of the time it will be of no consequence. PCR followed by sequencing still will show a simple match between this person and everyone else in the same maternal line (who does not exhibit what we can call this kind of "late-onset" somatic heteroplasmy). Likewise, two samples from the same individual normally will show the sequence that the individual received at conception from his or her mother—and her mothers, and all those mothers before her, going back to a germ-line mutation that actually took hold in the mother who founded a new egg-cell line of mtDNA for her descendants.

Sometimes, however, an mtDNA sequence that is a somatic variant will be amplified along with, or even instead of, the initial, maternal sequence. To account for the possibility of detectable heteroplasmy (whether inherited or somatic), the FBI does not declare an exclusion unless two samples differ at more than one point. But it also does not call a near-perfect match (a single-point difference) an inclusion. It is quite unusual for maternally unrelated individuals to have such closely related sequences, but it can happen. Therefore, the FBI merely says that the difference at only one point in the sequence is "inconclusive."

Shields, however, regarded these inconclusive near matches as if they were actual inclusions. He "testified that, because of the possibility of heteroplasmy, the FBI changed their matching criteria as to when two samples may be said to match, that is, when the donor of the known sample cannot be excluded as the source of the questioned sample" (*Pappas*, 776 A.2d at 1105). He complained that "while the new matching criteria reduce the probability of false negatives, they increase the likelihood of false positives, i.e., incorrectly including a known sample as a source of the questioned sample" (id.).

Designating a near match "inconclusive" does reduce the risk of a false negative because it avoids treating a detectably heteroplasmic individual as two different people. But as long as the analyst does not present a near match as positive proof of identity, the refusal to exclude the suspect does not increase the risk of a false positive, for no positive finding has been declared. Therefore, the FBI's unwillingness to exclude an individual who almost matches is not a defect in the protocol. To the contrary, it is a reasonable response to Shields's concern that hair heteroplasmy is universal.[4] Because heteroplasmy is common, it would be foolish to assume that a discrepancy in a single base pair means that the individual is part of a different maternal line that just happens to have an almost matching sequence. The two sequences could be from different maternal "clans," but heteroplasmy is the more probable explanation. Consequently, the individual who almost matches cannot be excluded.

The Supreme Court's Opinion

The trial court denied Pappas's motion to suppress the FBI's sequencing results. A jury found Pappas guilty of robbery and larceny. On appeal, Pappas argued that the trial court erred in admitting the mtDNA evidence because mitotyping did not satisfy Connecticut's version of *Daubert* and because, even if it did, it was substantially more prejudicial than probative.

The Connecticut Supreme Court devoted only a few sentences to the *Daubert* objection. It pointed to the 1996 NRC report and stated that "the procedures used to extract and chart the chemical bases of mtDNA—extraction, PCR amplification, capillary electrophoresis, and the use of an automated sequencing machine to generate a chromatograph [*sic*]—are scientifically valid and generally accepted in the scientific community" (id. at 1108).[5] Contamination could be an "important" issue "in a particular case," but it was not a sufficient reason to exclude PCR-based testing in its entirety (id.). Likewise, heteroplasmy could complicate the interpretation of results in some cases, but it was not a barrier to admissibility in general.

The court devoted considerably more attention to the FBI's attempt to estimate a random-match probability. This process should be a simple matter, but historically, statistical analysis of forensic evidence has not been the FBI's strong suit. In the first courtroom skirmish, in *Ware,* the FBI analyst did not try to present a random-match probability or a probability of not excluding an innocent suspect. In *Pappas,* however, Agent Wilson (or the prosecutor) was not content to say that "the sequence previously had not been observed" in the FBI's database of mtDNA types—at that point, a convenience sample of 1,657 sequences, 916 of which were from Caucasians. Extrapolating and overlooking the fact that almost matching individuals cannot be excluded, Wilson "concluded that approximately 99.75 percent of the Caucasian population could be excluded as the source of the mtDNA" (id. at 1104, note omitted). At trial, the Caucasian reference database had grown to 1,219 samples, and "[h]e testified that, at a 95 percent 'confidence interval,' 99.7 percent of the Caucasian population could be excluded as the source of the questioned sample" (id. at 1110, note omitted).

Shields was properly critical of this presentation, but his exposition of the point (as it was described by the court) also was flawed.[6] To dissect the issue cleanly, we need to see how the FBI produced its number for the proportion of the population that "could be excluded." Basically, it applied the 1992 NRC committee's "counting rule" (Chapter 5). Suppose that we want to know the percentage of the population that has the surname "Crosset." We dial telephone numbers randomly and find no Crossets whatsoever among the 1,219 people who answer. Maybe there are no Crossets to be found, but there might be a few. If the proportion p of Crossets were 0.25%, then the chance of not finding any Crossets in 1,219 random calls is $(1-p)^{1219} = (1-0.0025)^{1219} = (0.9975)^{1219} = 0.05$. In other words, as much as 0.25% of the population could be Crossets, and we still would have a 5% chance of drawing a sample with no Crossets. Therefore, "a 95 percent confidence interval" for the population proportion is between 0% (in which case we are certain to find no Crossets in the sample) and 0.25% (in which case we would still find samples with so few Crossets about one time in twenty). If the population proportion were much higher than 0.25%, then there would be even less of a chance than 0.05 of getting a sample devoid of Crossets. By this reasoning, we can be 95% "confident" that the population proportion lies between 0% and 0.25%.[7] Thus Agent Wilson could have reported that if the FBI's sample of Caucasians is representative of the general Caucasian population, then a plausible estimate of the proportion of that population with the mtDNA sequence in the hairs on the sweatshirt would go from 0% up to 0.25%. Had he stopped there, his testimony would have been quite defensible. Indeed, although Wilson did not provide the formula he used to arrive at the 0.25% figure, the Connecticut Supreme Court located a suitable approximation and verified that arithmetic.

But Wilson spoke of the proportion that "could be excluded." His testimony was "that at a 95 percent 'confidence interval,' 99.7% of the Caucasian population could be excluded as the source of the questioned sample" (id.). The problem with this formulation is that two groups of people cannot be excluded—those whose sequences match perfectly (the 0.25%) and those whose sequences are one base pair away from a perfect match. The FBI cannot have it both ways. If it does not exclude the very close matches, then those matches, as well as the perfect matches, need to be counted among the part of the population that would not be excluded. (If we regard single-letter variants of "Crosset" as too close to exclude, then the Crossits, Krossets, Clossets, and others in the sample of names have to added to the 0.25%.) Thus Shields corrected Wilson's omission. He reported that adding in the near matches "would exclude 99.3 percent of the Caucasian population" (id. at 1105, note omitted). So Wilson was wrong, but not by all that much.

In addition to the *Daubert* challenge, Pappas argued "that the trial court should have excluded that mtDNA evidence because its probative value was outweighed by undue prejudice" (id. at 1112). The Connecticut Supreme Court found "no merit to the defendant's claim" (id. at 1113). It noted:

> Wilson conceded that the frequency was an estimate based upon the FBI database, and that it is not representative of the New London population. Wilson also testified that mtDNA does not positively identify individuals because everyone within a given maternal line would have the same mtDNA type. . . .

[¶] During its charge to the jury, the court instructed the jury on the mtDNA evidence and carefully pointed out the differences between nuclear DNA results and mtDNA results. (Id. at 1112, note omitted)

As for the possible "psychological effect on the jury" (id. at 1113), the court observed that "mtDNA evidence is similar to other types of class evidence such as blood type, and microscopic hair analysis, that we traditionally have held is admissible in criminal trials" (id.). As in these situations, the court was willing to "presume that the jury followed the court's instructions," and it cited research mentioned in the 1996 NRC report to suggest "that, while jurors are influenced by DNA evidence, they are not overwhelmed by that evidence" (id. at 1113). After *Pappas,* the admissibility of mtDNA typing was assured.

An Experiment in Delaware

Was the *Pappas* court correct in assuming that jurors in mtDNA cases would follow the judge's cautions against confusing mitochondrial DNA typing with the nuclear DNA testing that was presented in the newspaper headlines, on radio talk shows, and in television dramas as offering almost magical solutions to criminal cases? Was the 1996 NRC committee correct in suggesting that the existing psychological literature did not support the conclusion that jurors would be overwhelmed by either type of DNA matches and the associated statistics? What would more psychological research of the type recommended in the report reveal about juror understanding of DNA and probabilities?

In Arizona, Maricopa County Superior Court judge Michael Dann had spearheaded a series of reforms to improve the performance of juries. Traditionally, jurors do not take notes, do not discuss the case among themselves before the formal deliberations begin, are not informed of the law that they are expected to apply until the evidence has been presented, and are not encouraged to ask questions. The Arizona reforms championed by Judge Dann changed this, and when he retired from the bench, he proposed testing whether these or other reforms actually improve jury fact finding and decision making. Teaming up with social psychologist Valerie Hans (then at the University of Delaware), he asked the National Institute of Justice to fund a study using mock juries and various decision aids in a case involving challenging scientific evidence. The idea was to present an abbreviated but realistic videotape of a mock trial to citizens who appeared for jury duty. Different groups of mock jurors would operate under different conditions. Some would be allowed to take notes and ask questions or would be given a notebook of information or a chart to structure their thinking.[8]

The researchers decided that DNA evidence would provide a good platform for investigating the reforms, as well as the broader question of juror comprehension of scientific and probabilistic evidence. They asked me to help develop the details of a case and a questionnaire to test the jurors' understanding of DNA evidence. I suggested that mitochondrial DNA would be of more interest than the largely settled nDNA issues, and that the *Pappas* case might provide the core of a factual scenario. Judge Dann condensed the trial transcript and altered the facts to weaken the case against the defendant, rendering the mtDNA evidence more salient. He also changed the names of the parties and witnesses and relocated the case to the hypothetical city of "Middletown," which had about 58,000 Caucasians, just under half of whom, or about 29,000, were males.[9]

We also modified the FBI agent's testimony to substitute a more up-to-date database of 5,071 Caucasians for the smaller one in *Pappas* and to replace the complicated 95% confidence interval with a simple $1/(n+1)$ estimate (Chapter 5). That is, the FBI agent (named Agent Jaye in the tradition of the film *Men in Black*) testified that only 1 in 5,072 Caucasian men had mtDNA types that matched that of hairs from the sweatshirt, meaning that 99.98% of that population would be excluded as a possible source of the hairs. On cross-examination, Jaye agreed that in Middletown, this would leave "six white males as the possible source of those hairs" (Kaye et al. 2007, 807).

Finally, we created the character of "Dr. Elizabeth Allison" for a defense expert.[10] This college professor of genetics questioned the FBI calculation on two primary grounds—that the FBI's database was not a random sample, and that Agent Jaye failed to account for possible heteroplasmy in its calculation of the proportion of the Caucasians who would not be definitively excluded. Most of her testimony was directed to the latter issue. Explaining this approach by the FBI, Allison observed that the random-match probability was affected by the number of near matches, as well as the number of exact matches. She found nine instances in which the crime-scene sample was one base pair off from a sample in the FBI database. She asserted that even if one granted the appropriateness of the FBI reference database, "some 10 out of every 5,072 white men in Middletown would have a DNA sequence that could not be excluded from that of the hairs in this case." Applying a nonexclusion proportion of 10/5,072, or 0.2%, to the town's white male population, she concluded that "about 57 Caucasian males in the Middletown metro area could not be excluded as the source of the hairs found on the sweatshirt" (ibid.).

The prosecutor argued in closing that

> [the defendant's] mtDNA profile is so rare that it has not been observed in the FBI's database of over 5,071 samples. Over 99.98 percent of white males can be excluded as the source of the hairs. Of course, there could be some other white males in this metro area with similar mtDNA. But the DNA evidence confirms what all the other evidence says—everything points to this defendant. (Ibid., 831)

Although heteroplasmy in the population means that near matches cannot be excluded, he added that "the defendant's DNA witness . . . agreed with the FBI analyst that the defendant is not heteroplasmic. Her conclusions and her calculations should be disregarded. They are simply not relevant to this case" (ibid., 832).

In his summation, the defense lawyer reminded the jury that "anyone in the defendant's family would have the very same mtDNA" (a slight overstatement). He disputed the 99.98% figure and emphasized the large number of men whose mitotypes might be within a single base pair of the sequence from the hairs and hence "could not be excluded under the FBI's criterion for exclusions":

> [A]bout 57 other Caucasian males in the metro Middletown area—not to mention the much larger number of white men in the state or who might have been visiting in town at the time of the bank robbery—had the same mtDNA profile as the defendant and could have been the source of the hair on the sweatshirt. The defendant is just 1 of these 57 or so men who could have left the hairs on the sweatshirt. That doesn't add up to proof beyond a reasonable doubt. It is not even close. (Ibid.)

As noted in Chapter 2, the jurors generally were not overly impressed by the mtDNA match. In filling out questionnaires, nine out of ten of them understood that mitotyping was less discriminating than nDNA analysis. Despite the exclusion probability of 99.98% (according to the prosecution) or 99.8% (according to the defense) and the lack of any clearly exculpatory evidence (such as an alibi), only half the jurors accepted as true the statement that the mtDNA evidence showed only about a 1% chance that someone else was the robber. On average, the jurors were only 75% confident that the defendant was the robber, and only about one-third of the jurors placed this source probability in the range of 0.90 or higher. Although the match surely was relevant, 40% agreed that it was "completely irrelevant because a substantial number of other people also could be the source of the hairs." In the end, nearly as many jurors voted to acquit as to convict.

These Delaware jurors were sensitive to the limitations in the database. In deliberations, some were highly skeptical of the extrapolation from the database to Middletown. For example, one juror pointed out:

> [T]he FBI got 5,072 different individuals, got their DNA samples. We don't know if they were out in San Francisco and got 5,072 DNA samples from Latin Americans only and then decided to use that database to characterize the population in Middletown, or if it was a random sample across the country. We have no idea whether that database that they have is actually representative of the population in Middletown or not. For all we know there could be three major families in Middletown that have been there for hundreds of years that have lots of family members in the area and, consequently, you see a lot of the DNA pattern out of that particular city. (Ibid., 809)

Likewise, although random sampling error was not mentioned in testimony or argument, some jurors worried that even if the national database were representative of Middletown, the match probability in the Middletown population could be larger than that in the FBI's national sample. For instance, one juror exclaimed, "A database of 5,000 people—how many people are in the United States? 5,000 is nothing" (ibid., 810). Another invoked his professional knowledge to declare that "I'm a research scientist by trade, and when they started talking about mitochondrial DNA, I thought, 'Oh boy this is good stuff!' It's very timely, you know, but I can tell you that their percentages could change if they measured DNA samples from 10,000 people rather than 5,000 people" (ibid.).

Most jurors succeeded in sorting out some of the fallacious arguments or suggestions in the case. Well over half of them knew, after hearing the experts and before deliberation, that mitochondria are found outside the nucleus of the cell, that the sequence of base pairs is important, that about 600 base pairs are analyzed, and that a match is the same mtDNA sequence in two samples. Of course, one might worry that the rest of them did not absorb this much basic biology, but two key features of mtDNA— its maternal inheritance and its similarity for all members of a maternal lineage—were well understood. Although the defense introduced evidence that the defendant's wayward half brother on his father's side lived in town at the time of the robbery, the presence of this paternal half brother did not confuse the jurors. Fully 90% correctly rejected after deliberation the suggestion that the mtDNA evidence could have come from the defendant's brother if the two had the same father but different mothers (Dann, Hans, and Kaye 2004).

The most challenging expert testimony probably involved heteroplasmy, an unfamiliar and intimidating term. Jurors did less well on recalling the meaning of the term, but about two-thirds of them were able to identify as correct a basic definition of heteroplasmy. More important, a similar percentage rejected the prosecution's misguided effort to ignore heteroplasmy in computing the proportion of the population "not excluded" and asserted that it was still necessary to consider heteroplasmy even though the defendant was not heteroplasmic.

My interpretation of these results is that juries can be trusted to handle mtDNA evidence as well as they do many other forms of evidence. Of course, no single study is a secure foundation for sweeping generalizations, but the results are broadly consistent with other mock-jury research with related evidence. It appears that neither the three letters DNA nor the less familiar Greek word "mitochondria" cast a spell over jurors. They can understand the important characteristics of mitotyping and the limitations on existing databases. The databases are small, and variations by region and subpopulation merit more attention (Adams 2005; Kaestle et al. 2006; Egeland and Salas 2008), but if the evidence is not oversold, it should be allowed to remain on the forensic market.

CHAPTER THIRTEEN

Learning from DNA

THERE WAS A TIME when genetic technology moved sedately from the laboratory to the courtroom. No longer. Within a year or two of the discovery of the polymerase chain reaction as a technique for amplifying DNA fragments, a criminal court was confronted with a PCR-based test for identity. Likewise, it took only two years or so for VNTR-RFLP testing to be brought from the laboratory bench to the judicial bench (see the timeline in the Appendix). Police agencies are even quicker to try out new technologies. Today's patrol car has a machine to measure breath alcohol concentration and a video camera. Tomorrow's may have a portable DNA analyzer (Liu et al. 2008).

To some, this pace may seem commendable. Why, they might ask, should the law hesitate to use science to do justice? To others, the pace seems frenetic. Particularly in criminal matters, they argue, sober reflection and study should precede the widespread adoption of a powerful new technology. A market-driven system, with commercial enterprises offering to supply police and prosecutors the evidence they need to investigate and prosecute cases, encourages speed. Companies like Cellmark and Lifecodes wasted no time in promoting and disseminating the newest forensic DNA technology (Aronson 2007, 17–21, 30–32, 96).

Of course, the profit motive is not the only force for rapid deployment. An inventor seeking fame or imbued with a saintly desire to make the world better may be no less motivated to propel his discovery into the judicial limelight.[1] Sir Alec Jeffreys was hardly shy in promoting "DNA fingerprinting" practically from the moment of its discovery, but he spurned an offer of venture capital to commercialize it (Aronson 2004, 63). In a simpler day, William Moulton Marston earnestly sought to publicize his lie-detection system (Golan 2004, 244). The problem, of course, is that the inventor's natural enthusiasm (or the commercial entrepreneur's desire to create and satisfy consumer demand) is no substitute for thorough testing of the invention.

The judicial response to this dynamic is the exclusion of scientific evidence—from psychology to physics, economics to engineering, geology to genetics—on the ground that it has not been shown to be scientifically valid or generally accepted as valid in the scientific community. Indeed, the *Frye* requirement of general acceptance emerged as a brake on Marston's unbridled enthusiasm for his new method of detecting deception. Since then, a wide array of scientific and pseudoscientific techniques have been excluded (at some point in at least some jurisdictions) as lacking scientific acceptance or validity. Specifically, *Frye* or *Daubert* has blocked the evidentiary use of graphology, polygraphy, hypnotic and drug-induced testimony, voice-stress analysis, voice spectrograms, various forms of spectroscopy, infrared sensing of aircraft, retesting of breath samples for alcohol content, psychological profiles of battered women and child abusers, post-traumatic stress disorder as indicating rape, penile plethysmography as indicating sexual deviancy, therapy to recover repressed memories, astronomical calculations, ear prints, and some forms of serological and DNA testing (*McCormick on Evidence*, 2006, 1:828). But these evidentiary standards are notoriously difficult to apply, and the history of DNA evidence illuminates some systemic problems with them.

The saga also connects with broader themes in the law governing expert testimony generally. It has been said that "[n]o human activity short of armed conflict or dogmatic religious controversy is more partisan than litigation" (Risinger and Saks 2003, 35). What problems afflict this adversary system as it applies expert knowledge to the resolution of legal disputes? How might these problems be ameliorated?

This chapter places the DNA experience in the context of these perennial questions. Because this book has focused on the special scrutiny courts have given to the capacity of the genetic theories and procedures brought before them to resolve questions of human identity, most of the analysis concerns the vetting of new methods or technologies. This discussion is followed by some remarks on the "production-line" problems of ensuring quality, accuracy, and even honesty in the routine operation of a forensic laboratory.

One category of problems that arose as courts tried to decide whether DNA evidence should be admissible was purely legal. The *Daubert* and *Frye* standards sound reasonable enough in the abstract. Who would favor using invalid or unreliable science? But the platitudinous agreement dissolves as soon as the rules are applied in concrete cases that inevitably spawn a host of subsidiary questions with less obvious answers. Examples are plentiful. Is there a viable distinction between the general acceptance or validity of an analytical method and the procedures that are (or should be) instituted to ensure that the method is properly employed? Defendants seeking to exclude DNA results assumed that there is not. They sought to transform the "moratorium" on DNA testing that one science reporter read into the 1992 NRC report into a judicial reality by construing laboratory accreditation, periodic proficiency testing, internal record keeping, or the like as integral components of the technology. The courts did not agree.

Does the methodology of DNA typing necessarily include quantitative rather than qualitative testimony? This is a question that the Arizona Supreme Court explicitly left unanswered in *State v. Bible*. It remains significant when one considers what jurors might be told about the proportion of a population that shares particular mitochondrial DNA sequences.

How much debate is compatible with "general acceptance"? We saw how the California Supreme Court's formulations left the issue wide open in *People v. Simpson,* and it emerges repeatedly in *Daubert,* as well as *Frye,* jurisdictions.

All these examples involve the content of the law. They are not fundamentally different from the open texture and frayed edges of legal doctrine in other domains. They fill the pages of law review articles and legal treatises. As judges incrementally supply the answers in the cases before them, choices are made, and a more definite, but never quite finished, pattern emerges.

The other category of problems with the admissibility standards for scientific evidence has been called epistemic. These problems involve the acquisition of reliable knowledge about social or scientific facts rather than the molding of amorphous legal doctrine. The nub of the problem is the difficulty in gaining a clear picture of a scientific disagreement solely from the testimony of party-selected experts (Hand 1901). Expert evidence fits uncomfortably into the normal trial process, which relies on judges or jurors who almost always lack any understanding of the field of knowledge on which the expert is drawing. Dissatisfaction with the system has an impressive pedigree. The Supreme Court of Michigan, for example, cautioned in 1892 that "[e]xpert evidence, while useful in many cases, is dangerous in all, and should be restricted . . . to those cases where its use is well nigh indispensable"(McNally v. Colwell, 52 N.W. at 73). Various grounds have been offered for this distrust of party-selected experts.

Payment

A common and enduring complaint about expert witnesses is that they are not to be trusted because they are (usually) paid by one party (e.g., Wigmore 1923, 1 § 563; Kaye, Bernstein, and Mnookin 2004, 339–341). Of course, ordinary witnesses often are biased for or against one party, such as a friend or family member. The law relies on cross-examination to expose these biases and presumes that the fact finder will discount the testimony accordingly. Opposing counsel can try to discredit a "hired-gun" expert for taking money for his testimony (or, rather, "for his time and effort"), but this lawyer may well have his own expert on his payroll (Bernstein 2008).

The early DNA admissibility hearings, however, do not fit the stereotype of highly paid, dueling professional experts. In the DNA cases, a coterie of Ph.D.'s crisscrossed the country, providing increasingly predictable and polarized testimony. On the prosecution side, many of the experts did not accept payment for their time. Likewise, some of the defense experts, especially in the second wave of cases that raised doubts about the quality of the early laboratory work and the foundations of the probability estimates, also did not seek compensation. To the chagrin of many expert witnesses, a standard line of impeachment is to ask about the witness's fees (e.g., Helpern and Knight 1977, 64, calling it "snide" and "below the belt"; Spain 1974, 91, characterizing it as "corny."). Yet throughout much of the life cycle of the admissibility debate, attempts by counsel to impeach experts by asking about substantial compensation fizzled. For instance, in an effort to block Bruce Weir's analyses of mixed stains in *People v. Simpson*, William Thompson tried hard to portray Weir as indebted to the DNA-testing establishment. Some of this cross-examination went as follows:

> *Q:* You mentioned in direct examination that you work as a consultant for Cellmark diagnostics?
> *A:* No. I said my university has a contract with Cellmark. . . .
> *Q:* How much money does it involve?
> *A:* $8,000 a year.
> *Q:* $8,000. And the money is granted to the university?
> *A:* The money is paid to the university.
> *Q:* Is it put in a fund that is at your disposal?
> *A:* I wish. . . .

Q: Okay. And you are being compensated for your time on this case?

A: No, I'm not. In fact, I'm losing money. The county is not paying enough for my meals each day.

Q: All right. So—so with regard to the work you did in preparing?

A: With regard from the beginning to the end, there will be not one dollar comes to me. . . . [I]t is flowing the other way. . . .

Q: Flowing the other direction. I think that is very public-spirited of you, sir. (Transcript of Examination of Bruce Weir, June 22, 1995, *44–45)

The state did slightly better, establishing that William Shields was earning $10,000 for his work in opposing Weir's calculations (Transcript of Examination of Bruce Weir and William Shields, June 22, 1995).

However, freedom from financial involvement simply opens the expert to being painted as an ideologue. One forensic pathologist likened the inquiry about compensation to the question, "And when did you stop beating your wife? . . . You're damned if you do and damned if you don't. If you answer yes, you're being paid, your questioner intimates you are doing it for the money. . . . If you answer no, he intimates that you have such an overriding personal interest in the outcome that you cannot give fair testimony" (Spain 1974, 91). Lewontin, who declined compensation for his expert witnessing, concluded that "your credibility is damaged no matter what you do" (Roberts 1992b, 735), and Weir confronted the dilemma in *Simpson*. When the court allowed him to testify about the mixed stains, the prosecution had him tell the jury that he was not being paid. To which defense counsel Peter Neufeld responded:

Q: Dr. Weir, a moment ago, you said that you weren't receiving any financial awards [*sic*] for testifying in this case; is that correct?

A: Yes. My travel expenses are paid.

Q: Okay. And would it be fair to say, Doctor, that sometimes people say something for money and sometimes people say something because they have a certain political agenda or certain philosophical statement they want to get across. Is that a fair statement?

A: As long as you don't apply it personally, it's a fair statement.

Q: Okay. And I believe you said a moment ago, sir, when you were describing yourself and the issues that have arisen in forensic DNA typing, you said that you've been very angry at the way these issues have been described in courts over the last several years; is that correct?

A: In some instances, yes. (Transcript of Examination of Bruce Weir and Denise Lewis, June 26, 1995, *29)

The fervor of some witnesses spoke for itself, as when Diane Lavett testified that "I don't want my beloved DNA molecule misused by courts of law" (United States v. Perry, No. CR 91-395-SC (D.N.M. Sept. 7, 1995)).

In time, as the debate over population genetics played itself out, some of the most scientifically prominent experts—including Hartl, Lewontin, and Weir—withdrew from the fray. After 1996, when the defenders of independence plainly had prevailed in the scientific arena, the efforts of prosecutors to tar and feather the remaining defense experts had far more impact. When science journalists stopped writing about a "raging debate," and when some consulting incomes reached six digits, courts took note of the "financial stake" of persistent witnesses (Chapter 8).

But what made the prosecution's attacks effective was not simply that these experts were pocketing a few (or many) dollars in exchange for their services in each case. It was the insidious suggestion that they no longer were in the game for the sake of what they held to be the truth. The truly worrisome "financial stake" lay not in the normal incentive of repeat players to be "good" (convincing) witnesses, but in the insinuation that they were clinging to an untenable position so that they could stay in the business of trying to be convincing. If the only scientists who testify to certain propositions are those who are being paid to present that particular view, courts may wonder whether the witnesses are beating a dead horse because, to them, it is a cash cow. This is not the usual payment problem that Wigmore saw in all expert testimony funded by one party to a case, but neither is it unique to the DNA experience. It can be seen in some "toxic tort" cases when a small number of experts persist in espousing views held by a dwindling number of their colleagues in epidemiology or toxicology.

Of course, the mere fact that a "financial stake" in the perpetuation of a controversy can contribute to the persistence of a minority view does not necessarily mean that those who accept payment are insincere or venal. There are less odious explanations for persistence. In particular, a second concern relates to the strategies the parties use to select their experts.

Selection and Cultivation

Even the most honest and neutral experts can hold different opinions. Typically, lawyers are forced to draw their ordinary fact witnesses from a very limited pool. If Pat and Jamie drive into each other at an intersection, there are only so many people who can give firsthand testimony about what happened. The attorneys are stuck with the testimonial limitations of these available fact witnesses. By contrast, in most circumstances the lawyer who needs an expert has a wider opportunity to shop for the witness with a pleasing courtroom manner and the willingness to endorse the attorney's theory of the case. Even an expert with idiosyncratic or eccentric views may be well credentialed and sincere, two attributes that judges especially appreciate. When there is a rich body of scientific opinion on a topic, it is likely to be clustered about some central position with outlying views in the tails of the curve that represents the distribution of individual opinions. The parties may draw their experts from the tails, giving the court a distorted picture of the general state of the field. This selection effect can be a reason to discount the testimony of experts. Over 130 years ago, an English judge explained that he had "the greatest possible distrust of scientific evidence" because

> [a] man may go . . . to half-a-dozen experts. . . . He takes their honest opinions, he finds three in his favor and three against him; he says to the three in his favor, "will you be kind enough to give evidence?" And he pays the three against him their fees and leaves them alone; the other side does the same. It may not be three out of six, it may be three out of fifty. I was told in one case, where a person wanted a certain thing done, that they went to sixty-eight people before they found one. (Mnookin 2008, 588)

In the first wave of DNA cases, however, this kind of selection strategy was not open to the defense. The commercial laboratories recruited distinguished academic scientists to review their operations and, it was hoped, give their blessing. The forensic technology was so new that defense counsel could not pick and choose from a thick tail of skeptical scientists. They had to go without or to make do with the likes of the public schoolteacher and private forensic scientist who questioned "radioactive technology" in *Kelly v. State* (Chapter 4). This phase in the introduction of a new forensic technology is hardly confined to DNA evidence. The desirability of a deeper pool of knowledgeable experts is one argument for the conservative strategy of waiting for general acceptance beyond a narrow group of scientists.

The defense bar responded by cultivating its own DNA experts. The tactics of the lawyers in neutralizing opposing experts and in enticing their own experts to express unequivocal or extreme positions produced a positive feedback effect. Annoyed with superficial and exaggerated arguments, upon returning from the courtroom, the academic scientists jousted in their journals. But their presentations sometimes obfuscated more than they clarified, as when treatments of population structure traded on an ambiguity between the impact of structure in a general population and the extent of variations across subpopulations (Chapter 6). The prominence of the experts on both sides of the growing divide caught the attention of journalists, who inevitably thrive on controversy. Scientists with opposing theories can become estranged, but the legal system encouraged the antagonists to inflict additional wounds on one another. Even in the calmest of times, lawyers will try to lure experts out on a limb, to have them avoid the normal academic disclaimers, and to present only those parts of their thinking that support the client's cause (Saks 1987; Kaye 1990). This may well have happened with DNA evidence. This phenomenon would account for the expert testimony or reports in *United States v. Yee* (Chapter 5) that criticized the FBI's match windows for being too wide even though narrower windows for declaring a match would have produced a match.

Exposure

A third concern with expert witnessing is that it often is isolated from normal professional peer scrutiny. An academic expert may be willing to say things in the courtroom that he would not dare put into print. In a run-of-the-mill case that no one ever hears about, such statements will not impede his professional career. As a young pathologist, Milton Helpern, the legendary medical examiner for the city of New York, testified in a civil case that "it is simply not possible for anyone to say with certainty exactly when [the insured] died. You can argue either way" (Houts 1967, 121). Yet, two eminent pathologists at least thirty years his senior had given definitive, flagrantly opposed opinions. At a medical meeting a few weeks later, Helpern asked one of them, "Would you have told your students in class the same thing that you testified to in that courtroom?" (ibid. 123). "Tut! Tut! Tut!, my boy," the venerable pathologist replied. "Don't ever confuse the courtroom with the classroom" (ibid.).

To reduce the divergence between conduct in the courtroom and behavior in the profession at large, the United States Supreme Court has spoken of the need for judges "to make certain that an expert, whether basing testimony upon professional studies or personal experience, employs in the courtroom the same level of intellectual rigor that characterizes the practice of an expert in the relevant field" (Kumho Tire Co., Ltd., v. Carmichael, 526 U.S. at 152). How? By excluding testimony that lacks sufficient "intellectual rigor."

Carmichael was a personal injury suit. A "rear tire of a minivan . . . blew out," and "[i]n the accident that followed, one of the passengers died, and others were severely injured" (id. at 142). "[A]n expert in tire failure analysis" was prepared to testify that a defect in the tire caused the accident (id.). Applying *Daubert* "flexibly" (id. at 145), the federal district court excluded the proposed testimony and granted judgment for the manufacturer because there was no other adequate evidence that the company was responsible. The court of appeals reversed on the ground that "'a *Daubert* analysis' applies only where an expert relies 'on the application of scientific principles,' rather than 'on skill- or experience-based observation'" (id. at 146). The Supreme Court held that the trial judge had the discretion to apply the factors listed in *Daubert* to determine if the expert's analysis of the particular tire was sufficiently rigorous. In this particular case, the Court decided that the district court was within its "lawful discretion" in concluding that the expert's "failure to satisfy either *Daubert*'s factors or any other set of reasonable reliability criteria" justified excluding his testimony (id. at 158). The Carmichaels lost their case.

This inquiry into "intellectual rigor" is easily adapted to apply to a number of issues with DNA evidence. Was Lifecodes' interpretation of all the bands in *Castro* truly rigorous, or was it, as Lander and others claimed, a departure from normal laboratory work with RFLPs? Was the company's use of monomorphic probes to correct for band shifting reasonable?

At the same time, to decide the general question whether a methodology or its application possesses enough rigor, the court needs intellectually rigorous testimony about the methodology that is being challenged as lacking in rigor. The separation of professional reputation from courtroom behavior removes the normal incentive for the expert addressing this metaquestion to be intellectually rigorous. The lack of professional accountability for courtroom performances has led some professional societies to promulgate standards and (occasionally) to discipline their members. An extreme example is the determination (reversed in the courts) of the British General Medical Council that the eminent pediatrician Sir Roy Meadow was "unfit to practise" as a result of his flawed testimony in the prosecution of Sally Clark that the probability of two cases of sudden infant death syndrome in the same family was only one in 73 million (General Medical Council v. Meadow; Marshall 2005). In the same vein, some commentators and courts have flirted with the idea of a tort action for expert-witness malpractice (Hanson 1996). These remedies can be of some value at the margins, but as a practical matter, they are limited to the most egregious cases.

Would some mechanism for the exposure of an academic expert's work as a consultant or witness have more of a disciplining effect? On the basis of the saga of DNA evidence, it is hard to say. One persistent critic of the basic product rule was Seymour Geisser, a statistician with illustrious credentials and both feet firmly planted in the university campus. Yet Geisser lost credibility with the Supreme Court of Washington when it became apparent that he was testifying without keeping abreast of the literature (State v. Copeland, 922 P.2d at 1318 n.5). Would his testimony have been any different had a panel of his fellow statisticians been sitting in the courtroom? Considering his presentations and publications on DNA evidence after the statistical debate was winding down (e.g., Geisser 1996), I have my doubts. Even more clearly, there was no lack of professional transparency in the presentations of some of the most prominent experts in the population-genetics debates. They put their ideas into print and never modulated the amplitude of their rhetoric. (See, for example, the discussion of Lander (1991a), Lewontin and Hartl (1991), and Hartl and Lewontin (1993) in Chapter 6, and Hartl (1994) in Chapter 7.)[2] Similar remarks apply to several of the witnesses who appeared for the government. Critics in the academic world responded to the excessive claims of forensic labs. The excesses of the critics, in turn, generated predictable reactions from their targets. Positions hardened and rhetoric escalated, leaving some observers to wonder why we cannot all play nicely.

Another response to the lack of peer accountability for courtroom opinions is to look at whether a theory or method of analysis was developed specifically for litigation. Courts can be more comfortable with a theory or method developed and used outside the courtroom than with material created in response to a party's needs or desires. Determining that the method is the same, however, can require care. Thin-multigel electrophoresis of degraded samples was not equivalent to the well-established types of electrophoresis previously used in biochemical research (Chapter 1). Similarly, Southern blotting and autoradiography of diallelic RFLPs were applied outside criminal investigations, but the interpretation of VNTR autoradiographs from biological fluids from crime scenes was more complex (Chapter 3). Yet the idealized view that because "scientists developed DNA testing in academic and industrial settings for reasons having nothing to do with solving crimes," its "potential limitations" were "already investigated and identified" continues to be advanced (Cooley 2007, 337).

Neutral Expertise

The standard criticisms of adversary expert testimony—payment, selection, and cultivation—argue in favor of supplementing, if not supplanting, the system with more neutral experts (Joiner v. General Electric Company, 522 U.S. at 149–150; Gross 1991; Breyer 1998; Erzinçlioglu 1998; D. E. Bernstein 2008). Courts have an inherent power to appoint expert advisers, special masters, and testifying witnesses. Federal Rule of Evidence 706 confirms this last power, and proposals to make more use of court-appointed experts or to create specialized, expert judges have a long history (Sink 1956; Kaye, Bernstein, and Mnookin 2004, 347; Mnookin 2008, 598–599).

Nonetheless, court-appointed experts often will deliver less than one naively might hope for. "More neutral" does not mean free from all biases or agendas (e.g., Deason 1998). The scientists who know the most about a controversial subject are likely to have staked out a position and to have their own pet theories. We interpret new information in light of our prior, strongly held beliefs. Cognitive dissonance affects scientists, just as it does the rest of the human race. A joke among lawyers concerns the advocate who grew weary of his retained expert's desire to qualify every statement with "on the other hand," so he decided to seek out a one-handed expert. Equally rare, perhaps, is the leading researcher on a subject who is truly open minded on the issue. On the other hand, if the search for appointed experts is confined to those who have not published on the subject, and if the issue is complex and the literature extensive, the start-up costs of educating a previously uninformed expert will be substantial. But even if total neutrality is not a feasible goal, the eradication of bias due to the affiliation with an adversarial party to a legal dispute is practicable.

Judges appointed their own experts in some DNA cases (e.g., *United States v. Yee*), but the impact of the appointed experts was not obvious. Better results were obtained, albeit at great expense, in litigation over silicone-gel breast implants. Court-appointed expert panels were able to inform the judiciary of what was already apparent in the literature—that no causal link to immune diseases had been established despite substantial research (Sanders and Kaye 1997; D. E. Bernstein 1999; National Research Council Committee on the Safety of Silicone Breast Implants 1999). Of course, in cases in which scientific opinion is more balanced, the courts' experts should not make the controversy disappear (Kaye, Bernstein, and Mnookin 2004, 357–358; Mnookin 2008, 599), but they can provide a more accurate determination of the extent and nature of the disagreement in the broader scientific community than is likely to emerge from the posturing or juxtapositioning of purely partisan experts.

Courts also sought opinions about the scientific status of forensic DNA methods from outside the courtroom. Some appellate opinions quoted from population-genetics textbooks and other scientific references. Indeed, if judges were skilled in reading the scientific literature, if the literature were rich enough, and if it were free from distorting influences of funding sources with interests in the outcomes, then judges would not need to rely on live witnesses in pretrial admissibility hearings. But this is a bit like saying that if pigs had wings, they could fly. The underlying sources of the disagreement may not be clear early in the life cycle of a scientific controversy. Not much will have been published, and disagreements may arise among honest scientists because they have different standards of what it takes to prove that a technology or procedure is valid and reliable. In the early population-genetics debate, for example, everyone was an empiricist, but the defenders of independence did not expect to see much population structure for VNTRs. The population structuralists were holding out for more studies of the specific loci. On the basis of the extant literature, generalist judges could discern that more studies were desirable, but they were hard pressed to tell whether they were necessary.[3]

Despite these obstacles, courts and the lawyers who brief them should not avoid the scientific literature. Some insight into a controversy usually can be gained by inspecting the writings that are intended to convince other scientists. Yet DNA cases like *Bible* (Chapter 6) and *Johnson* (Chapter 7) demonstrate how judges at the highest level of the court system can err when they try to read the literature (or even the trial record) for themselves. To avoid technical errors in opinions that have top-heavy discussions of science, courts could even use outside experts as sounding boards for drafts of their opinions. Although this idea proved shocking to one court (State v. Hummert, 933 P.2d at 1195), appellate courts already obtain prepublication expert review of the discussions of legal rules and cases by deciding as panels rather than individuals. The preparation of a joint opinion requires the circulation of drafts among the judges and their clerks, producing a kind of expert peer review of the legal research and reasoning. As long as suitable procedures, such as notice to the parties, are followed, no rule of law forces a judge's written explanation of science and technology to be insulated from prepublication review by scientific experts of the court's choosing (Kaye 1997a; 2001a, 53 n.66; cf. Cheng 2007).

In addition to appointing a small set of experts in specific cases or inspecting the research literature for themselves, courts sometimes can look to government-sponsored reviews. These are more common in the public health field. The National Research Council, for example, has produced studies of the health effects of vaccines, electromagnetic fields, silicone breast implants, and dioxins (National Research Council Committee to Review the Adverse Consequences of Pertussis and Rubella Vaccines 1991; National Research Council Committee on the Possible Effects of Electromagnetic Fields on Biologic Systems 1997; National Research Council Committee on the Safety of Silicone Breast Implants 1999; National Research Council Committee on EPA's Exposure and Human Health Reassessment of TCDD and Related Compounds 2006). Forensic science also has produced its share of technical controversies that have been examined in this way. The first such study, in 1979, was on the spectrographic comparison of voices (National Research Council Committee on Evaluation of

Sound Spectrograms 1979); the latest, in 2004, examined the compositional analysis of bullet-lead evidence (National Research Council Committee on Scientific Assessment of Bullet Lead Elemental Composition Comparison 2004). Both these NRC reports contributed to curbing the use of these technologies.[4]

Between the studies of sound spectrograms and bullet-lead analysis came the 1992 and 1996 reports on DNA evidence. For a combination of reasons, these two reports were less useful to the courts than they might have been. The 1992 DNA report was the victim of changing circumstances (Chapter 6). Although the charge that it was totally devoid of expertise in statistics and population genetics was overdrawn, there was an element of truth in the accusation: The committee was constituted to deal primarily with different issues, but uncertainty and confusion over population structure were emerging as the most potent objections to admissibility, and it was the aspect of genetic knowledge that was the most engaging to the high-powered, testifying experts. Consequently, the committee's recommendations on improving laboratory performance were all but lost in the reaction to its endorsement of the policy-driven ceiling methods for expressing the significance of DNA matches. Courts seeking a statement of what the burgeoning research actually showed about the likely effect of population structure on random-match probabilities (either within the general population or across subpopulations) could not find it in this report. The report's apparent insistence on a sharply quantified presentation of the significance of a match made this gap even more treacherous, and the rapid decision to empanel a second committee to clarify the science without dispensing such legal prescriptions weakened the authoritativeness of this first effort.

The second report also was less effective than it could have been (Chapter 7). The problem was delay. The attempt at a fast-track report failed because it took too long to secure funding, to assemble the group, to digest the literature, to prepare a polished product, and to move it through the NAS's review and final editing process. In the main, the report largely reinforced the clear trend back toward admissibility. However, the report may have squelched any efforts to manipulate the ceiling methods by searching though studies of small samples from different regions to find large allele frequencies to multiply. By treating population structure in a more technically correct way and showing that usually this makes little difference, the report also may have shortened the life span of the interim ceiling procedure in those jurisdictions that had approved of or required it. Its endorsement of PCR-based procedures smoothed the way for the admission of the next generations of typing methods. The outcomes probably would have been the same even if there had been no report, but there might have been more zigzagging on the way there.

A variation on the often-belated appointment of ad hoc committees might be a freestanding, national forensic science institute. An independent body that was neither run by nor beholden to law-enforcement officials and had sufficient funding and expertise might be able to identify and address emerging technological issues more nimbly and expeditiously than the current procedure of waiting for Congress or a federal agency to resort to a blue-ribbon committee in response to an apparent crisis. The National Institute of Forensic Science proposed by the National Research Council Committee on Identifying the Needs of the Forensic Science Community (2009, S-14; hereinafter cited as NRC 2009) would be an obvious organization to conduct or commission such assessments proactively.

Other Reforms

Traditionally, the format for pretrial hearings on the admissibility of scientific testimony evidence is the same as that for trial testimony from ordinary witnesses. In a criminal trial, the government presents its witnesses, who are subject to cross-examination, one at a time. The defense presents its witnesses, if any. The witnesses do not talk to each other. This method of grouping the testimony and isolating the witnesses works well enough with fact witnesses in a reasonably short trial. It works less well in complex litigation with many issues and witnesses. In that situation, the information relating to the issues will emerge in a disjointed fashion.

Disjointed testimony from experts is especially difficult to evaluate, and there is little reason to keep the experts isolated from one another. Courts have discretion to depart from the usual script for the order and mode of testimony, and doing so at a pretrial hearing on admissibility can be productive. Rather than simply listen to experts tell lawyers what the lawyers want to hear, judges could ask opposing experts to talk to each other rather than past each other. They might be required to produce a single, jointly authored report that lists the issues on which they agree, the points of disagreement, and the reasons for the disagreement. More radically, the experts might be required to produce the joint report away from the supervision of (and without revision by) the lawyers for the parties. This report could be submitted to an independent expert selected by the court, if desired (cf. Nesson and Demers 1998). At the hearing (or even at trial), the experts could engage in a more direct exchange, something closer to a panel discussion to amplify or clarify the points raised in the joint report.

The judge or the appointed expert could keep the discussion on track and prevent the joint testimony from degenerating into mere bickering. The procedure has achieved some popularity in Australia, where it is referred to as expert "hot tubbing" (Liptak 2008).

Unfortunately, the cooperative model will not always succeed. When the experts truly are hired guns or are infected with personal animosity, pretrial cooperation is unlikely. But at worst, the court can revert to the current standard operating procedure, and at best, a joint statement will produce greater clarity and more consensus. A major conversion such as that in *People v. Castro* (Chapter 4) is obviously exceptional, but a less adversarial dialogue among retained experts who are not consciously biased or dogmatic generally should be possible.

Thus far, I have been considering the creation and acquisition of information·for the courts on the status of a scientific theory or procedure. The quotidian production of case-specific evidence pursuant to an accepted technology also could be improved. A persistent issue in struggles over the admissibility of DNA identification is the perceived need for centralized review and regulation. From the ad hoc consensus statement in *People v. Castro* to the two NRC book-length reports, scientists have pleaded for new institutions to evaluate new developments, for detailed protocols for testing, and for internal and external reviews of laboratory performance— in short, for more regulation. Despite the high drama of the confrontations between the population structuralists and the defenders of independence, most of the problems with DNA testing have not been errors of high theory or the introduction of methods that did not work. Rather, they have been production-line problems in the generation of the evidence for specific cases—from overselling a particular interpretation of a mixture (and failing to disclose data that might raise a question about that interpretation) to outright lies by analysts about their bench work. (Such cases are collected in, e.g., Giannelli and McMunigal 2007; see also W. C. Thompson 2008; Thompson and Dioso-Villa 2008.) These kinds of abuses are not restricted to DNA typing but date back to the bad old days of forensic serology (Kaye 1997c; 14–15; Giannelli and McMunigal 2007).

The criminal justice system is not very good at detecting these errors. Prosecutors do not usually seek out errors or questionable judgments by laboratories, and government laboratories work hand in hand with law-enforcement authorities, who often are the source of or conduit for their funding (NRC 2009, 2-2-2-4). As the earlier anecdote about Milton Helpern indicated, the conversion or perversion of expertise into expert advocacy is well known to the medicolegal world. Helpern further observed that "P.G.H. Brouardel, a famous, late nineteenth century French medico-legalist . . . said something which is as true now as when he wrote it, if not more so: 'If the law has made you a witness, remain a man of science. You have no victim to avenge, no guilty or innocent person to convict or save—you must bear testimony within the limits of science' " (Helpern and Knight 1977, 65). If Helpern had his way, this quotation would "be framed and hung in every pathologist's office, as well as over every witness stand in every court" (ibid.). From the standpoint of many police and prosecutors, however, criminalists and DNA analysts are there to develop and present part of the case against their suspects. A 2006 report of a crime-scene specialist called to an automobile crash scene in Scottsdale, Arizona, reads as follows:

> Detective Thompson requested that I swab and impound both airbags in vehicle 1. . . . I noted red stains . . . on both the driver and the front passenger airbags. . . . I swabbed a representative spot of the bloodstains on the driver's airbag. . . . I cut and collected the driver's airbag. . . . I then swabbed a representative spot on the front passenger airbag. . . . . . . At the direction of Detective Thompson no further action was taken.

One senses that the investigating officer, not the crime-scene specialist, is calling the shots here. The police department's written request to the laboratory does not ask for an independent and comprehensive assessment of the forensic evidence but requests an analysis of only one, specific swab and directs the laboratory to "[p]lease check [this] airbag swab against blood sample DNA to prove suspect 1 is the driver." Of course, the laboratory is not going to come back with a report that states that suspect 1 is the driver if its STR testing plainly excludes the suspect as the source of material swabbed from the driver's airbag, but what might additional swabs from both airbags show? The culture and institutional arrangements in which police laboratories operate thus can be difficult to reconcile with Brouardel's admonition that "[y]ou have no victim to avenge, no guilty or innocent person to convict or save." With the right leadership and oversight of casework and testimony, forensic scientists may be able to follow rather than channel the scientific evidence even when it leads in an unexpected or unwanted direction. But the current system creates incentives or temptations for the scientists and technicians to meet the demands or desires of the police and prosecutors, who may have their suspect firmly in place.

One can hope that any leanings in this direction are subject to correction by the adversary system, but an overburdened adversary system does not exert a strong corrective force. Too many criminal defense lawyers, stretched and stressed by large caseloads, remain ignorant of possible infirmities in laboratory work and are unable to understand and counter statistics that may be tendentiously presented. Most of their clients are too poor to hire consulting experts who might uncover a possible problem, and courts, with limited budgets, usually demand a strong showing that there is a problem before they will authorize payment for a defense expert (Giannelli 2004). Appellate courts, in turn, are loath to overturn convictions even when they sense that the science was oversold. Instead, one reads opinions such as the one in *Commonwealth v. Teixeira,* where the Massachusetts Appeals Court wrote in 1996 that"[i]t may have been scientifically and statistically unsound for [FBI technician John] Quill to claim a laboratory error rate of zero, but that proposition is not so ordinary and obvious that a judge should be held to take judicial notice of it" (662 N.E.2d at 729).

The most promising line of attack on these problems is the creation of incentives for the forensic scientists or technicians working in the laboratory and testifying in the courtroom to get it right in the first place and to avoid slipping into the role of advocates. Lawyers and scholars therefore have touted the value of institutional reform—separating laboratories from the police; fostering competition among laboratories; requiring auditing, proficiency testing, and regulation of laboratories; supplying more expert services to defendants; and demanding more awareness and action to discourage prosecutors from using dubious scientific evidence whenever it seems convenient (Giannelli and McMunigal 2007; D. E. Bernstein 2008, 460; see also House of Commons Science and Technology Committee 2005; NRC 2009). Like the perennial calls for more neutral expertise, legislating these changes will not be a panacea, but industrial-strength quality control is not too much to demand. The great DNA wars over admissibility are over.[5] The courts have discerned the essential features of the structure of the DNA molecule and the biochemical techniques for analyzing the characteristics of particular DNA samples. Although skirmishes will spring up as new DNA loci and technologies come on line, the forensic DNA scientists have learned the major lessons about the need for validation and publication. The more enduring challenge for the legal and the forensic science communities resides in how the evidence is generated and presented.

Timeline of Selected Developments in or Affecting Forensic Genetics

Abbreviations: HLA, human leukocyte antigen; mtDNA, mitochondrial DNA; NRC, National Research Council; QA, quality assurance; QC, quality control; RFLP, restriction fragment length polymorphism; STR, short tandem repeat; VNTR, variable number tandem repeat

1900	Human ABO blood group discovered
1915	First antibody test (for ABO) developed
1920s–1950s	Other blood groups and serum proteins discovered
1924	Genetics of the ABO system elucidated
1930s	ABO typing introduced in courts
1953	Chemical structure of DNA described
1954–1980s	Electrophoresis for separating protein and DNA variants developed
1964	Test for HLA (tissue types) developed
1970	Restriction enzymes discovered
1970s–1980s	HLA and extended red blood cell (RBC) testing used in courts

1975   Southern blotting developed

1977   DNA sequencing methods developed

1980   RFLP mapping described; first VNTR discovered

1980s   Electrophoretic variants of RBC enzymes used in courts

1984   Multilocus VNTR probe discovered

1985   Multilocus VNTR profiling presented in an immigration proceeding in England (*Sarbah*)

            PCR discovered

1986   Single-locus VNTR profiling used in a criminal investigation in England (*Pitchfork*)

            PCR-*DQA* analysis admitted in an American case (*Pestinikis*)

1988   Single-locus VNTR profiling upheld in a Florida appellate court (*Andrews*)

            First commercial kit for PCR-HLA, dot-blot, and oligonucleotide hybridization

            First DNA data-banking law enacted (Colorado)

1989   Single-locus VNTR profiling excluded in a New York trial court (*Castro*)

            Lander testifies in *Castro* and publishes in *Nature*

            First convict exonerated by DNA (United States)

1991   Lewontin and Hartl testify in *Yee* and publish in *Science*

            First useful polymorphic human STRs characterized

1992   First NRC report calls for improved QC and QA and presents the "ceiling principle"

            First commercial forensic STR profiling kits developed

            First Y-STR described and used in casework (Germany)

            First use of mtDNA in casework (United Kingdom)

1994	O. J. Simpson moves to exclude DNA evidence before his murder trial begins
	"DNA Fingerprinting Dispute Laid to Rest" by Lander and Budowle in *Nature*
1995	O. J. Simpson withdraws motion to exclude DNA evidence
	O. J. Simpson moves to exclude DNA evidence (again)
	First national DNA data bank for law enforcement established (United Kingdom)
1996	Second NRC report calls for improved QC and QA and rejects the ceiling methods
1997	STR profiling from touched objects and single cells demonstrated
1999	Tennessee appellate court upholds mtDNA testing (*Ware*)
2007	200th convict is exonerated by DNA testing (United States)
2009	NRC report on strengthening forensic science presents DNA evidence as a model

Notes

1. Before DNA

1. Individuals with type B blood have B antigens and anti-A antibodies in their blood serum. In addition to these two types of people, there are type AB individuals, who have both A and B antigens and no antibodies at all. Then there are Type O individuals, who lack both the A and B antigens but have both antibodies. The O cells can pass unnoticed, making type O people universal donors (they can donate blood to anyone), but their antibodies mean that they are not universal recipients.

2. The A allele codes for an enzyme that adds one kind of sugar molecule to the end of the basic ABO blood-group protein (known as the H antigen). The enzyme determined by the B allele adds a different sort of sugar to the end. The O allele produces a protein with no enzymatic activity, leaving the H antigen unchanged.

3. Both sides promptly declared that they had secured a victory in the Supreme Court because the new standard for scientific evidence would produce the result they desired. In the end, the court of appeals adhered to its original decision on other grounds. Daubert v. Merrell Dow Pharms., Inc., 43 F.3d 1311 (9th Cir. 1995) (even if unpublished epidemiological reanalyses of data used in previous studies were admissible, they would not show a substantial enough association between Bendectin and limb-reduction defects to overcome a motion for summary judgment).

4. For instance, Schiff (1929, 992) wrote in the *Lancet* that "the leading representatives of forensic medicine, serology, and the science of inheritance, unanimously expressed the opinion that the examination of the blood groups, when

performed by an expert, 'is a reliable method which may be used advantageously for forensic purposes, especially for the exclusion of paternity.' " "Bernstein's formula of inheritance," he added, "is generally accepted as the principle underlying the method."

5. Because the gels were made of a starch, the conflict over the admissibility of electrophoresis of serum proteins sometimes was called the "starch wars" (Aronson 2006).

6. When it came time to submit a final report to the funding agency, the federal Law Enforcement Assistance Administration (LEAA), Grunbaum charged that Wraxall had manipulated and falsified the results. Although an ad hoc committee appointed by the LEAA to review Grunbaum's charges held that the allegations were unfounded, the government declined to publish any final report. In a compromise of sorts, it supplied a booklet to any forensic scientist or laboratory that requested a copy (Aronson 2006, 64).

7. E.g., People v. Reilly, 242 Cal. Rptr. 496, 503 (Cal. Ct. App. 1987); State v. Fenney, 448 N.W.2d 54, 60 (Minn. 1989). However, in holding that it was error to find general acceptance of electrophoresis of serum proteins on the basis only of the acceptance of ABO typing and the testimony of Stolorow that electrophoresis was widely used in crime laboratories, the Illinois Appellate Court observed that "that use in crime labs alone [cannot] justify admission of evidence in the face of a bona fide scientific dispute" and called for "cautious consideration of this evidence." People v. Harbold, 464 N.E.2d 734, 747 (Ill. App. Ct. 1984).

8. Various occupations have been ascribed to Juricek. Compare People v. Morris, 245 Cal. Rptr. 52, 58 n.1 (Ct. App. 1988) ("a genetic counselor [with] no experience with bloodstain genetic marker detection"); People v. Marlow, 41 Cal. Rptr. 2d 5, 17 n.21 (Ct. App. 1995), *rev. granted,* 899 P.2d 65 (Cal. 1995), *rev. dismissed,* 987 P.2d 695 (Cal. 1999) ("Dr. [Diane Juricek] Lavett completed her Ph.D. and post doctoral research in genetics at Emory University. She is an associate professor at the State University of New York at Cortland and is the author of the 'Student Companion with Complete Solutions for an Introduction to Genetic Analysis' "); and People v. Axell, 1 Cal. Rptr. 2d 411, 419 (Ct. App. 1991) ("Doctor Diane [Juricek] Lavett, a Professor of Genetics at State University of New York and at Emory University"). She published her misgivings in Juricek (1984).

9. E.g., People v. Stoughton, 460 N.W.2d 591 (Mich. Ct. App. 1990); State v. Thomas, 421 S.E.2d 227, 231 (W.Va. 1992). However, it has been said that no credence should be given to this study because "the authors did not release the underlying data from their research for others to review." *Stoughton,* 460 N.W.2d at 593 (describing one expert's testimony).

2. Trial by Mathematics

1. People v. Mountain, 486 N.E.2d at 805. The court added that "where the defendant can show that the potential prejudice outweighs the probative value the court may in its discretion exclude the evidence." Id. at 806.

2. The Illinois Appellate Court has continued to apply the rule in a small class of cases. See People v. Schulz, 506 N.E.2d 1343 (Ill. App. Ct. 1987).
3. Some people call such reasoning "the defense attorney's fallacy" (see, e.g., United States v. Chischilly, 30 F.3d 1144, 1157 (9th Cir. 1994); Berger 1997, 1107). However, I cannot see any fallacy in it. A possible fallacy that favors the defense is the jump to the conclusion that evidence of a match is irrelevant or weak just because a large number of people in the general population also are expected to match.
4. Bayes's rule is commonly used in paternity litigation (with an arbitrary assumption that, before considering the genetic evidence, a man named as the father is 50% likely to be the actual father).
5. State v. Carlson, 267 N.W.2d at 172. As noted in Kaye (1997c, 27–28), Gaudette's statistical studies soon received severe criticism.
6. State v. Boyd, 331 N.W.2d at 481. The figure of 1,121 for the "mean number of men" excluded is not consistent with an inclusion probability of 0.03 to 0.06. The latter estimate means that in testing the entire population, between 100/3 and 100/6 (about 17 to 33) men would be excluded, on average, for every man who would be included. The logic that leads to the 1,121 figure for the paternity index and the related estimate of 99.911% for the probability of paternity is discussed in, e.g., *McCormick on Evidence* (2006, 1:924–926).
7. This experiment is discussed in greater detail in Chapter 12.
8. In fairness, it should be noted that Lopez did not make a *Frye* objection to Moorman's testimony. In fact, it was the defense that asked her about statistics:

> Even though Moorman's credentials as a statistics expert were questionable, defendant opened the door on cross-examination when he questioned Moorman about the frequency of Type O blood in the general population. There had been no mention of percentages or frequency in the State's direct examination. During the redirect examination, the State laid a proper foundation to inquire into the frequency of the four enzymes and Type O blood in the population.
>
> Where a defendant opens the door to an issue, he interjects the issue into his case, and cannot then contend that bringing it to the jury's attention was an error. . . . Moreover, defendant presented his own expert testimony on the statistics. (People v. Lopez, 593 N.E.2d at 656–657 (citations omitted))

3. The Dawn of DNA Typing

1. Although variations at a (different) single locus lead to each disease, there are far more than two possible alleles at each locus. Tay-Sachs disease is a recessive, neurodegenerative disorder that, in the classic infantile form, is usually fatal by age two or three years. It is caused by a mutation in the hexosaminidase A gene (*HEXA*). Seventy-eight mutations in the *HEXA* gene have been described, including sixty-five single-base substitutions, one large and ten small deletions,

and two small insertions (Myerowitz 1997). However, the most prevalent forms of *HEXA* code for an enzyme that functions normally. The disease occurs when certain mutant alleles are inherited from both parents, resulting in proteins that are dysfunctional. Thus the simple Mendelian model ignores variations in the disease related to different mutant alleles, and it treats the loci as having one dominant allele *A* and another recessive allele *a* (which really designates a class of alleles associated with the disease).

Likewise, "sickle-cell disease" refers to a collection of genetic disorders. At least 476 beta globin gene variants exist. Individuals with "sickle-cell anemia" have two copies of a particular allele. Individuals with other types of sickle-cell disease possess one copy of this allele and one copy of another beta globin gene variant (Ashley-Koch, Yang, and Olney 2000).

2. "Each cell contains thousands of different proteins: enzymes that make new molecules and catalyze nearly all chemical processes in cells; structural components that give cells their shape and help them move; hormones that transmit signals throughout the body; antibodies that recognize foreign molecules; and transport molecules that carry oxygen" (National Center for Biotechnology Information 2004).

3. It takes a "word" of three base pairs to code for a particular one of the twenty amino acids. A two-letter word is too short—only $4 \times 4 = 16$ such words can be made with four letters. Because there are $4 \times 4 \times 4 = 64$ possible triplet code words, in some cases the same amino acid is coded by several different three-letter DNA words (codons). The mapping of the triplets of DNA base pairs to the twenty amino acids is the genetic code. For example, the triplet TTT leads to an amino acid known as phenylalanine. Some additional triplets, such as TAG, do not code for an amino acid but stop the addition of new amino acids to the protein chain.

It follows that the structure of a protein with, say, 100 amino acids can be encoded by 300 DNA base pairs. This would be the protein-coding portion of the gene for this protein. However, cells do not "read" the coding DNA directly. Instead, proteins "transcribe" a copy of the coding part of the gene into messenger RNA, a single-stranded molecule with four nucleotides that are almost the same as those in the DNA. (In the messenger RNA, the base uracil is substituted for the thymine in the DNA.) The messenger RNA then carries the code words for the protein from the cell nucleus to the surrounding cellular material. Here, other molecules working at a kind of assembly bench translate the messenger RNA into new proteins by stringing together the amino acids in the order dictated by the messenger RNA.

4. A broader and more contemporary definition is that a gene is "a complete chromosomal segment responsible for making a functional product" (Snyder and Gerstein 2003, 258). "This definition has several logical components: the expression of a gene product, the requirement that it be functional, and the inclusion of both coding and regulatory regions" (ibid.). Because some "RNA has structural, catalytic, and even regulatory properties," this definition encompasses DNA sequences that lead not only to proteins but also to those that lead to some RNA transcripts (ibid.).

Indeed, from the standpoint of what is transcribed, a gene need not be confined to a single region on one chromosome. A process known as alternative splicing of the messenger RNA enables a cell to select different combinations of protein-coding sequences (exons) within a single region to make different "mature" RNA transcripts. It can even recruit exons from more than one contiguous region that formerly would have been designated a "gene" (Seringhaus and Gerstein 2008; Zimmer 2008).

5. Some introns may serve as sites for recombination.

6. This example is adapted from Odelberg and White (1989).

7. The second figure reflects an arithmetic error. Under the analysis in the article, it should be 5×10^{-21} (Cohen 1990, 362).

8. This particular error was the use of $\bar{p}$, the mean proportion of shared bands, to estimate $\bar{P}$, the mean probability of a random match. Cohen pointed out that even granting all the other assumptions, one should calculate $P_i = p_i^n$ for each individual i and average these numbers to obtain the mean random-match probability. This "arithmetric mean" will be smaller than Jeffreys's estimate of the "geometric mean" $(\bar{P})^n$. Although Cohen wrote that the effect was substantial, when Jeffreys computed the arithmetic mean for his data, he found that the mean random-match probability was 8.4×10^{-17}—not very different from the original estimate of 5×10^{-19} (Cohen 1990, 366; see also Jeffreys, Turner, and Debenham 1991, 837).

9. Jeffreys, Turner, and Debenham (1991, 824) state that the multilocus "probes termed 33.6 and 33.15 have been developed which have been extensively applied to civil and criminal casework." The two references they cite, however, do not refer to any criminal cases in which the probes were used. Thompson and Ford (1989, 49) also claim that "[t]ests based on [multilocus] probes have been admitted in evidence in Great Britain in about twenty criminal cases, with convictions resulting in most." This statement came from a vice president of the U.S. subsidiary of Cellmark Diagnostics, the company that Jeffreys licensed to develop his probes for forensic use, and it does not specify the types of probes used in these cases. Contrary to the citation in Thompson and Ford, the *Pitchfork* case is not an example of the admission of a multilocus-probe test. As discussed later in the text, that case involved single-locus testing.

One case in which the multilocus probes might have been used is the prosecution of Robert Melias in November 1987, in Bristol Crown Court, for robbing and raping a crippled woman. Reports about the case state that the Home Office used some form of DNA testing that gave a random-match probability of 1 in 4 million (Wilson and Wilson 2003, 304). Whatever probes were used, it is not obvious that the judge considered the admissibility of the results, because Melias pled guilty (McCarthy 1987; McCartney 2006, 92). Yet, the case often is presented as if it were the first conviction in which a court relied directly on DNA evidence (e.g., Belair 1991, 4 n.8).

Two cases of convictions based in part on multiple locus probes are *R. v. Deen* (see Lynch et al. 2008, 165, 168) and *R. v. Doheny*. However, it is not clear that objections were lodged to the use of these probes. Instead, the convictions were reversed because of errors in the computation or presentation of the random-match probabilities. In *Deen* the trial judge slipped into the transposition fallacy, prompting a reversal (ibid. 179). In *Doheny* transposition again occurred, but the Court of Appeals reversed on the ground that the reported match probability of 1 in 40 million was calculated incorrectly. The difficulty, according to testimony from the geneticist Paul Debenham and the statistician Peter Donnelly, was that this figure was obtained by multiplying the probabilities for overlapping multiple locus and single-locus matches.

10. According to Aronson (2005, 129), "[T]he police officers had good reason to question the validity and reliability of this novel technique." The reason given for this judgment, however, is misguided. Aronson writes that "in a 1996 interview, Jeffreys admitted that the single-locus probe technique had not been fully developed: 'We were basically flying by the seat of our pants', he said. Had there been a match, it would have been necessary to estimate the probability of it being a chance occurrence; whereas the mismatch between the DNA from the crime scene and the suspect meant that the kitchen porter could be excluded from the investigation without any need for complex statistics. 'In a way it was good that we had an exclusion', said Jeffreys." This "admission" by Jeffreys simply means that Jeffreys was not in a position to estimate the probability of a random match with the newly minted single-locus probe. That fact did not give the police any reason to doubt the validity and reliability of an exclusion. In fact, the reliability of the exclusion was shown by subsequent tests with the same probes conducted by the Forensic Science Service (Forensic Science Service n.d.; Newton 2004b). Even an inclusion would have had probative value, although this value would have been difficult to quantify at the time.

11. DNA testing for paternity was admitted at the Old Bailey in a case "of unlawful intercourse with a fourteen-year-old mentally handicapped girl who had given birth to [a] baby" (Wambaugh 1989, 316).

4. The Emergence of VNTR Profiling

1. To be more precise, there are three core sequences for this VNTR: GGGGGTGTT, GGGGTTGTT, and GGGGTGTT (Holmlunda and Lindblom 1998).

2. Jeffreys's group contributed to this approach, describing the individual loci that were responding to the "fingerprint probes" and deriving new "locus-specific" probes for them (Wong et al. 1987).

3. *Andrews* often is miscited as the first criminal case in the United States to admit DNA evidence (e.g., Lander 1992, 193; Mnookin 2006, 214; Aronson 2007, 35; cf. Sheindlin 1996, front cover (describing Judge Sheindlin as "[t]he first judge to rule on DNA evidence in the courtroom")). That distinction belongs to an unreported 1986 Pennsylvania case, *Commonwealth v. Pestinikis,* which involved a less discriminating form of testing, namely, an early form of *DQA* analysis (discussed in Chapter 9).

4. For discussion of the history of these companies, see Aronson (2007, 15, 17–21, 30–32). On their operating structure, see Daemmrich (1998).

5. Currently, Robin Cotton is an associate professor and director of the Biomedical Forensic Sciences Program at the Boston University School of Medicine.

6. *Cobey v. State,* 559 A.2d at 398. The state produced five experts who testified to general acceptance. Id. at 392.

7. A 1998 advertising manual titled *DNA Fingerprinting*^{SM} states that "[i]f the two patterns match exactly, then it is virtually certain that the accused left that specimen at the crime scene" (Cellmark Diagnostics 1988, 8). The accompanying figures show a multilocus VNTR profile. Single-locus probes are "in most cases not as discriminating as the multi-locus probes," but, according to Cellmark, they "can differentiate individuals to the degree of excluding the world's population" (ibid., 9).

8. For an earlier example, see People v. Wesley, 533 N.Y.S.2d 643 (Albany County Ct. 1988).

9. Kelly v. State, 792 S.W.2d 579, 583 (Tex. Ct. App. 1990), *aff'd*, 824 S.W.2d 568 (Tex. Ct. Crim. App. 1992). Radioactive isotopes are used in producing the bands in the autoradiograph.

10. Sheindlin (1996, 24) gives larger numbers: twenty-eight and sixty-seven, respectively.

11. The opinion on the admissibility of the evidence does not state the estimated population frequency. However, in Sheindlin (1996, 66), the judge, misconstruing the frequency as the probability "that the blood could belong to another person beside Vilma," recites "[o]ne in 189 million!" See also Lander (1989a), and Aronson (2007, 64). R. Lewin (1989), Coleman and Swenson (1994, 6), and Mnookin (2006) give the figure 1/189,200,000. D. Thompson (1989) and Lander (1992, 197) refer to 1/100,000,000.

12. According to Balazs et al. (1989, 183), up to six fragments can occur in an individual because each chromosome contains between one and three closely spaced regions of VNTR separated by a spacer region up to a few thousand base pairs long, and the restriction site for the enzyme used to analyze the polymorphism is located in the spacer region. According to Aronson (2007, 64), "Howard Cooke, who had invented the probe used by Lifecodes to target the DXYS14 locus, and David Page, of MIT, testified that . . . the probe in question could produce banding patterns with anywhere from one to six bands." However, Lander (1992, 198) states that as many as eight bands can arise at this locus.

13. Lander (1992) refers to Roberts as an organizer of the meeting. Aronson (2004, 199–200), citing uncatalogued records of the Cold Spring Harbor Laboratory, reports that the only organizers were "Jan Witkowski, a molecular biologist from Cold Spring Harbor Laboratory," "Jack Ballantyne, a well-respected forensic scientist," and George Sensabaugh, a forensic scientist on the faculty of the University of California at Berkeley. Efforts "to bring David Housman on as a fourth organizer" were "in vein [*sic*]."

14. Other writers relate slightly different accounts. Judge Sheindlin believed that the meeting occurred in "Lander's hotel room . . . [o]ver the course of a long evening" (Sheindlin 1996, 73). Mnookin (2006, 226) reports that "eight of the ten witnesses who were contacted liked the idea of meeting, [but] only four were able to fit the meeting into their schedules." Accord, Parloff (1989); Schmeck (1989). Sheindlin recounts that the statement was "signed by all of

the scientists except Dr. Baird" (Sheindlin 1996, 73). Finally, although Sheindlin speaks of a "secret meeting," Parloff (1989) reports that it "was arranged with the knowledge of the lawyers on both sides."

15. People v. Castro, 545 N.Y.S.2d at 996.

16. The entire pattern of bands appeared to be shifted, and Baird sought to correct for this in a way that seemed ad hoc or inadequately validated to other scientists (Norman 1989). Lifecodes' senior vice president defended the adjustment as a "conservative choice" in response to the "normal defense tactic" of "[m]uddying the waters" (Winkler 1990).

17. Levy (1996, 53–55) paints a much rosier picture of these efforts than does Aronson (2007).

18. United States v. Jakobetz, 747 F. Supp. at 258. In affirming, the Second Circuit determined that the district court correctly applied its reliability standard for scientific evidence rather than the general acceptance standard of *Frye v. United States*. Beyond that, in dictum, this court observed that the testimony to the district court and the scientific literature established that VNTR profiling as practiced by the FBI also satisfied *Frye*. It wrote that "the general theories of genetics which support DNA profiling are unanimously accepted within the scientific community," and "the specific techniques used by the FBI lab in RFLP analysis are commonly used by scientists in microbiology and genetics research" (955 F.2d at 799). The court did not discuss the details of the FBI's procedures, but it also remarked that although *Castro*'s neoteric "third prong" was not necessary for admissibility, it too had been met. Giving a green light to DNA evidence, the court of appeals advised that "in the future, we do not think that such extensive hearings and findings should be conducted in every case" (id.). Rather, "[t]he district court should focus on whether accepted protocol was adequately followed in a specific case, but the court, in exercising its discretion, should be mindful that this issue should go more to the weight than to the admissibility of the evidence. Rarely should such a factual determination be excluded from jury consideration" (id. at 800).

5. The Intensifying Debate over Probability and Population Genetics

1. See State v. Kromah, 657 N.W.2d 564 (Minn. 2003) (*Kim* not overruled, and there is a need for caution with probabilities); State v. Alt, 505 N.W.2d 72 (Minn. 1993) ("the only DNA frequency evidence to be admitted at trial is the population frequency evidence of the individual bands. State v. Johnson, 498 N.W.2d 10 (Minn. 1993); State v. Jobe, 486 N.W.2d 407 (Minn. 1992); State v. Kim, 398 N.W.2d 544 (Minn. 1987)").

2. E.g., United States v. Morrow, 374 F. Supp. 2d 51, 65 (D.D.C. 2005) ("careful oversight by the district court and proper explanation can easily thwart this issue").

3. It favors the defendant in the mathematical sense that the proper transposition via Bayes's rule would have produced even larger posterior odds. The rule states that

posterior odds = likelihood ratio × prior odds.

For example, as indicated in Chapter 2, a random-match probability of 1 in 1,000,000 produces a likelihood ratio of 1,000,000. If the prior odds were 2:1, then the formula gives posterior odds of 2,000,000:1—an even more compelling number than the 1,000,000:1 odds that come from a naive transposition. Of course, this calculation of the likelihood ratio ignores other hypotheses (such as a close relative or a false-positive match due to laboratory error or police mishandling) that are consistent with innocence. To the extent that transposition prevents the jury from considering these matters, it is not a good thing for the defendant.

4. A small correlation in the errors arises because of band shifting.
5. But see People v. Venegas, 954 P.2d 525 (Cal. 1998) (upholding the reversal of a rape conviction because the FBI used floating bins whose width was smaller than the match window).
6. *Jakobetz,* 747 F. Supp. at 258. For details of the FBI's procedures, which were less conservative than those later recommended by the NRC committees, see NRC (1996, 143–144).
7. Kaye 1996. For examples of this error, see State v. Bible, 858 P.2d 1152, 1185–1186, 1186 n.25 (Ariz. 1993); People v. Castro, 545 N.Y.S.2d 985, 992 (Sup. Ct. 1989); State v. Pennell, 584 A.2d 513, 517 (Del. Super. Ct. 1989); State v. Cauthron, 846 P.2d 502, 513 (Wash. 1993); Thompson and Ford (1989, 82).
8. The Hardy–Weinberg equilibrium proportions will occur in a closed, infinite population when mating, viability, and survival are uncorrelated with the gene in question.
9. Lewontin also is known for his social activism and particularly for resigning from the National Academy of Sciences in 1971 when he "discovered that among other things, the National Academy, through its operating arm, the National Research Council, had committees doing secret war research" (Kreisler 2003).
10. However, the court noted that Anderson's "more conservative figure" of 1/250,000 could be admitted (*Caldwell,* 393 S.E.2d at 444). Beyond stating that he used "the database itself, and not 'any population theory,'" the court did not explain how Anderson arrived at this number (id.).
11. In contradiction to Roberts and Koshland, Peter Neufeld (1993, 193), alleged, without specifying the source of his information, that "the content of the article" was altered.
12. For details, see Roberts (1991, 1722); Levy (1996, 115–116); and Derksen (2000, 821–822). Although Derksen maintains that "Koshland was also contacted by officials from the FBI, who told him the article was 'devastating,'" she cites no source. Koshland himself insisted that "he heard from no one in the FBI or in the government" (Roberts 1991, 1722).
13. Wooley explained that he was perplexed that Hartl had written the article because Hartl had said derogatory things about defense counsel after the pretrial hearing and had been badly misled by those who had asked him to become involved in DNA litigation (Levy 1996, 113).

14. In an article on the *Castro* case and broader issues, Lander (1989a, 505) reiterated the call for a "National Academy of Sciences committee . . . report on guidelines for DNA fingerprinting." With some chagrin, he noted that "[a]n academy study on DNA fingerprinting had been planned for last year, but was postponed indefinitely when the National Institute of Justice [the research arm of the Department of Justice] would not finance it. As one justice official told me, the study was unwelcome: scientists had done their part by discovering DNA; it was not their job to tell forensic labs how to use it."

15. Despite signing on to the statement that "with the relatively small number of loci used and the available population data," an expert should "avoid assertions in court that a particular genotype is unique in the population" (NRC 1992, 92), Mary-Claire King "essentially testified that a 'match' over three probes has always meant that the compared samples came from the same individual or from an identical twin." State v. Hummert, 905 P.2d 493, 499 (Ariz. Ct. App. 1994), *vacated,* 933 P.2d 1187 (Ariz. 1997). In a Washington case, "Dr. King went further and stated 'In my view, there's absolutely no doubt those two samples came from the same human being.'" State v. Buckner, 890 P.2d 460 (Wash. 1995). Her testimony prompted appellate courts to reverse the convictions in these cases.

16. Compare Balding (2005, 95) (describing a similar "pseudo-count" procedure of adding both the defendant's alleles and those from the crime scene in estimating allele frequencies). Applying this procedure to the genotype count would give a frequency of $(x+2)/(N+2)$.

17. For a random-match probability p between 0% and 1.3%, it would not be too surprising to find zero matches in a sample of 225 unrelated individuals. When $p = 1.32\%$, the chance of not seeing the genotype in 225 random draws is $(1-p)^{225} = (1-0.0132)^{225} = 0.05$. For larger p, the chance of no matches would be smaller. Hence $p = 0.013$ or so emerges as the upper limit of a "confidence interval" that has a "confidence coefficient" of $1 - 0.05 = 0.95$.

    More generally, there are various methods for estimating the probability of an event that has occurred zero times in a sample of n observations—the "zero-numerator" problem (Winkler, Smith, and Frynack 2002, 1). The NRC's counting method simply produces an approximate upper bound of $3/n$ on the probability (see Jovanovic and Levy 1997, 137).

18. National Academy of Sciences officials declined to forward Hicks's letter to committee members (Sherman 1992, 1). McKusick reported that "the main changes in the final version 'came out of the review process, not Hicks' letter'" (Anderson 1992a). Lander warned that "the defense community should not use [allegations of attempted meddling] to undermine the report when it didn't happen" (Sherman 1992, 1). Yet Barry Scheck still worried that "the FBI's pressure" might be responsible for any differences in "the two reports" (ibid.).

19. Kolata's (1994) suggestion that "the ceiling principle was devised two years ago by the experts whom the National Research Council had called together" overlooks this longer history.

20. Before the NRC report was released, the FBI had decided to acquire convenience samples from forensic laboratories in five European countries (Anderson 1992b). Although this was less ambitious than the committee's proposed fifteen to twenty samples, it might have been consistent with the committee's idea of "random samples." The report suggested what sounds suspiciously like convenience sampling—going to "blood banks in the appropriate country" or "sampling recent immigrants" (NRC 1992, 84).

21. In a refinement, the committee recommended vaulting the ceiling slightly but using the upper end of a 95% confidence interval for the allele frequency (NRC 1992, 92). The reasons the committee offered do not withstand analysis (Kaye 1993, 171–172). To reflect random sampling error, rather than raising the individual allele frequencies, it would be more sensible to calculate a single confidence interval after estimating the full genotype frequency with the unadjusted allele ceilings (Weir 1993).

22. According to Aronson (2007, 164), "[S]omeone close to the defense community" had "leaked" the report's "main conclusions" to Kolata "several days before the report was due to be released."

23. Professor William Thompson, who was misidentified as a "lawyer and molecular biologist" (he holds degrees in law and psychology and taught in the Criminology Department at the University of California at Irvine), "said that the report 'raises serious concerns about whether DNA tests as they currently are done meet the standards for admissibility in court'" (Kolata 1992c, A6)." In her confession-of-error story, Kolata insisted that "[s]everal law professors, including Paul Giannelli of Case Western Reserve University, Edward Imwinkelried of the University of California at Davis, and Randolph Jonakait of New York University [in actuality, at the New York Law School, which has no affiliation with NYU], said that in their view these recommendations were tantamount to saying that DNA evidence should not be admissible at this time" (Kolata 1992a, A19). Defense lawyer Peter Neufeld was adamant: "There's no question that it says that" (ibid.). Unless the law professors had the time to study the book-length report and to offer considered legal advice, however, Kolata's effort to attribute her interpretation of the report to them seems strained.

24. The new prefatory material appears to have been something of a compromise. A final paragraph adds that "accreditation and proficiency testing [are] essential," so that "[a]fter a sufficient time for implementation of quality-assurance programs has passed, courts should view quality control as necessary for general acceptance" (NRC 1992, x).

25. The sentence reads: "Because it is impossible or impractical to draw a large enough population to test directly calculated frequencies of any particular profile much below 1 in 1,000, there is not a sufficient body of empirical data on which to base a claim that such frequency calculations are reliable or valid" (Lewontin 1992, 39–40).

6. The Initial Reaction to the 1992 NRC Report

1. Although no systematic studies of the frequencies of such hairs have been undertaken, the criminalist believed that "the majority of the population would not have the same range of hair characteristics" (*Axell*, 1 Cal. Rptr. 2d at 415).

2. Succumbing to the transposition fallacy, the court wrote that "Cellmark reported that the frequency of that DNA banding pattern in the Hispanic population is approximately 1 in 6 billion. . . . Simply put, Cellmark's analysis meant that the chance that anyone else but appellant left the unknown hairs at the scene of the crime is 6 billion to 1" (*Axell*, 1 Cal. Rptr. 2d at 415).

3. Even before these developments, another division of the court of appeal, relying largely on Lewontin's testimony in *Yee*, held that the state failed to establish general acceptance of the FBI's basic-product-rule estimates. *People v. Pizarro,* 12 Cal. Rptr. 2d 436 (Ct. App. 1992), *on remand,* 3 Cal. Rptr. 3d 21 (Ct. App. 2003).

4. For a more complete analysis of this opinion, see Kaye (1997a).

5. Judge Sheindlin thought that the court reversed the conviction, which "stunned the prosecutors in the case. There were so many pieces of evidence linking Bible to the crime. How could this happen?" (Sheindlin, 1996, 105). The short answer is that it did not. The concluding words of the opinion are "we affirm Defendant's convictions and sentences" (*Bible,* 858 P.2d at 1191).

6. See also United States v. Porter, 618 A.2d 629 (D.C. 1992).

7. See also State v. Alt, 505 N.W.2d 72 (Minn. 1993) (affirming the holding at 504 N.W.2d 38 (Minn. Ct. App. 1993), that interim-ceiling computations at each locus are admissible under the general-acceptance standard).

8. One population geneticist, Jerry Coyne of the University of Chicago, described the report as "the consensus document of the scientific community" (Kolata 1992c, A6).

9. Lander (1993a) acknowledged that the interim ceiling product rule "involve[d] some elements of judgment," but he insisted that they were neither "illogical" nor "arbitrary." He explained that "[t]he committee prescribed an upper bound of 50:1 for the contribution of each genetic locus to the overall odds on the basis of quantitative estimates (of the effects of sample error and genetic drift) that indicated this would make adequate allowance for fluctuations among population groups." Oddly, the report omitted the "quantitative estimates," making it all but impossible to evaluate the committee's undisclosed reasoning.

10. Thompson informed me that her criticism was directed not just at the ceiling principle but also at various other attempts by both sides to adjust the uses of the ceiling principle to their perceived goals. The trial court excluded the evidence, but the court of appeals held that this ruling was error because Thompson's position was not widely shared. State v. Hollis, No. 33007-1-I, 1997 WL 306432 (Wash. Ct. App. June 9, 1997) (unpublished opinion).

11. Lempert's support dropped still more after publication when he realized that "in most forensic situations the problem the ceiling principle was designed to

resolve—the possibility that forensic data bases would by ignoring population substructure substantially underestimate relevant allele frequencies—hardly ever exists because the proper reference population for estimating allele frequencies is typically a mixed population fairly represented by the data bases now in use" (Lempert 1993a, 43). Nonetheless, he continued to support the procedure as a "second-best" method for compensating for the possibilities that the laboratory had committed a false-positive error and that the source of the crime-scene DNA was a close relative of the defendant.

12. The differences between the population structuralists and the defenders of independence were evaluated in the 1996 NRC report, which largely sided with the defenders and repudiated parts of the 1992 report. I served on the committee that wrote the later report (see Chapter 7).

13. In the face of such criticism, the editor of the *American Journal of Human Genetics*, who had invited Lander to write an article, felt compelled to offer "a few words about editorial policy" (Epstein 1991, 697). He responded to "the concern . . . that a 'peer-reviewed journal' has permitted the publication of an invited editorial which is perceived as both flawed in reasoning and potentially injurious to the use, in legal settings, of evidence obtained by molecular genetic techniques" by explaining that "unless stated otherwise, [an invited editorial is not] subjected to formal peer review. An invited editorial represents the opinion of its author alone and does not constitute an expression of the position of either the editor who solicited its preparation, of the *Journal*, or of The American Society of Human Genetics. . . . [A]n opinion is just that—an opinion, hopefully informed—and should not be taken as being more than that" (ibid.).

14. Budowle et al. (1991) also suggested that an excess of single bands in the FBI databases had resulted from "null alleles"—fragments that either were so small that they ran off the gel or that had too few repeats to be detected.

15. Statistically sophisticated database testers responded to the low-power argument with demonstrations that their tests had the power to detect substantial departures from independence (e.g., Devlin, Risch, and Roeder 1991; Weir 1992a, 878–881).

16. See, e.g., Buckleton, Walsh, and Harbison (2001, 81–84). Statistical tests for correlations still had a place in the validation of DNA identification. Weir (1992a, 877) cogently explained that even though excess heterozygosity and population structure were distinct phenomena, it remained important to test whether loci used in genotype-frequency computations were in Hardy-Weinberg equilibrium: "[T]he issue is not whether the conditions that will lead to HW frequencies are satisfied in a population, but whether the HW relations . . . hold. Instead of asking whether or not a set of biological conditions are met, the issue is whether or not there is a particular relation between allelic and genotype frequencies in a population. If the relation holds then allelic frequencies can be used to generate genotypic frequencies."

17. Compare Lewontin 2006 ("Of the remaining 15% of human variation, between a quarter and a half is between local populations within classically defined human 'races,' between the French and the Ukrainians, between the

Kikuyu and the Ewe, between the Japanese and the Koreans"). Outside the forensic context, some anthropologists, striving to make the valid point that most genetic variability lies within rather than across "races," continue to present Lewontin's 1972 work with its "new statistical analysis" as if it were the last word on the subject of the extent of "racial" variation (Marks 2002, 81–82).

18. Lander (1991b, 901) dismissed these studies as irrelevant: "Ironically, the very reason that VNTRs are so much better for identification than are conventional genetics markers is also the reason why their population genetic properties cannot simply be inferred by analogy with conventional markers. Specifically, VNTRs are highly polymorphic and highly mutable. Because they have allele frequencies that are much lower and mutation rates that may be 1,000-fold higher than those of conventional genetic markers, genetic drift can have much greater proportional effects on allele frequencies. (Note that proportional effects are the relevant matter when one is computing odds.) Accordingly, studies of broad racial groups such as Caucasians, blacks, and Orientals may mask considerable heterogeneity among ethnic subpopulations." The defenders of independence, however, reasoned that because the VNTRs were not subject to natural selection (because they had no phenotypes), the forces that maintained a concentration of particular alleles in a population were inoperative. They emphasized that "VNTR loci are likely to be noncoding genetic markers and thus are expected to behave independently, as has been found for almost all noncoding and neutral human genetic markers examined to date" (Devlin, Risch, and Roeder 1992, 347; see also Weir 1992a, 873 ("surveys of neutral variants in human populations rarely reveal departures from Hardy-Weinberg equilibrium or linkage equilibrium between unlinked loci")).

19. Brookfield (1991) also employed a standard model of population structure to examine the effect of structure on the random-match probability. For "realistic values," he concluded that "[i]t is impossible to find a case in which a likelihood ratio constituting strong evidence against a subject is converted to one in which the evidence has become weak" (ibid., 100).

20. Hartl and Lewontin (1993) convey the opposite impression. Relying on the study by Krane et al. (1992) discussed in the next note in this chapter, they wrote that "the methods currently used in court are not conservative—they are systematically prejudiced against the defendant—and no amount of argument will make them conservative." This is not what Krane et al. actually found. The underestimation reported in that article (but questioned in subsequent commentary) arose from using the allele frequencies for one subpopulation to characterize the significance of a match between two members of a different subpopulation. In a general-population case, however, the relevant comparison is between the estimate for a population ignoring all subpopulations and the estimate for a population accounting for the stratification within that population.

21. An early study of the type prescribed by the structuralists suggested that the errors due to using a population database rather than a subpopulation database

in a subpopulation case would tend to favor the prosecution but that the estimated and true values usually would be within one order of magnitude. Daniel Krane and his colleagues (Krane et al. 1992) acquired blood samples from Finns in Helsinki, Italians in Milan, and Caucasians in St. Louis. The most useful part of the study compared the three-locus genotype frequencies using the allele frequencies in the population database of St. Louis Caucasians with the genotype frequencies obtained by using the data within the European subpopulations. In other words, the researchers asked what would have happened if the population of plausible suspects were Finns or Italians rather than Caucasians but the laboratory used the available database for all Caucasians. They found that 20% of the estimated match probabilities were too large, meaning that the use of the general-population database benefited the defendant. Another 62% were too small, but the discrepancy was within a power of 10. For example, a reported probability of 1 in 10 million might have been as high as 1 in 1 million. In only 18% of the cases did the error exceed a factor of 10 in favor of the prosecution, and then it was almost always within a factor of 100 (ibid., 10586).

7. Ending the Debate over Population Genetics

1. The affinal model was well known among population geneticists and had been brought to bear on the estimation of forensic VNTR frequencies in Brookfield (1991), Nichols and Balding (1991), and Morton (1992). For later discussions, see Crow and Dennison (1993); Balding and Nichols (1994, 1995); Weir (1994); Evett and Weir (1998); Balding (2005); and Buckleton (2005)
2. Lander did not name his three critics, but among the committee's list of "participants" in its work (NRC 1992, 177), the three scientists who most visibly regarded population structure as a vastly overblown concern were Bruce Budowle, Kenneth Kidd, and Bruce Weir (see Chapter 6).
3. Lempert (1993b, 1) referred to ceiling calculations as "scientifically problematic," but he explained that "[w]hen the ceiling principle was first suggested, it was far more defensible scientifically than it was when the report finally appeared, and when the report appeared, it was more defensible than it appears today."
4. In comments sent to me in July 2008, Weir stated that he attended "a couple of meetings at Quantico in the early 1990s," but that he would not characterize his interactions with the FBI as those of a consultant. Another peculiar description of Weir comes from Lynch et al. (2008, 62), who portray him as "a prominent forensic analyst testifying for the prosecution." Weir had no experience as a forensic analyst. A pioneer in the field of statistical genetics, he was then a professor of statistics and genetics at North Carolina State University.
5. I participated as one of several lawyers on the ad hoc committee, expressing some doubt about the FBI's suggestion that the 1992 report was causing large numbers of appellate courts to reject DNA evidence. I pointed out that it was hard to know how much the report had accelerated the trend toward restricting DNA testimony because of statistical issues. Cf. infra note 7. I supported a new

study, however, because the first one left too many questions unanswered, and the courts did not always perceive the report's limitations.

6. The National Institute of Justice (NIJ) (the research arm of the Department of Justice) was the principal funding agency, with much smaller contributions from the State Justice Institute, the National Science Foundation, the National Institutes of Health, and the Department of Energy.

7. A more complete description of the law as of October 18, 1993, would be that the defense prevailed on the issue of the admissibility of basic-product-rule computations in four appellate cases citing the 1992 report and involving DNA testing by the FBI laboratory. These were United States v. Porter, 618 A.2d 629 (D.C. 1992) (remanding to consider the admissibility of the interim-ceiling rule); Commonwealth v. Lanigan, 596 N.E.2d 311 (Mass. 1992); State v. Vandebogart, 616 A.2d 483 (N.H. 1992), and State v. Anderson, 853 P.2d 135 (N.M. Ct. App. 1993), rev'd, 881 P.2d 29 (N.M. 1994). With respect to the same issue, the defense also prevailed in two cases out of five citing the NRC report and involving the commercial laboratories. Compare People v. Barney, 10 Cal. Rptr. 2d 731 (Ct. App. 1992) (Cellmark), and State v. Bible, 858 P.2d 1152 (Ariz. 1993) (Cellmark's use of the basic product rule did not satisfy the general-acceptance standard when the court misread the expert's testimony as conceding that the population was not in Hardy-Weinberg equilibrium), with People v. Wesley, 589 N.Y.S.2d 197 (App. Div. 1992) (Lifecodes); State v. Pierce, 597 N.E.2d 107 (Ohio 1992) (Cellmark); and State v. Futch, 860 P.2d 264 (Or. Ct. App. 1993) (Lifecodes).

8. The task assigned to the committee was to "perform a study updating the previous NRC report, *DNA Technology in Forensic Science.* The study will emphasize statistical and population genetics issues in the use of DNA evidence. The committee will review relevant studies and data, especially those that have accumulated since the previous report. It will seek input from appropriate experts, including those in the legal and forensics communities, and will encourage the submission of cases from the courts. Among the issues examined will be the extent of population subdivision and the degree to which this information can or should be taken into account in the calculation of probabilities or likelihood ratios. The committee will review and explain the major alternative approaches to statistical evaluation of DNA evidence, along with their assumptions, merits, and limitations. It will also specifically rectify those statements regarding statistical and population genetics issues in the previous report that have been seriously misinterpreted or led to unintended procedures" (NRC 1996, 49).

A preliminary draft of the project proposal similarly spoke of reviewing alternative statistical approaches "to access and describe the degree of certainty of DNA evidence in ways useful to the courts; consider sources of errors such as differences among laboratories and techniques; and lay out alternative perspectives about the assumptions and assessments of uncertainty and describe their implications in court cases" (Aronson 2007, 170). According to Aronson, the FBI balked at this description, and "[b]y the time the final proposal had been prepared, the NAS had made almost all the changes requested by the FBI regarding which issues were salient and the relationship between science and the law" (ibid.). Yet both formulations of the committee's task required it to examine all the sources of uncertainty in and the assumptions behind the DNA probabilities, and that is what the committee did.

9. At the same time, the lack of previous usage of the ceiling methods by forensic laboratories did not deter the committee from inquiring into—and advocating—that procedure, at least as a way to avoid an impasse.

10. At a briefing about the report for the sponsoring organizations, the FBI's Bruce Budowle suggested that this recommendation was undesirable. See also Marshall (1996, 804) (reporting that California prosecutor Rockne Harmon expressed concern that the recommendation "will prompt defense attorneys to file appeals on grounds that statistics are misunderstood"). In contrast, Richard Rau, representing the NIJ, stated that the institute favored such research.

11. This recommendation (4.1) specified that for single-banded VNTR loci, the overstated frequency of $2p$ should be used in preference to the Hardy-Weinberg p^2 proportion, but that for loci in which homozygosity was less ambiguous, a proportion closer to p^2 could be used.

12. See supra note 1. In one version of the model that applies to selectively neutral alleles (those that are unrelated to the number of offspring that people have), the proportion of homozygous individuals in the subgroup is not simply p_1^2, as it is in the general population, but the larger number $p_1^2 + p_1(1 - p_1)F_{ST}$ (Crow and Dennison 1993). F_{ST} is a measure of the extent of population subdivision, and the NRC report suggests that 1% is a conservative estimate of its value. For a VNTR allele with a population frequency of, say, 5%, the estimated effect of population structure is to increase the homozygote frequency from 0.25% to 0.2975%. For VNTRs, however, the FBI already used the far larger quantity $2p_1 = 10\%$ to estimate the homozygote frequency. Thus even the small, theoretical adjustment required to estimate the frequency of a homozygous locus in a subpopulation is a moot point.

For heterozygotes in the subpopulation, the frequency is reduced by a factor of $1 - F_{ST}$ from the value it would have if the whole population were mating at random. The basic product rule would evaluate the chance of a random match at a single, heterozygous locus in the whole population as $2p_1 p_2$. With the correction based on F_{ST}, the chance of a random match at this one locus within the subpopulation would be $2p_1 p_2 (1 - F_{ST})$. Because this quantity is smaller than the population estimate of $2p_1 p_2$, the basic product rule (with the use of $2p$ for homozygotes) always overestimates single-locus frequencies in subpopulations.

From the perspective of subpopulation theory, then, it seems that the entire debate over Hardy-Weinberg equilibrium and population structure was a red herring. This theory suggests that for single-locus genotypes, the basic product rule already benefits defendants. The situation for multilocus genotypes, however, is different. F_{ST} is not constant from subpopulation to subpopulation, and an additional term involving the variance of F_{ST} increases the probability of a multilocus match. In the end, the estimated probability for the multilocus subpopulation match will be larger than the estimate that comes from the basic product rule for a homogeneous population (ibid.).

13. For studies of the relative degree of overestimation of the various procedures, see Buckleton (2005). The NRC report also considered a third situation. Recommendation 4.3 states that "[i]f the person who contributed the evidence sample is from a group or tribe for which no adequate database exists, data from several other groups or tribes thought to be closely related to it should be used" in basic-product-rule calculations for each group or tribe (NRC 1996, 123).

14. See United States v. Gaines, 979 F. Supp. 1429 (S.D. Fla. 1997); and United States v. Shea, 957 F. Supp. 331, 343 (D.N.H. 1997), *aff'd,* 159 F.3d 37 (1st Cir. 1998). More recent cases include United States v. Morrow, 374 F. Supp. 2d 51 (D.D.C. 2005); United States v. Ewell, 252 F. Supp. 2d 104, 108–109 (D.N.J. 2003); United States v. Trala, 162 F. Supp. 2d 336, 343 (D. Del. 2001); People v. Brown, 110 Cal. Rptr. 2d 750 (Ct. App. 2001); and Commonwealth v. Gaynor, 820 N.E.2d 233, 254 (Mass. 2005).

15. In an interview conducted in 1998 as part of a "multi-sited ethnographic study" (Lynch et al. 2008, xiv), Lewontin elaborated on this theme, stating that the report "has ended the controversy because the establishment has spoken. . . . [A]nybody who is engaged over a long time in such struggles knows that usually you lose. . . . I mean the power of the state has been marshaled, uh, to end the affair" (ibid., 227).

16. In a 1997 wrongful-death action, Simpson "was found liable and ordered to pay damages to the victims' families totaling $33.5 million," most of which has not been collected (Friess 2008).

17. As if all this were not enough, Simpson recently returned to the news and the justice system. On October 3, 2008, "13 years to the day after he was acquitted in the killings of his ex-wife, Nicole Brown Simpson, and her friend Ronald L. Goldman," a "Las Vegas jury of nine women and three men found Mr. Simpson guilty of storming into a room at the [Palace Station] hotel casino . . . with a group of five friends, at least two of whom carried guns, and seizing memorabilia worth thousands of dollars from the dealers, Bruce L. Fromong and Alfred Beardsley." Declaring that "[t]he evidence was overwhelming," the judge sentenced Simpson to serve between nine and thirty-three years in prison (Friess 2008).

18. At the time of the verdict, I suggested that although O. J. probably was the murderer—a view shared by some of the jurors who have given interviews or written on the subject—the defense may have succeeded in inducing a kind of "global skepticism" that created a reasonable doubt in the minds of jurors sincerely trying to apply the law to the evidence, and that "the defense may have created just enough doubt, ambiguity, or confusion to liberate the jurors from the evidence—to allow them to decide the case, if only subconsciously, on the basis of prejudices or attitudes that would not have been decisive had the case against Simpson been clearer" (Kaye 1995d). In response, Barry Scheck wrote me, objecting that he and his cocounsel had proved Simpson

innocent to a reasonable doubt and that my speculations undermined the American jury system.

19. Yet another DNA expert who was involved in Simpson's defense was Laurence Mueller, "a professor in the Department of Ecology and Evolutionary Biology at the University of California, Irvine.... [T]he trio became known as the 'combine from Irvine'" (Coleman and Swenson 1994, 103).

20. Major sections of the memorandum were taken, verbatim and without acknowledgment, from W. C. Thompson (1993). The prosecution memorandum complained that "[w]hile there is nothing wrong with excerpting from this type of material, it is somewhat troubling that the brief makes numerous references to the article as some sort of legal authority, when the brief and the article are one and the same in many areas."

21. The memorandum may have had a less ambitious goal as well—to convince the judge that the DNA findings, to the extent that they were admissible, would have to be on terms favorable to the defense. For example, the memorandum questioned the admission of random-match probabilities when the risk of laboratory error was orders of magnitude larger than the probability of a true but coincidental match. Even if the court refused to exclude the DNA results lock, stock, and barrel, it might limit the expert to presenting the finding of a match between samples and the probability that such a match would arise because of a mistake at the laboratory as estimated by proficiency tests. Or, even if it approved the admission of a random-match probability, it might insist on a ceiling estimate.

22. The discussion in this section is adapted from Kaye (1994b).

23. In a review of the FBI development process, Aronson (2007, 89–119) writes that the FBI was not open to outside scientific advice except for that from a community of friendly experts supported by grants that were more akin to contracts. Among other things, he suggests that it is wrong for federal agencies to fund research on topics that are not chosen by "the normal peer-review mechanisms set up to determine which research the scientific community considered worth doing" (ibid., 111); that it is unethical for scientists to propose and undertake funded research directed at courtroom controversies when "there [is] little doubt that the results would benefit the prosecution and not the defense"; and that the National Institute of Justice, which is the scientific research arm of the Justice Department, is the FBI's "parent agency" (ibid., 113). The claims about the ethics and funding of applied research seem dubious. The statement about the organization of the Department of Justice is simply wrong (Boyd 1995).

24. The article states that the "specific comments about the working and intent of the NRC committee itself . . . necessarily are based on E. S. L.'s recollections" (Lander and Budowle 1994, 735).

25. Lander's explanation for not speaking out earlier was that "the committee members had agreed to let the report speak for itself"—a decision that he now saw as probably "unwise . . . because it has allowed a minor academic debate to snowball" (Lander and Budowle 1994, 737). However, rather than

simply let the report speak for itself, Lander (1993a) solidified the impression that the ceiling product rule was advanced as a distinct alternative to the basic product rule when he wrote that "[c]ritics are welcome to try to achieve 'general acceptance' of a looser standard for DNA fingerprinting. However, this may be slow in coming. . . . [A] looser rule . . . will likely provoke continued litigation that will hamper the use of this important and powerful criminalistic tool." Surely, any rule looser than the total expulsion of the basic product method would have "provoked continued litigation" and failed to meet the expressed objective of eliminating all "significant controversy about the validity of the method" (ibid.).

26. Furthermore, it is worth noting that significance testing never has been "universally accepted" (e.g., Savage 1954; Morrison and Henkel 1970; Oakes 1986; Royall 1997; Nickerson 2000), and its "seeming objectivity is illusory" (Kaye 1986b, 1334). One of the many difficulties is that the choice of any particular significance level is somewhat arbitrary.

27. Rather late in the game, the California Supreme Court did the same in *People v. Venegas,* 954 P.2d 525 (Cal. 1998).

28. He also provided figures for other populations and databases. For the Orange County, California databases, they were 1 in 38 million (Caucasian), 1 in 807 million (black), and 1 in 177 million (Vietnamese). For the FBI's databases, they were 1 in 55 million (Southwest Hispanic—Texas), 1 in 2.3 billion (Southeast Hispanic—Florida), 1 in 2.4 billion (black—United States), and 1 in 3 billion (Caucasian—United States) (*Soto,* 981 P.2d at 971).

8. Moving Back to Errors and Relatives

1. In 2003, Krane took the more moderate position that "empirical research has . . . allayed much of the concern" about population structure (Thompson and Krane 2003, 11-60).

2. With apologies for mixed metaphors, it should be added that the lowly fruit fly has been the outstanding workhorse for genetic research since 1908. Thousands of "fly people, as biologists who study the laboratory fruit fly call themselves," are attempting to "take the fly apart and put it back together by thorough analysis of its genes" (Wade 1997). "Medical research agencies willingly support fruit fly research because its relevance to humans has turned out to be surprisingly direct" (ibid.). Thus, when a colleague of mine testified as an expert witness for the defendant in a paternity case and was asked if the only DNA she studied was in fruit flies, she informed plaintiff's counsel that "it is the same DNA as yours."

3. For additional discussion of the case law, see Kaye, Bernstein, and Mnookin 2004, 459.

4. The prosecution had feared that something like this might happen. It had asked the judge to question Simpson to ensure that he was "willing to give up [the] right to challenge the admissibility of the DNA test results—including both results showing that samples match as well as the statistical significance of those matches—and to allow the DNA test results to be presented to the

jury." Scheck objected that "[i]t is highly inappropriate to go through this kind
of allocution of waiver." He offered a more ambiguous alternative—that the
court could "ask Mr. Simpson if he has read the notice of withdrawal and
understands it." Simpson then confirmed that "I have full confidence in my
lawyers and we withdraw the previous motion. I understand the waiver."
Judge Lance Ito was satisfied. He announced, "All right. I think that is a sufficient waiver," and went on to other matters (Transcript, Jan. 4, 1995). There
was, however, no record of what Simpson's counsel told Simpson the consequences of the motion were, and, in particular, whether they were congruent
with the prosecution's view that the waiver meant that "the evidence of
matching DNA and accompanying statistics would be admitted into
evidence."

5. See, e.g., Kelly v. State, 792 S.W. 2d 579 (Tex. Ct. App. 1990), *aff'd,* 824
S.W.2d 568 (Tex. Ct. Crim. App. 1992 (discussed in Chapter 4); cf. Commonwealth v. Teixeira, 662 N.E.2d 726, 729 (Mass. App. Ct. 1996) ("It may have
been scientifically and statistically unsound for [FBI technician John] Quill to
claim a laboratory error rate of zero, but that proposition is not so ordinary
and obvious that a judge should be held to take judicial notice of it").

6. The act required laboratories that request federal funds, as many state DNA
laboratories do, to be accredited within two years "by a nonprofit professional association of persons actively involved in forensic science that is na-
tionally recognized within the forensic science community" and to "undergo
external audits, not less than once every 2 years, that demonstrate compliance
with standards established by the Director of the Federal Bureau of Investigation." 42 U.S.C. § 14132(b)(2) (2006).

7. Contrary to the intimations of some observers (Marshall 1996, 803, quoting
Lewontin; Aronson 2007), the committee took this perspective not because
the FBI told it to, but because its members understood this to be the normal
role of a scientific advisory committee.

8. The meaning of "double-blind" here is obscure. In clinical trials of a drug,
both clinicians and patients can be blinded as to whether a patient is receiving
a placebo or the drug. This double-blinding controls for the placebo effect,
and it prevents any bias on the part of the clinicians evaluating the outcome
for a patient. Proficiency tests are "single-blinded." Only the laboratory needs
to be kept in the dark as to whether it is dealing with a simulated case or a
real one.

9. Indeed, there is even a legal doctrine that deems evidence of error on other
occasions inadmissible without some special link to the case at bar (Imwinkelried and Kaye 2001, 462–463).

10. Presumably, Benjamin testified that the chance that two siblings of parents
who are heterozygous at the same locus and whose alleles at this locus are
distinct would contribute the same single-locus genotype to two zygotes was
1/4. Or he might have been referring to the mean probability of two full siblings matching at a given locus, which is somewhat larger than 1/4 (NRC
1996, 113). The chance of a multilocus match in these circumstances will be
much smaller (ibid.).

11. Lempert (1997, 461) proposed that an exception to this requirement apply when "the state excludes each named relative through DNA testing or exculpatory non-DNA evidence or by showing that substantial non-DNA evidence of a sort unlikely to implicate the relatives implicates the defendant."

12. Lewontin had made and continued to make the broadly similar argument that as a practical matter, subpopulations and relatives always are a dominant concern in considering the probability of a match in an innocent suspect (e.g., Lewontin 1994a, 1997).

13. On the basis of state-by-state data, the report did suggest that "individual within-race profile frequencies from different geographic areas in the United States usually differ by less than a factor of 10 in either direction" (NRC 1996, 150–151).

14. Of course, if the defense were to stipulate that the source of the crime-scene sample is not an unrelated individual, then the balance of probative value and prejudicial impact might warrant the exclusion of the statistic. Cf. Old Chief v. United States, 519 U.S. 172 (1997).

15. See, e.g., Dawid (2002). The formula could be extended to deal with the hypothesis that one of Troy's brothers is the source, but that is not a feature of the transposition fallacy that worried the court.

16. The probability that the five-locus profile of two full siblings is the same is larger than $1/1,024$ because $1/4$ captures only the chance of identical alleles at each locus by descent. It does not consider the possibility that both parents might share some alleles. If 10% is used as a rough estimate of the mean allele frequency (see National Commission on the Future of DNA Evidence 2000, 41, and formula (4.9b) of the NRC report (1996, 113)), a better estimate of the probability of identical alleles at each locus is 0.305 rather than Romero's 0.25. The chance of a five-locus match between two siblings then is $0.305^5 = 1/379$ rather than $1/1,024$. A still-better estimate would use the NRC formulas with the frequencies for the specific alleles that were present in the crime-scene sample. In the habeas proceedings, Troy supplied a report from Laurence Mueller that found this number to be $1/263$ (Brown v. Farwell, 525 F.3d at 800).

17. The numbers in the text are based on the estimate of $1/1,024$ for the chance that any two full siblings will match at five loci. The chance that Brother A will match Troy is then $P(A) = 1/1,024$. The chance that Brother B will match is $P(B) = 1/1,024$. The chance that A or B will match is $P(A \text{ or } B) = P(A) + P(B) - P(A \& B) = 1/1,024 + 1/1,024 - (1/1,024)^2 < 1/512$. By an extension of this reasoning, $P(A \text{ or } B \text{ or } C \text{ or } D) < P(A) + P(B) + P(C) + P(D) = 4/1,024 = 1/256$.

    Using the specific allele frequencies in the case to arrive at $1/263$ rather than $1/1,024$ for "the chance of a single sibling matching Troy Brown's DNA profile," the report from Mueller arrived at the following figures: $1/132$ for "the chance that among two brothers, one or more would match"; and $1/66$ for the chance that at least one of the other four brothers would match (525 F.3d at 800).

18. With Mueller's large figure of $1/66$ for a match to at least one brother, the DNA match, standing alone, might not seem so overwhelming, but Judge

O'Scannlain pointed to various other items of evidence making Troy far and away the most likely one of the Brown brothers to have been the rapist.

19. Of course, the whole inquiry into probabilities concerning siblings could have been avoided had the brothers been tested. Given the evidence making Troy the most likely culprit, it is a good bet that none of the four brothers matched, but we should not have to guess.

9. Moving on to Short Tandem Repeat Loci

1. Some time elapsed between Mullis's often-told story of his 1983 epiphany during a Friday night drive through the California woods (Mullis 1997) and the reality of a workable PCR system (Fore, Wiechers, and Cook-Deegan 2006).

2. If the evidence was obviously flawed because of contamination, however, the judge would have discretion to exclude it under the normal unfair-prejudice standard. (See Chapter 4, discussing the "third prong" of *People v. Castro*.)

3. For an earlier claim of uniqueness from Lewontin's colleague, Daniel Hartl, based on a nine-locus VNTR match, see *State v. Bloom,* discussed in Chapter 5 and in Aronson (2007, 187).

4. Although the report remained an internal document for the commission, it was circulated and its contents were discussed at a commission meeting. The working group's chair, Professor Michael Smith, finessed the impasse in the group by obtaining agreement to transmit it to the full commission as a "Report to the Working Group." However, the NIJ never disseminated it along with the other commission work products, and the government never published or distributed it. Its contents appear as Imwinkelried and Kaye (2001) and Kaye (2001b).

5. VNTRs are also called minisatellites. For an explanation of the origin of this nomenclature, see, for example, Brown (2002, 217).

6. Some variations on this simple theme are known (Butler and Reeder 2008).

7. Gel electrophoresis with fluorescent tags also has been used (J. M. Butler 2005, 361–367). Thompson and Dioso-Villa (2008) discuss how this system, as applied to small and mixed samples, may have been manipulated to provide results to suit the prosecution's case in the 1999 trial of Robin Lovitt for the murder of a pool manager in Virginia.

8. Applied Biosystems makes this kit. The amelogenin gene, which is found on the X and the Y chromosomes, codes for a protein that is a major component of tooth enamel matrix. The copy on the X chromosome is 112 bp long. The copy on the Y chromosome has a string of six base pairs deleted, making it slightly shorter (106 bp) (Goodwin, Linacre, and Hade 2007, 53). A female (XX) will have one peak at 112 bp. A male (XY) will have two peaks (at 106 and 112 bp).

9. In a recent California case stemming from the rape and murder of a five-year-old girl, Libby was able to inflate the prosecution's random-match probability of 1 in 600 million by a factor of 100 million to conclude that it could be as large as 1 in 6. The prosecution's cross-examination "brought out . . . that

Libby has never been a tenured professor, that . . . he has done consulting work for defense attorneys since 1988 [and] that 70 to 80 percent of his personal income comes from consultations in defense cases" (Hall 2005).

10. Transcending Race and Unscrambling Mixed Stains

1. In *People v. Prince,* 36 Cal. Rptr. 3d 300 (Cal. Ct. App. 2005), *rev. dismissed,* 142 P.3d 1184 (Cal. 2006), however, the court of appeal applied this conditional-relevance reasoning to a case in which the state presented statistics for a range of racial and ethnic groups, thus avoiding the *Pizarro* error. The *Prince* court nevertheless insisted that "[t]he probative value (hence, the relevancy) of a [DNA] profile's frequency in an ethnic population depends on proof that the perpetrator belongs to that ethnic group" (36 Cal. Rptr. 3d at 304). *People v. Wilson,* 136 P.3d 864 (Cal. 2006), put an end to the grotesque *Pizarro* line of cases. In an incisive opinion by Justice Ming Chin—the author of the *Barney* opinion that so roiled prosecutors and the FBI at the high point of the population-structure debate—the California Supreme Court recognized the logical difficulties with the dictum in *Pizarro.* Further analysis is available in Kaye (2007a, 2008).
2. This evidence included fingerprints in the girl's bedroom, hairs in Wilson's car, and a DNA test of an anal swab that "showed a DNA profile matching appellant's DNA profile [for which] only 1 in 2083 Hispanics would match" (Wilson v. State, 185 S.W. 3d at 485 n.6). "What is more," the court wrote, "appellant has not explained how the newer more discerning DNA test would cast doubt on the reliability of the earlier test" (id.). The explanation is obvious. The original testing was incriminating as far as it went, but it was not conclusive. If testing at newer STR loci were to exclude Wilson, then he would be among the 1 in every 2,000 or so innocent Hispanic men expected to match at the loci analyzed in the original testing.
3. See Wilson v. State, 185 S.W. 3d at 485 ("In none of appellant's various direct and post-conviction appeals, in neither of his trials, nor here, does appellant claim that the State was prosecuting the wrong man. . . . Rather, he avers he is entitled to new testing or retesting of evidence because there may have been an additional man involved in the abduction, rape, and murder of Maggie").
4. This analysis contradicts Texas Code Crim. Proc. Ann. art. 64.03(a)(2)(A) (Vernon 2006), which merely requires proof "by a preponderance of the evidence, that he 'would not have been convicted if exculpatory results had been obtained through DNA testing' " (Wilson v. State, 185 S.W. 3d at 484). Conclusive proof that another, named individual was responsible is not and should not be required.
5. For an opinion written ten years after the second NCR report and contemporaneously with a slew of textbooks on forensic genetics and statistics, the suggestion that the procedures for computing random-match probabilities are "new and only marginally understood" is puzzling. The opinion proceeds to present a variety of other misconceptions about the technology of DNA profiling and statistics. For example, it confuses recessive traits with statistical dependence,

suggests that mitochondrial DNA is more polymorphic than nuclear DNA, misapprehends the nature of an expected value, and mistakes a functional relationship for a statistical one.

6. Tiger's father, Earl Woods, is "half-black, one-quarter American Indian, [and] one-quarter white" (Kamiya 1997). His mother is "half Thai, a quarter Chinese, and a quarter white" (Nordlinger 2002).

7. Weigand, Schurenkamp, and Schutte (1992); cf. Elliot et al. (2003) (describing another procedure for differential extraction that works with smaller samples).

8. For a more advanced treatment of the issue and the statistical literature, see Clayton and Buckleton (2005).

9. Strangely, none of the country's leading DNA lawyers or other experts in the case seemed to know this until Weir called it to their attention (Transcript, Examination of Bruce Weir, June 22, 1995).

10. After Thompson asked about the meaning of words like "permissible," "wrong," and "reasonable" in statistical practice, the court reminded him that "[t]he issue is which of these two methodologies am I going to adopt for use in front of this jury. Which makes sense both in terms of the science that produces these results and the math that is applied to it?" (Transcript, Examination of Bruce Weir, June 22, 1995, 38). When Thompson asked another general question about "an additional meaning of that term 'bootstrapping' in scientific circles," Judge Ito had had it:

> *The Court:* Counsel, this is not real helpful to me.
> *Mr. Thompson:* Well, I—
> *The Court:* It is really not. It is a complete waste of my time so far, this whole discussion about bootstrapping.
> *Mr. Thompson:* I was. I think I can tie it in, your honor, if you will give me some leeway.
> *The Court:* No. (Ibid., June 22, 1995, 39)

11. Weir spent a long night in his hotel room coming up with the new figures. Occupied with this task, he did not reexamine calculations that he had finished the previous day and transmitted to the defense without double-checking them. The defense waited for him to testify before the jury, and then, on cross-examination, Neufeld pounced. A chastened Weir agreed that some of his figures understated by a factor of two or so the conditional probabilities and thus were "biased" in a statistical sense. (He did not state, and the defense apparently did not notice, that other figures in his original report were, for a different reason, "biased" in favor of Simpson by a factor of two.)

12. For a straightforward presentation of the usual, somewhat subjective approach, see J. M. Butler (2005, 158–167). Perlin (2006) describes a procedure that could be used to verify the reliability of these judgments. In *Roberts v. United States,* 916 A.2d 922 (D.C. 2007), the District of Columbia Court of Appeals held that mixture analysis is admissible even without such studies.

13. Some X-Y crossing over does occur at the tips of the Y chromosome. In fact, observations of this recombination are part of the evidence that led to the

realization that X and Y chromosomes form one of the twenty-three pairs of human chromosomes (Buckleton, Walsh, and Harbison 2005, 318–319). The loci used in forensic identification are located in the nonrecombining region that constitutes 95 percent of the Y chromosome.

14. Using family names to study the extent of inbreeding is a well-established mode of inquiry in population genetics (Smith 2002). "Marital-isonomy" analysis relies on the fact that a man and woman who had the same surname before their marriage are more likely to be related than a couple with different names (Crow 1996).

15. The laboratory analyst who performed the testing in the case also testified.

11. Outside the Nucleus

1. Portions of this section are adapted from D. Butler (2007). See also Owen and Hamilton (2006).

2. According to Hofreiter et al. (2004, 407), "the finger showed a mixture of mitochondrial DNA (mtDNA) sequences from different individuals, and in two of four amplifications showed a minority sequence that matched a rare sequence motif shared by Prince Phillip [sic] and Alexandra."

12. Stray Hairs

1. Houck and Budowle (2002, 4) cautioned "that microscopy is not a 'screening test' and mtDNA analysis is not a 'confirmatory test.' Both methods, or either, can provide probative information to an investigation: one is not superior to another as both analyze different characteristics." Although the prosecution should not be permitted to introduce a microscopic match if it has not performed mtDNA tests, arguments that mtDNA should be a "replacement for traditional microscopic hair analysis" except when the DNA results are inconclusive (Cheng 2005, 118) seem too simplistic. Exclusions based on microscopy may be quite convincing even without DNA sequencing, and inclusions may be admissible in response to arguments that deficiencies in existing mtDNA population databases make it impossible to draw useful inferences from mitotyping.

2. The defense perspective on this issue is presented in Kaestle et al. (2006).

3. Many researchers believe that such accumulated mutations contribute to the aging process, and a variety of degenerative diseases are related to mtDNA changes (Lin and Beal 2006). However, the medically important sequence variations are in the mitochondrial genes rather than the D-loop sequenced for identification (but see Melton 2004, 4–5, citing reports of some disease associations).

4. This exchange of expert views in *Pappas* is reminiscent of the arguments about match windows in gel electrophoresis. In *Jakobetz,* for instance, Nadeau's theory that the FBI should have used smaller windows made no real difference when the match fell within the smaller windows he arbitrarily preferred (Chapter 5). Here too, the more stringent criteria for an inclusion—a match at every base pair—was met.

5. Technically, "chromatography" refers to a set of laboratory techniques that pass a liquid or gas carrying a mixture over a stationary material that attracts the different components of the mixture in varying degrees. For example, the stationary material might be a thin film of viscous liquid coated on the surface of solid particles packed into a column made of glass or metal. The components of the fluid mixture with a strong attraction to the coated particles will move more slowly than those with weak attraction. After the sample is flushed from the stationary material, the different components will emerge from the column at different times. The electrophoretic separation and detection system to which the *Pappas* court referred does not rely on such absorption and elution. Rather, it uses an electric field to separate the DNA fragments of different lengths, and it produces an electropherogram rather than a chromatogram.

6. Shields "stated that because the FBI would not exclude as a match two samples that differed by one chemical base, other samples in the database that differ by one such base should be included in the estimated mtDNA type frequency" (*Pappas,* 776 A.2d at 1105). This statement is incorrect because the frequency of the sequence actually detected has to be based on that one sequence, not on other nearby sequences.

7. Using the value of the confidence coefficient for an interval estimate as the probability that the population proportion lies within the interval from 0% to 0.25% is a version of the transposition fallacy, which has been encountered in earlier chapters.

8. For details of the different experimental conditions, see Dann, Hans, and Kaye (2004, 2007).

9. As it turns out, "Middletown" corresponds to an actual city in Delaware—an unintended fact that came up in several jury deliberations. Fortunately, no jurors noticed the difference in the size of the populations in our hypothetical town and the real one, and these juries were not concerned with particular attributes of the real Middletown, Delaware.

10. Allison's testimony was presented by the nearly eponymous Lizabeth Allison, the Dorman Family Term Distinguished Professor of Biology at the College of William and Mary in Virginia.

13. Learning from DNA

1. Governments also can innovate, as the FBI laboratory did, but their constituencies and incentives are different and may diverge from those of private or social entrepreneurs (see Aronson 2007).

2. This is not to say that they were wrong to stick to their guns, but it does suggest that some experts are more uninhibited or less easily intimidated than others and that lawyers can locate these experts.

3. Appropriately enough, forensic scientists have come to realize that even if their discipline has accepted a technique without published research demonstrating its validity, the judicial system may not be satisfied (e.g., Bradley et al. 2006, 504). However, they do not necessarily help their cause when, rather

than simply presenting their research findings to their colleagues, they opine on whether their work satisfies the legal standards for admitting scientific evidence. In one report on "the validity of using posterior-anterior radiographs of the hand to make positive identifications of unknown human remains," for instance, the researchers tested the ability of examiners "to attempt to match 10 simulated postmortem radiographs of skeletonized hands to 40 simulated antemortem radiographs of fleshed cadaver hands." On the basis of good results with a grand total of twelve examiners, they assured their readers that the "method appears to satisfy the requirements of *Daubert*'s guidelines" (Koot, Sauer, and Fenton 2005, 267). Amateur lawyers are no better than amateur scientists.

4. The scope of the two reports is strikingly different. The voice-identification report was largely confined to describing scientific principles and studies. The bullet-lead report included legal arguments that seemed to be in tension with the chapters on the statistical issues. I have suggested that NRC reports on particular forms of scientific evidence are more likely to succeed when they assemble and assess the existing research literature and then outline, but do not try to prescribe, policy choices (Kaye 2006). More recently, the NRC released a sweeping (if somewhat cursory) survey of many fields of forensic science, calling in the strongest terms for major institutional changes (National Research Council Committee on Identifying the Needs of the Forensic Science Community 2009) [hereafter NRC 2009]. This report generally avoids overt conclusions on what the legal standards for admitting scientific evidence imply about evidence from particular fields of forensic science.

5. The 2009 NRC committee regarded "DNA typing [as] the standard against which many other forensic individualization techniques are judged" (NRC 2009, 5-3). The report indicates that the only sources of error in nuclear DNA analysis are "fraud or an error in labeling or handling" or a duplication of the profile somewhere else in the population (ibid.). This presentation overlooks the lack of easily applied standards and procedures for the interpretation of mixtures and highly degraded samples (Chapter 10).

References

Abbey, David M. 1999. "The Thomas Jefferson Paternity Case." *Nature* 397: 32.

Abbott, Alison. 1992. "FBI Attaches Strings to Its DNA Database." *Nature* 357: 618.

ABC Radio International. 2002. "The Science Show." Sept. 21. Available at www.abc.net.au/rn/scienceshow/stories/2002/684602.htm (accessed Mar. 12, 2007) (transcript of interview with Sir Alec Jeffreys).

Abele, Doris. 2002. "Toxic Oxygen." *Nature* 420: 27.

Adams, Julian. 2005. "Nuclear and Mitochondrial DNA in the Courtroom." *Journal of Law and Policy* 13: 69–97.

Alberts, Bruce, Dennis Bray, Julian Lewis, Martin Raff, Keith Roberts, and James D. Watson. 1983. *Molecular Biology of the Cell.* New York: Garland.

Aldhous, Peter. 1993. "Geneticists Attack NRC Report as Scientifically Flawed." *Science* 259: 755–756.

Allen-Mills, Tony. 2008. "Clara Rojas Hides Truth of Her Jungle Baby." *Sunday Times,* Jan. 20.

American Association of Blood Banks. 2006. "Highlights of Transfusion Medicine History." Apr. 12. Available at www.aabb.org/Content/About_Blood/Highlights_of_Transfusion_Medicine_History/ (accessed Feb. 28, 2007).

Anderson, Christopher. 1992a. "Academy Approves, Critics Still Cry Foul." *Nature* 356: 552.

———. 1992b. "FBI Gives In on Genetics." *Nature* 355: 663.

Annas, George J. 1993. "Privacy Rules for DNA Databanks: Protecting Coded 'Future Diaries.'" *Journal of the American Medical Association* 270: 2346–2350.

Aronson, Jay D. 2004. "The Introduction, Contestation, and Regulation of Forensic DNA Analysis in the American Legal System (1984–1994)." Ph.D. diss., University of Minnesota.

———. 2005. "DNA Fingerprinting on Trial: The Dramatic Early History of a New Forensic Technique." *Endeavor* 29: 126–131.

———. 2006. "The 'Starch Wars' and the Early History of DNA Profiling." *Forensic Science Review* 18: 59–72.

———. 2007. *Genetic Witness: Science, Law, and Controversy in the Making of DNA Profiling.* New Brunswick, N.J.: Rutgers University Press.

Ashley-Koch, Allison, Quanhe Yang, and Richard S. Olney. 2000. "Sickle Hemoglobin (Hb S) Allele and Sickle Cell Disease: A HuGE Review." *American Journal of Epidemiology* 151: 839–845.

Attewill, Fred. 2007. "Remains of Tsar's Heir May Have Been Found." *Guardian,* Aug. 24. Available at www.guardian.co.uk/world/2007/aug/24/russia (accessed May 5, 2009).

Avery, Oswald T., Colin M. MacLeod, and Maclyn McCarty. 1944. "Studies on the Chemical Nature of the Substance Inducing Transformation of Pneumococcal Types." *Journal of Experimental Medicine* 79: 137–158.

Ayala, Francisco J. 1992. "DNA Law." *Journal of Molecular Evolution* 35: 273–276.

———. 1993. "Junk Science and DNA Typing in the Courtroom." *Contention* 2: 45–60.

Balazs, Ivan, Michael Baird, Mindy Clyne, and Ellie Meade. 1989. "Human Population Genetic Studies of Five Hypervariable DNA Loci." *American Journal of Human Genetics* 44: 182–190.

Balding, David J. 1997. "Errors and Misunderstandings in the Second NRC Report." *Jurimetrics: The Journal of Law, Science, and Technology* 37: 469–476.

———. 2005. *Weight of Evidence for Forensic DNA Profiles.* Chichester: John Wiley & Sons.

Balding, D. J., and R. A. Nichols. 1994. "DNA Profile Match Probability Calculations: How to Allow for Population Stratification, Relatedness, Database Selection and Single Bands." *Forensic Science International* 64: 125–140.

———. 1995. "A Method for Quantifying Differentiation between Populations at Multi-allelic Loci and Its Implications for Investigating Identity and Paternity." *Genetica* 96: 3–12.

Ballantyne, Jack, George Sensabaugh, and Jan Witkowski, eds. 1989. *DNA Technology and Forensic Science.* Cold Spring Harbor, N.Y.: Cold Spring Harbor Laboratory Press.

Bamshad M., S. Wooding, B. A. Salisbury, and J. C. Stephens. 2004. "Deconstructing the Relationship between Genetics and Race." *Nature Reviews Genetics* 5: 598–609.

BBC News. 2007. "Lost Romanov Bones 'Identified.'" Sept. 28. Available at http://news.bbc.co.uk/2/hi/europe/7018503.stm (accessed Mar. 27, 2008).

Belair, Robert R. 1991. "Forensic DNA Analysis: Issues." Washington, D.C.: Bureau of Justice Statistics. Available at www.ncjrs.gov/pdffiles1/pr/128567.pdf (accessed May 21, 2009).

Belin, Thomas R., David W. Gjertson, and Hu Ming-yi. 1997. "Summarizing DNA Evidence When Relatives Are Possible Suspects." *Journal of the American Statistical Association* 92: 706–716.

Berger, Margaret. 1997. "Laboratory Error Seen through the Lens of Science and Policy." *University of California at Davis Law Review* 30: 1081–1109.

Bernstein, David E. 1999. "The Breast Implant Fiasco." *California Law Review* 87: 457–510.

———. 2008. "Expert Witnesses, Adversarial Bias, and the (Partial) Failure of the *Daubert* Revolution." *Iowa Law Review* 93: 451–489.

Bernstein, Felix. 1924. "Ergebnisse einer biostatistichen zusammenfassenden Betrachtung über die erblichen Blutstrukturen des Menschen." *Klinische Wochenschrift* 3: 1495–1497.

———. 1925. "Zusammenfassende Betrachtungen über die erblichen Blutstrukturen des Menschen." *Zeitschrift fur Induktive Abstammungs und Vererbungslehre* 37: 237–370.

Berry, Donald A. 1992. "Statistical Issues in DNA Identification." In Paul R. Billings, ed., *DNA on Trial: Genetic Identification and Criminal Justice*, 91–108. Cold Spring Harbor, N.Y.: Cold Spring Harbor Laboratory Press.

———. 1994. "Comment." *Statistical Science* 9: 252–255.

———. 2005. "Seymour Geisser, 1929–2004." *Journal of the Royal Statistical Society: Series A (Statistics in Society)* 168: 245–246.

Berry, D. A., I. W. Evett, and R. Pinchin. 1992. "Statistical Inference in Crime Investigations Using Deoxyribonucleic Acid Profiling." *Applied Statistics* 41: 499–531.

Biesecker, Leslie G., Joan E. Bailey-Wilson, Jack Ballantyne, Howard Baum, Frederick R. Bieber, Charles Brenner, et al. 2005. "DNA Identifications after the 9/11 World Trade Center Attack." *Science* 310: 1122–1123.

Blake, E., J. Mihalovich, R. Higuchi, P.S. Walsh, and H. Erlich. 1992. "Polymerase Chain Reaction (PCR) Amplification and Human Leukocyte Antigen (HLA)-DQ-a Oligonucleotide Typing on Biological Evidence Samples: Casework Experience." *Journal of Forensic Sciences* 37: 700–726.

Blanchetot, A., V. Wilson, D. Wood, and A. J. Jeffreys. 1983. "The Seal Myoglobin Gene: An Unusually Long Globin Gene." *Nature* 301: 732–734.

Botstein, D., R. L. White, M. Skolnick, Raymond L. White, Mark Skolnick, and Ronald W. Davis. 1980. "Construction of a Genetic Linkage Map in Man Using Restriction Fragment Length Polymorphisms." *American Journal of Human Genetics* 32: 314–331.

Bowler, Peter J. 1989. *The Mendelian Revolution: The Emergence of Hereditarian Concepts in Modern Science and Society*. Baltimore: Johns Hopkins University Press.

Boyd, David G. 1995. "More on DNA Typing Controversy." *Nature* 373: 98–99 (letter).

Bradley, Maureen J., Roger L. Keagy, Preston C. Lowe, Roger L. Keagy, Preston C. Lowe, Michael P. Rickenbach, Diana M. Wright, et al. 2006. "A Validation Study for Duct Tape End Matches." *Journal of Forensic Sciences* 51: 504–508.

Breyer, Stephen. 1998. "The Interdependence of Science and Law." *Science* 280: 537–538.

Brodzinsky, Sibylla, and Rory Carroll. 2008. "Reborn: Ex-hostage Reunited with Son." *Guardian,* Jan. 15, 17.

Brookfield, John. 1991. "The Effect of Population Subdivision on Estimates of the Likelihood Ratio in Criminal Cases Using Single-Locus Probes." *Heredity* 69: 97–100.

Brown, Ryan A., and George J. Armelagos. 2001. "Apportionment of Racial Diversity: A Review." *Evolutionary Anthropology* 10: 34–40.

Brown, T. 2002. *Genomes.* 2d ed. Oxford: BIOS Scientific Publishers.

Buckleton, John S. 2005. "Population Genetic Models." In John S. Buckleton, Christopher M. Triggs, and Simon J. Walsh, eds., *Forensic DNA Evidence Interpretation,* 65–122. Boca Raton, Fla.: CRC Press.

Buckleton, John, and James Curran. 2008. "A Discussion of the Merits of Random Man Not Excluded and Likelihood Ratios." *Forensic Science International: Genetics* 2: 343–348.

Buckleton, John, and Peter Gill. 2005. "Low Copy Number." In John S. Buckleton, Christopher M. Triggs, and Simon J. Walsh, eds., *Forensic DNA Evidence Interpretation,* 275–297. Boca Raton, Fla.: CRC Press.

Buckleton, J. S., K. A. J. Walsh, and I. W. Evett. 1991. "Who Is the Random Man?" *Journal of the Forensic Science Society* 31: 463–468.

Buckleton, John S., Simon Walsh, and Sallyann Harbison. 2001. "The Fallacy of Independence Testing and the Use of the Product Rule." *Science and Justice* 41: 81–84.

———. "Nonautosomal Forensic Markers." 2005. In John S. Buckleton, Christopher M. Triggs, and Simon J. Walsh, eds., *Forensic DNA Evidence Interpretation,* 299–340. Boca Raton, Fla.: CRC Press.

Budowle, B., and R. C. Allen. 1987. "Electrophoresis Reliability: I. The Contaminant Issue." *Journal of Forensic Sciences* 32: 1537–1550.

Budowle, B., A. M. Giusti, J. S. Waye, F. S. Buechtel, R. M. Fourney, Dwight E. Adams, Lawrence H. Presley, et al. 1991. "Fixed-Bin Analysis for Statistical Evaluation of Continuous Distributions of Allelic Data from VNTR Loci, for Use in Forensic Comparisons." *American Journal of Human Genetics* 48: 841–855.

Budowle, B., K. L. Monson, and J. R. Wooley. 1992. "Reliability of Statistical Estimates in Forensic DNA Typing." In Paul R. Billings, ed., *DNA on Trial: Genetic Identification and Criminal Justice,* 79–90. Cold Spring Harbor, N.Y.: Cold Spring Harbor Laboratory Press.

Butler, Declan. 2007. "Ghost Buster." *Nature* 445: 811.

Butler, John M. 2005. *Forensic DNA Typing: Biology, Technology, and Genetics of STR Markers.* 2d ed. Amsterdam and Boston: Elsevier Academic Press.

———. 2006. "Genetics and Genomics of Core Short Tandem Repeat Loci Used in Human Identity Testing." *Journal of Forensic Sciences* 51: 253–265.

Butler, John M., Eric Buel, Federica Crivellente, and Bruce R. McCord. 2004. "Forensic DNA Typing by Capillary Electrophoresis Using the ABI Prism 310 and 3100 Genetic Analyzers for STR Analysis." *Electrophoresis* 25: 1397–1412.

Butler, John M., and Dennis J. Reeder. 2008. "Overview of STR Fact Sheets." In *Short Tandem Repeat DNA Internet DataBase*. Available at http://cstl.nist .gov/div831/strbase/str_fact.htm (accessed Jan. 22, 2008).

Caskey, C. Thomas, and Holly Hammond. 1989. "DNA-based Identification: Disease and Criminals." In Jack Ballantyne, George Sensabaugh, and Jan Witkowski, eds., *DNA Technology and Forensic Science*, 127–135. Cold Spring Harbor, N.Y.: Cold Spring Harbor Laboratory Press.

CBS News. 2004. "The O. J. Case 10 Years Later: Multiple Questions Linger, Simpson Still Adamant." June 11. Available at www.cbsnews.com/stories/2004/06/11/national/main622502.shtml (accessed Dec. 28, 2007).

Cellmark Diagnostics. 1988. *DNA Fingerprinting*. Germantown, Md.: Cellmark Diagnostics.

Cerri, N., S. Manzoni, A. Verzeletti, and F. De Ferrari. 2004. "Population Data on D16S539, D2S1338 and D19S433 Loci in a Population Sample from Brescia (Italy)." *Progress in Forensic Genetics* 10: 210–212.

Chakraborty, Ranajit. 1991. "Statistical Interpretation of DNA Typing Data." *American Journal of Human Genetics* 49: 895–897.

———. 1993. "NRC Report on DNA Typing." *Science* 260: 1059.

Chakraborty, Ranajit, and Kenneth K. Kidd. 1991. "The Utility of DNA Typing in Forensic Work." *Science* 254: 1735–1739.

Chakraborty, Ranajit, M. R. Srinivasan, L. Jin, and M. de Andrade. 1992. "Effects of Population Subdivision and Allele Frequency Differences on Interpretation of DNA Typing Data for Human Identification." In *Proceedings of the 3rd International Symposium on Human Identification*, 205–222. Madison, Wis.: Promega Corporation.

Cheng, Edward K. 2005. "Mitochondrial DNA: Emerging Legal Issues." *Journal of Law and Policy* 13: 99–118.

———. 2007. "Independent Judicial Research in the *Daubert* Age." *Duke Law Journal* 56: 1263–1316.

Clayton, Tim, and John Buckleton. 2005. "Mixtures." In John S. Buckleton, Christopher M. Triggs, and Simon J. Walsh, eds., *Forensic DNA Evidence Interpretation*, 217–274. Boca Raton, Fla.: CRC Press.

Cohen, Joel E. 1990. "DNA Fingerprinting for Forensic Identification: Potential Effects on Data Interpretation of Subpopulation Heterogeneity and Band Number Variability." *American Journal of Human Genetics* 46: 358–368.

———. 1992. "The Ceiling Principle Is Not Always Conservative in Assigning Genotype Frequencies for Forensic DNA Testing." *American Journal of Human Genetics* 51: 1165–1167.

Cold Spring Harbor Laboratory. n.d. "Genetic Origins." Available at www.genetic-origins.org/mito/theory.html (Mar. 14, 2008).

Coleman, Howard, and Eric Swenson. 1994. *DNA in the Courtroom: A Trial Watcher's Guide*. Seattle: GeneLex.

Cooley, Amanda, Carrie Bess, and Marsha Rubin-Jackson. 1996. *Madame Foreman: A Rush to Judgment*. Beverly Hills, Calif.: Dove Books.

Cooley, Craig M. 2004. "Reforming the Forensic Science Community to Avert the Ultimate Injustice." *Stanford Law and Policy Review* 15: 381–446.

————. 2007. "Forensic Science and Capital Punishment Reform: An 'Intellectually Honest' Assessment." *George Mason University Civil Rights Law Journal* 17: 299–422.

Cowell, R., S. Lauritzen, and J. Mortera. 2007. "Identification and Separation of DNA Mixtures Using Peak Area Information." *Forensic Science International* 166: 28–34.

Cramér, Harald. 1946. *Mathematical Methods of Statistics*. Princeton, N.J.: Princeton University Press.

Cree, Lynsey M., David C. Samuels, Susana Chuva de Sousa Lopes, Harsha Karur Rajasimha, Passorn Wonnapinij, Jeffrey R. Mann, Hans-Henrik M. Dahl, et al. 2008. "A Reduction of Mitochondrial DNA Molecules during Embryogenesis Explains the Rapid Segregation of Genotypes." *Nature Genetics* 40: 249–254.

Crow, James F. 1993. "Felix Bernstein and the First Human Marker Locus." *Genetics* 133: 4–7.

————. 1996. "Isonomy: A Thirty Year Retrospective." *Rivista di Antropologia* 74: 25–34.

Crow, J. F., and C. Denniston. 1993. "Population Genetics as It Relates to Human Identification." In *Proceedings of the 4th International Symposium on Human Identification,* 31–36. Madison, Wis.: Promega Corporation.

Curriden, Mark. 1996. "A New Evidence Tool: First Use of Mitochondrial DNA Test in a U.S. Criminal Trial." *American Bar Association Journal,* Nov., 18.

Daemmrich, Arthur. 1998. "The Evidence Does Not Speak for Itself: Expert Witnesses and the Organization of DNA-Typing Companies." *Social Studies of Science* 28: 741–772.

Dann, B. Michael, Valerie P. Hans, and David H. Kaye. 2004. *Testing the Effects of Selected Jury Trial Innovations on Juror Comprehension of Contested mtDNA Evidence: Final Technical Report*. Washington, D.C.: National Institute of Justice.

————. 2007. "Can Jury Trial Innovations Improve Juror Understanding of DNA Evidence?" *Judicature* 90: 152–156.

Davidson, Joe. 1992. "Study Supports Use in Court of DNA Tests." *Wall Street Journal,* Apr. 15, B1, B4.

Dawid, A. P. 2002. "Bayes's Theorem and Weighing Evidence by Juries." In Richard Swinburne, ed., *Bayes's Theorem,* 71–90. Oxford: Oxford University Press.

Deason, Ellen E. 1998. "Court-Appointed Expert Witnesses: Scientific Positivism Meets Bias and Deference." *Oregon Law Review* 77: 59–156.

Dennett, Daniel C. 1995. *Darwin's Dangerous Idea*. New York: Simon & Schuster.

Derksen, Linda. 2000. "Toward a Sociology of Measurement: The Meaning of Measurement Error in the Case of DNA Profiling." *Social Studies of Science* 30: 803–845.

Devlin, Bernard, and Neil Risch. 1992a. "Ethnic Differentiation at VNTR Loci, with Special Reference to Forensic Applications." *American Journal of Human Genetics* 51: 534–548.

———. 1992b. "A Note on Hardy-Weinberg Equilibrium of VNTR Data Using the FBI's Fixed-Bin Method." *American Journal of Human Genetics* 51: 549–553.

———. 1993. "Physical Properties of VNTR Data and Their Impact on Tests of Allelic Independence." *American Journal of Human Genetics* 53: 324–329.

Devlin, Bernard, Neil Risch, and Kathryn Roeder. 1990. "No Excess of Homozygosity at DNA Fingerprint Loci." *Science* 249: 1416–1420.

———. 1991. "Forensic DNA Tests and Hardy-Weinberg Equilibrium: Reply." *Science* 253: 1039–1041.

———. 1992. "Forensic Inference from DNA Fingerprints." *Journal of the American Statistical Association* 87: 337–350.

———. 1993a. "NRC Report on DNA Typing." *Science* 260: 1057–1058 (letter).

———. 1993b. "Statistical Evaluation of DNA Fingerprinting: A Critique of the NRC's Report." *Science* 259: 748–749, 837.

Devlin, Bernard, and Kathryn Roeder. 1997. "Population Genetic Issues in the Forensic Use of DNA." In David L. Faigman, David H. Kaye, Michael J. Saks, and Joseph Sanders eds., *Modern Scientific Evidence: The Law and Science of Expert Testimony,* 1: 710–747. St. Paul, Minn.: West Publishing.

Donnelly, Peter. 1995. "Nonindependence of Matches at Different Loci in DNA Profiles: Quantifying the Effect of Close Relatives on the Match Probability." *Heredity* 75: 26–34.

Downer, Joanna. 2002. "McKusick, 'Father of Genetic Medicine,' to Get National Medal of Science." Available at www.hopkinsmedicine.org/press/2002/MAY/020509.htm (accessed Apr. 3, 2007).

Edwards, A., A. Civitello, H. A. Hammond, and C. T. Caskey. 1991. "DNA Typing and Genetic Mapping with Trimeric and Tetrameric Tandem Repeats." *American Journal of Human Genetics* 49: 746–756.

Edwards, A., H. A. Hammond, L. Jin, C. T. Caskey, and R. Chakraborty. 1992. "Genetic Variation at Five Trimeric and Tetrameric Tandem Repeat Loci in Four Human Population Groups." *Genomics* 12: 241–253.

Edwards, A. W. F. 1972. *Likelihood: An Account of the Statistical Concept of Likelihood and its Application to Scientific Inference.* Cambridge: Cambridge University Press.

Egeland, T., and A. Salas. 2008. "Statistical Evaluation of Haploid Genetic Evidence." *Open Forensic Science Journal* 1: 4–11.

Elliott, K., D. S. Hill, C. Lambert, T. R. Burroughes, and P. Gill. 2003. "Use of Laser Microdissection Greatly Improves the Recovery of DNA from Sperm on Microscope Slides." *Forensic Science International* 137: 28–36.

Ellman, Ira M., and D. H. Kaye. 1979. "Probabilities and Proof: Can HLA and Blood Test Evidence Prove Paternity?" *New York University Law Review* 55: 1131–1162.

Elmer-DeWitt, Philip. 2004. "Eric Lander: Unraveling the Threads of Life." *Time,* Apr. 26, 103.

Epstein, C. J. 1991. "The Forensic Applications of Molecular Genetics—The Journal's Responsibilities." *American Journal of Human Genetics* 49: 697–698.

Erzinçlioglu, Zakaria. 1998. "British Forensic Science in the Dock." *Nature* 392: 859–860.

———. 2002. *Forensics: True Crime Scene Investigations.* New York: Barnes & Noble.

Evett, Ian W. 1992. "DNA Statistics: Putting the Problems into Perspective," *Jurimetrics. The Journal of Law, Science, and Technology* 33: 139–145.

Evett, Ian W., Cecilia Buffery, Geoffrey Willott, and D. Stoney. 1991. "A Guide to Interpreting Single Locus Profiles of DNA Mixtures in Forensic Cases." *Journal of the Forensic Science Society* 31: 41–47.

Evett, Ian W., and Peter Gill. 1991. "A Discussion of the Robustness of Methods for Assessing the Evidential Value of DNA Single Locus Profiles in Crime Investigations." *Electrophoresis* 12: 226–230.

Evett, I. W., and R. Pinchin. 1991. "DNA Single-Locus Profiles: Tests for Robustness of Statistical Procedures within the Context of Forensic Science." *International Journal of Law and Medicine* 104: 267–272.

Evett, I. W., J. Scranage, and R. Pinchin. 1993. "An Illustration of Efficient Statistical Methods for RFLP Analysis in Forensic Science." *American Journal of Human Genetics* 52: 498–505.

Evett, Ian W., and Bruce S. Weir. 1991. "Flawed Reasoning in Court." *New Directions in Statistics and Computing* 4(4): 19–21.

———. 1992. "Whose DNA?" *American Journal of Human Genetics* 50: 869.

———. 1998. *Interpreting DNA Evidence: Statistical Genetics for Forensic Scientists.* Sunderland, Mass.: Sinauer.

Ezzell, C. 1992. "Panel OKs DNA Fingerprints in Court Cases." *Science News* 141: 261.

Faigman, David L., David H. Kaye, Michael J. Saks, and Edward K. Cheng. 2006. *Modern Scientific Evidence: The Law and Science of Expert Testimony.* Eagan, Minn.: Thomson/West.

Finkelstein, Michael O., and William B. Fairley. 1970. "A Bayesian Approach to Identification Evidence." *Harvard Law Review* 83: 489–517.

Fischer, Eric A. 1993. Memorandum to Planning Group for Proposed DNA Forensics. June 9.

Fisher, Deborah, Mitchell M. Holland, Lloyd Mitchell, Paul S. Sledzik, Allison Webb Wilcox, Mark Wadhams, and Victor W. Weedn. 1993. "Extraction, Evaluation, and Amplification of DNA from Decalcified and Undecalcified United States Civil War Bone." *Journal of Forensic Sciences* 38: 60–68.

Fisher, George. 2006. "Green Felt Jungle: The Story of *People v. Collins.*" In Richard Lempert, ed., *Evidence Stories,* 7–27. New York: Foundation Press.

Fore, Joe, Jr., Ilse R. Wiechers, and Robert Cook-Deegan. 2006. "The Effects of Business Practices, Licensing, and Intellectual Property on Development and Dissemination of the Polymerase Chain Reaction: Case Study." *Journal of Biomedical Discovery and Collaboration* 1: 7. Available at www.j-biomed-discovery.com/content/1/1/7 (accessed Jan. 22, 2008).

Forensic Science Service. n.d. "Colin Pitchfork." Available at www.forensic.gov.uk/html/media/case-studies/f-18.html (accessed Feb. 22, 2009).

Forero, Juan. 2008. "DNA Links Boy to FARC Hostage: Colombian Rebels Appear to Have Deceived Venezuela's Chavez." *Washington Post,* Jan. 5, A14.

Foster, E. A., M. A. Jobling, and P. G. Taylor. 1998. "Jefferson Fathered Slave's Last Child." *Nature* 396: 27–28.

Friess, Steve. 2008. "After Apologies, Simpson Is Sentenced to at Least 9 Years for Armed Robbery." *New York Times*, Dec. 6, A9.

Frudakis, T., K. Venkateswarlu, M. J. Thomas, Z. Gaskin, S. Ginjupalli, S. Gunturi, V. Ponnuswamy, et al. 2003. "A Classifier for the SNP-Based Inference of Ancestry." *Journal of Forensic Sciences* 48: 771–782.

Garratty, George, Simone A. Glynn, and Robin McEntire. 2004. "ABO and Rh(D) Phenotype Frequencies of Different Racial/Ethnic Groups in the United States." *Transfusion* 44: 703–706.

Garrett, Brandon. 2008a. "Claiming Innocence." *Minnesota Law Review* 92: 1629–1723.

———. 2008b. "Judging Innocence." *Columbia Law Review* 108: 55–142.

Geisser, Seymour. 1992. "Some Statistical Issues in Medicine and Forensics." *Journal of the American Statistical Association* 87: 607–614.

———. 1996. "Some Statistical Issues in Forensic DNA Profiling." In Jack C. Lee, Wesley O. Johnson, and Arnold Zellner, eds., *Modeling and Prediction: Honoring Seymour Geisser*, 3–18. New York: Springer-Verlag.

Giannelli, Paul C. 1997a. "The DNA Story: an Alternative View." *Journal of Criminal Law and Criminology* 88: 380–422.

——— 1997b. "The Abuse of Scientific Evidence in Criminal Cases: The Need for Independent Crime Laboratories." *Virginia Journal of Social Policy and Law* 4: 439–478.

———. 2004. "*Ake v. Oklahoma*: The Right to Expert Assistance in a Post-*Daubert*, Post-DNA World." *Cornell Law Review* 89: 1305–1419.

———. 2007a. "Regulating Crime Laboratories: The Impact of DNA Evidence." *Journal of Law and Policy* 15: 59–92.

———. 2007b. "Wrongful Convictions and Forensic Science: The Need to Regulate Crime Labs." *North Carolina Law Review* 86: 163–236.

Giannelli, Paul C., and Edward J. Imwinkelried. 1993. *Scientific Evidence*. 2d ed. Vol. 1. Charlottesville, Va.: Michie.

Giannelli, Paul C., and Kevin C. McMunigal. 2007. "Prosecutors, Ethics, and Expert Witnesses." *Fordham Law Review* 76: 1493–1537.

Gibbons, Ann. 1998. "Calibrating the Mitochondrial Clock." *Science* 279: 28–29.

Gilbert, M., P. Thomas, Lynn P. Tomsho, Snjezana Rendulic, Michael Packard, Daniela I. Drautz, Andrei Sher, et al. 2007. "Whole-Genome Shotgun Sequencing of Mitochondria from Ancient Hair Shafts." *Science* 317: 1927–1930.

Gilder, J., R. Koppl, I. Kornfield, D. Krane, L. Mueller, and W. Thompson. 2008. "Comments on the Review of Low Copy Number Testing." *International Journal of Legal Medicine*, Sept. 18 online, DOI 10.1007/s00414-008-0281-z, Available at www.springerlink.com/content/x5g816283n6u6q71/full text.pdf (accessed May 19, 2009) (letter).

Gill, P., C. H. Brenner, J. S. Buckleton, A. Carracedo, M. Krawczak, W.R. Mayr, N. Morling, et al. 2006. "DNA Commission of the International Society of

Forensic Genetics: Recommendations on the Interpretation of Mixtures." *Forensic Science International* 160: 90–101.

Gill, Peter, and Erika Hagelberg. 2004. "Letter." *Science* 306: 408.

Gill, Peter, Pavel L. Ivanov, Colin Kimpton, Romelle Piercy, Nicola Benson, Gillian Tully, Ian Evett, et al. 1994. "Identification of the Remains of the Romanov Family by DNA Analysis." *Nature Genetics* 6: 130–135.

Gill, Peter, Alec J. Jeffreys, and David J. Werrett. 1985. "Forensic Applications of DNA 'Fingerprints'." *Nature* 318: 577–579.

Gill, Peter, and David J. Werrett. 1987. "Exclusion of a Man Charged with Murder by DNA Fingerprinting." *Forensic Science International* 35: 145–148.

Gitschier, Jane. 2005. "The Whole Side of It—An Interview with Neil Risch." *PLoS Genetics* 1: e14. Available at http://genetics.plosjournals.org/perlserv/?request =get-document&doi=10.1371/journal.pgen.0010014&ct=1 (accessed Aug. 20, 2007).

Golan, Tal. 2004. *Laws of Men and Laws of Nature: The History of Scientific Expert Testimony in England and America*. Cambridge, Mass.: Harvard University Press.

Gonzalez, Jenny Carolina. 2008. "Hostage Boy's Story Emerges." *Miami Herald,* Jan. 8, A1.

González-Andrade, Fabricio, and Dora Sánchez. 2004. "Genetic Profile of the Kichwas (Quichuas) from Ecuador by Analysis of STR Loci." *Human Biology* 76: 723–730.

Goodwin, William, Adrian Linacre, and Sibte Hade. 2007. *An Introduction to Forensic Genetics*. Chichester: John Wiley & Sons.

Green, P., and E. S. Lander. 1991. "Forensic DNA Tests and Hardy-Weinberg Equilibrium." *Science* 253: 1038–1039.

Gross, Samuel R. 1991. "Expert Evidence." *Wisconsin Law Review* 1991: 1113–1232.

Guillen, Gonzalo, and Jenny Carolina Gonzalez. 2008. "Colombian Hostage's Health Is Failing." *Miami Herald,* Mar. 28.

Gusella, J. F., Nancy S. Wexler, P. Michael Conneally, Susan L. Naylor, Mary Anne Anderson, Rudolph E. Tanzi, Paul C. Watkins, et al. 1983. "A Polymorphic DNA Marker Genetically Linked to Huntington's Disease." *Nature* 306: 234–238.

Haas-Rochholz, H., and G. Weiler. 1997. "Additional Primer Sets for an Amelogenin Gene PCR-Based DNA-Sex Test." *International Journal of Legal Medicine* 110: 312–315.

Haflon, Saul. 1998. "Collecting, Testing, and Convincing: Forensic DNA Experts in the Courts." *Social Studies of Science* 28: 801–828.

Hall, John. 2005. "Defense Continues Hammering at DNA in Avila Case." *North County Times (The Californian),* Apr. 20. Available at www.nctimes.com/ articles/2005/04/20/news/californian/0_11_194_20_05.txt (accessed Jan. 23, 2008).

Hand, Learned. 1901. "Historical and Practical Considerations Regarding Expert Testimony." *Harvard Law Review* 15: 40–58.

Hanson, Randall K. 1996. "Witness Immunity under Attack: Disarming 'Hired Guns'." Wake Forest Law Review 31: 497–511.

Hartl, Daniel L. 1994. "DNA Forensic Typing Dispute." *Nature* 372: 398–399 (letter).

Hartl, Daniel L., and Richard C. Lewontin. 1993. "DNA Fingerprinting Report." *Science* 260: 473–474.

———. 1994. "DNA Fingerprinting." *Science* 266: 201 (letter).

Hassman, Phillip E. 1977. "Annotation: Admissibility of Expert Medical Testimony as to Future Consequences of Injury as Affected by Expression in Terms of Probability or Possibility." *American Law Reports*, 3d Series, 75: 9–124.

Hayden, Erika Check. 2008. "International Genome Project Launched: Three-Year Study Will Capture Variation in 1,000 People." *Nature* 451: 378–379.

Helpern, Milton, and Bernard Knight. 1977. *Autopsy: The Memoirs of Milton Helpern, the World's Greatest Medical Detective.* New York: St. Martin's Press.

Hendricks, Melissa. 2000. "The Man Who Put Genetics on the Map." *Johns Hopkins Magazine,* Apr. Available at www.jhu.edu/~jhumag/0400web/38.html (accessed Apr. 3, 2007).

Herrin, George, Jr. 1993. "Probability of Matching RFLP Patterns from Unrelated Individuals." *American Journal of Human Genetics* 52: 491–497.

Highfield, Roger. 2004. "Scientists Reopen the Romanov Mystery." *Telegraph,* Dec. 17. Available at www.telegraph.co.uk/news/main.jhtml?xml=/news/2004/07/12/wtsar12.xml&sSheet=/news/2004/07/12/ixworld.html (accessed Mar. 27, 2008).

Hofreiter, Michael, Odile Loreille, Deborah Ferriola, and Thomas J. Parsons. 2004. "Ongoing Controversy over Romanov Remains." *Science* 306: 407–408 (letter).

Holland, Mitchell M., Deborah L. Fisher, Lloyd G. Mitchell, W. C. Rodriquez, J. J. Canik, C. R. Merril, and V. W. Weedn. 1993. "Mitochondrial DNA Sequence Analysis of Human Skeletal Remains: Identification of Remains from the Vietnam War." *Journal of Forensic Sciences* 38: 542–553.

Holland, M. M., and T. J. Parsons. 1999. "Mitochondrial DNA Sequence Analysis—Validation and Use for Forensic Casework." *Forensic Science Review* 11: 21–50.

Holmlunda, G., and B. Lindblom. 1998. "Different Ancestor Alleles: A Reason for the Bimodal Fragment Size Distribution in the Minisatellite D2S44 (YNH24)." *European Journal of Human Genetics* 6: 597–602.

Hopkin, Karen. 2007. "Eric S. Lander, Ph.D." Available at www.hhmi.org/biointeractive/genomics/lander.html (accessed Mar. 29, 2007).

Houck, Max M., and Bruce Budowle. 2002. "Correlation of Microscopic and Mitochondrial DNA Hair Comparisons." *Journal of Forensic Sciences* 47: 1–4.

House of Commons Science and Technology Committee. 2005. *Forensic Science on Trial.* London: Stationery Office Limited (seventh report of the session 2004–2005, HC 96-1, Mar. 29).

Houts, Marshall. 1967. *Where Death Delights: The Story of Milton Helpern and Forensic Medicine.* New York: Coward-McCann.

Imwinkelried, Edward J., and D. H. Kaye. 2001. "DNA Typing: Emerging or Neglected Issues." *Washington Law Review* 76: 413–474.

Ivanov, Pavel L., Mark J. Wadhams, Rhonda K. Roby, Mitchell M. Holland, Victor W. Weedn, and Thomas J. Parsons. 1996. "Mitochondrial DNA Sequence Heteroplasmy in the Grand Duke of Russia Establishes the Authenticity of the Remains of Tsar Nicholas II." *Nature Genetics* 12: 417–420.

James, Ian. 2008. "Ex-captive Describes Years with Guerrillas." *Washington Post,* Jan. 13, A21.

Jaroff, Leon. 1989. "The Gene Hunt." *Time,* Mar. 20. Available at www.time.com/ time/printout/0,8816,957263,00.html (accessed Dec. 27, 2008).

Jasanoff, Sheila. 1995. *Science at the Bar: Law, Science, and Technology in America.* Cambridge, Mass.: Harvard University Press.

Jeffreys, Alec J. 1987. "Highly Variable Minisatellites and DNA Fingerprints." *Biochemical Society Transactions* 15: 309–317.

———. 2005. "Genetic Fingerprinting." *Nature Medicine* 11: 1035–1039.

Jeffreys, Alec J., John F. Y. Brookfield, and Robert Semeonoff. 1985. "Positive Identification of an Immigrant Test-Case Using Human DNA Fingerprints." *Nature* 317: 818–819.

Jeffreys, A. J., M. Turner, and P. Debenham. 1991. "The Efficiency of Multilocus DNA Fingerprint Probes for Individualization and Establishment of Family Relationships, Determined from Extensive Casework." *American Journal of Human Genetics* 48: 824–840.

Jeffreys, Alec J., V. Wilson, A. Blanchetot, P. Weller, A. Geurts van Kessel, N. Spurr, E. Solomon, et al. 1984. "The Human Myoglobin Gene: A Third Dispersed Globin Locus in the Human Genome." *Nucleic Acids Research* 12: 3235–3243.

Jeffreys, Alec J., V. Wilson, and S. L. Thein. 1985a. "Hypervariable 'Minisatellite' Regions in Human DNA." *Nature* 314: 67–73.

———. 1985b. "Individual-Specific 'Fingerprints' of Human DNA." *Nature* 316: 76–79.

Jobling, Mark A., and Peter Gill. 2004. "Encoded Evidence: DNA in Forensic Analysis." *Nature Reviews Genetics* 5: 739–752.

Johannsen, Wilhelm. 1909. *Elemente der exakten Erblichkeitslehre.* Jena, Germany: Gustav Fischer.

Jonakait, Randolph N. 1982. "Will Blood Tell? Genetic Markers in Criminal Cases." *Emory Law Journal* 31: 833–911.

Jovanovic, B. D., and P. S. Levy. 1997. "A Look at the Rule of Three." *American Statistician* 51(2): 137–139.

Juengst, Eric T. 2004. "FACE Facts: Why Human Genetics Will Always Provoke Bioethics." *Journal of Law, Medicine and Ethics* 32: 267–275.

Juricek, D. K. 1984. "The Misapplication of Genetic Analysis in Forensic Science." *Journal of Forensic Sciences* 29: 8–16.

Kaestle, Frederika A., Ricky A. Kittles, Andrea L. Roth, and Edward J. Ungvarsky. 2006. "Database Limitations on the Evidentiary Value of Forensic Mitochondrial DNA Evidence." *American Criminal Law Review* 43: 53–88.

Kamiya, Gary. 1997. "Cablinasian like Me." *Salon,* Apr. Available at www.salon .com/april97/tiger970430.html (accessed Nov. 29, 2008).

Kaye, David H. 1986. "Is Proof of Statistical Significance Relevant?" *Washington Law Review* 61: 1333–1365.

———. 1990. "Improving Legal Statistics." *Law and Society Review* 24: 1255–1275.

———. 1993. "DNA Evidence: Probability, Population Genetics, and the Courts." *Harvard Journal of Law and Technology* 7: 101–172.

———. 1994a. "The DNA Chronicles: Is Simpson Really Collins?" 1994 WL 592117.

———. 1994b. "The DNA Chronicles: The Meaning of General Acceptance." 1994 WL 595559.

———. 1995a. "The DNA Chronicles: Is Simpson Really Cella?" 1995 WL 234887.

———. 1995b. "The Forensic Debut of the NRC's DNA Report: Population Structure, Ceiling Frequencies, and the Need for Numbers." *Genetica* 96: 99–105.

———. 1995c. "The Relevance of 'Matching' DNA: Is the Window Half Open or Half Shut?" *Journal of Criminal Law and Criminology* 85: 676–695.

———. 1995d. "Was the Verdict Wrong?" *Arizona State University Law Forum,* Winter, 12–15. (Also available on Westlaw at 1996 WL 1218).

———. 1996. "Cross-Examining Science." *Jurimetrics: The Journal of Law, Science, and Technology* 36: vii–x.

———. 1997a. "*Bible* Reading: DNA Evidence in Arizona." *Arizona State Law Journal* 28: 1035–1078.

———. 1997b. *Science in Evidence.* Cincinnati: Anderson Publishing.

———. 2001a. "Choice and Boundary Problems in *Logerquist, Hummert,* and *Kumho Tire.*" *Arizona State Law Journal* 33: 41–74.

———. 2001b. "The Constitutionality of DNA Sampling on Arrest." *Cornell Journal of Law and Public Policy* 10: 455–509.

———. 2003. "Questioning a Courtroom Proof of the Uniqueness of Fingerprints." *International Statistical Review* 71: 521–533.

———. 2004. "Logical Relevance: Problems with the Reference Population and DNA Mixtures in *People v. Pizarro.*" *Law, Probability, and Risk* 3: 211–220.

———. 2006. "The Current State of Bullet-Lead Evidence." *Jurimetrics: The Journal of Law, Science, and Technology* 46: 99–114.

———. 2007a. "DNA Probabilities in *People v. Prince:* When Are Racial and Ethnic Statistics Relevant?" In Terrence Speed, ed., *Probability and Statistics: Essays in Honor of David A. Freedman,* 289–301. Beachwood, Ohio: Institute of Mathematical Statistics.

———. 2007b. "Please, Let's Bury the Junk: The CODIS Loci and the Revelation of Private Information." *Northwestern University Law Review Colloquy* 102: 70–81.

———. 2008. "The Role of Race in DNA Evidence: What Experts Say, What California Courts Allow." *Southwestern Law Review* 37: 303–322.

———. 2009a. "Rounding Up the Usual Suspects: A Legal and Logical Analysis of DNA Database Trawls." *North Carolina Law Review* 87: 425–503.

———. 2009b. "Trawling DNA Databases for Partial Matches." *Cornell Journal of Law and Public Policy* 19 (in press).

Kaye, David H., David Bernstein, and Jennifer Mnookin. 2004. *The New Wigmore: A Treatise on Evidence; Expert Evidence.* New York: Aspen.

Kaye, David H., Valerie Hans, B. Michael Dann, Erin Farley, and Stephanie Albertson. 2007. "Statistics in the Jury Box: Do Jurors Understand Mitochondrial DNA Match Probabilities?" *Journal of Empirical Legal Studies* 4: 797–834.

Kaye, David H., and George F. Sensabaugh. 2006. "DNA Typing." In David L. Faigman David H. Kaye, Michael J Saks, and Joseph Sanders, eds., *Modern Scientific Evidence: The Law and Science of Expert Testimony,* § 32. Eagan, Minn.: Thomson/West.

Keynes, J. M. 1921. *A Treatise on Probability.* London: Macmillan.

Khrapko, Konstantin. 2008. "Two Ways to Make an mtDNA Bottleneck." *Nature Genetics* 40: 134–135.

King, Turi E., Stéphane J. Ballereau, Kevin E. Schürer, and Mark A. Jobling. 2006. "Genetic Signatures of Coancestry within Surnames." *Current Biology* 16: 384–388.

Knight, Alec, Lev A. Zhivotovsky, David H. Kass, Daryl E. Litwin, Lance D. Green, and P. Scott White. 2004. "Response." *Science* 306: 409.

Koehler, Jonathan J. 1993. "DNA Matches and Statistics: Important Questions, Surprising Answers." *Judicature* 76: 222–229.

———. 1997. "Why DNA Likelihood Ratios Should Account for Error (Even When a National Research Council Report Says They Should Not)." *Jurimetrics: The Journal of Law, Science, and Technology* 37: 425–437.

———. 2008. "Fingerprint Error Rates and Proficiency Tests: What They Are and Why They Matter." *Hastings Law Journal* 59: 1077–1098.

Koehler, Jonathan J., Audrey Chia, and Samuel Lindsey. 1995. "The Random Match Probability (RMP) in DNA Evidence: Irrelevant and Prejudicial?" *Jurimetrics: The Journal of Law, Science, and Technology* 35: 201–219.

Kolata, Gina. 1990. "Some Scientists Doubt the Value of 'DNA Fingerprint' Evidence." *New York Times*, Jan. 29, A1.

———. 1992a. "Chief Says Panel Backs Court's Use of a Genetic Test." *New York Times,* Apr. 15, A1.

———. 1992b. "DNA Fingerprinting: Built-in Conflict." *New York Times,* Apr. 17, A13.

———. 1992c. "U.S. Panel Seeking Restriction on Use of DNA in Courts: Labs' Standards Faulted, Judges Are Asked to Bar Genetic 'Fingerprinting' Until Basis in Science Is Stronger." *New York Times,* Apr. 14, A1, A6.

———. 1994. "Two Chief Rivals in the Battle over DNA Evidence Now Agree on Its Use." *New York Times,* Oct. 27, B14.

Koot, Michael G., Norman J. Sauer, and Todd W. Fenton. 2005. "Radiographic Human Identification Using Bones of the Hand: A Validation Study." *Journal of Forensic Sciences* 50: 263–267.

Koshland, Daniel E. 1994. "Response." *Science* 266: 202–203.

Krane, Dan. 2002. As quoted in "DNA for the Defense: New Wright State University Software Interprets DNA Evidence." Press release, Wright State University Communications and Marketing, Oct. 22, 2002. Available at www.wright.edu/cgi-bin/cm/news.cgi?action=news_item&id=387 (accessed Dec. 30, 2007).

Krane, Dan E., Robert W. Allen, Stanley A. Sawyer, Dmitri A. Petrov, and Daniel L. Hartl. 1992. "Genetic Differences at Four DNA Typing Loci in Finnish, Italian, and Mixed Caucasian Populations." *Proceedings of the National Academy of Sciences (USA)* 89: 10583–10587.

Kreisler, Harry. 2003. "Conversation with Richard Lewontin." Nov. 20. Available at http://globetrotter.berkeley.edu/people3/Lewontin/lewontin-con2.html (accessed July 21, 2007).

Krings, M., C. Capelli, F. Tschentscher, Matthias Krings, Cristian Capelli, Frank Tschentscher, Helga Geisert, et al. 2000. "A View of Neandertal Genetic Diversity." *Nature Genetics* 26: 144–146.

Krings, M., H. Geisert, R. W. Schmitz, Heike Krainitzki, and Svante Pääbo. 1999. "DNA Sequence of the Mitochondrial Hypervariable Region II from the Neandertal Type Specimen." *Proceedings of the National Academy of Sciences (USA)* 96: 5581–5585.

Krings, Matthias, Anne Stone, Ralf W. Schmitz, Heike Krainitzki, Mark Stoneking, and Svante Pääbo. 1997. "Neandertal DNA Sequences and the Origin of Modern Humans." *Cell* 90: 19–30.

Krishnan, Kim J., Amy K. Reeve, David C. Samuels, Patrick F. Chinnery, John K. Blackwood, Robert W. Taylor, Sjoerd Wanrooij, et al. 2008. "What Causes Mitochondrial DNA Deletions in Human Cells?" *Nature Genetics* 40: 275–279.

Kulish, Nicholas. 2009. "Telling Twins Apart Takes New Meaning in Berlin Jewelry Heist." *New York Times*, Feb. 21, A6.

Kurland, C. G., and S. G. E. Andersson. 2000. "Origin and Evolution of the Mitochondrial Proteome." *Microbiology and Molecular Biology* 64: 786–820.

Lancia, M., A. Coletti, G. Margiotta, E. Carnevali, and M. Bacci. 2006. "Allele Frequencies of 15 STR Loci in an Italian Population." *Progress in Forensic Genetics* 11: 340–342.

Lander, Eric S. 1989a. "DNA Fingerprinting on Trial." *Nature* 331: 501–505.

———. 1989b. "Population Genetic Considerations in the Forensic Use of DNA Typing." In Jack Ballantyne, George Sensabaugh, and Jan Witkowski, eds., *DNA Technology and Forensic Science,* 146–154. Cold Spring Harbor, N.Y.: Cold Spring Harbor Laboratory Press.

———. 1991a. "Invited Editorial: Research on DNA Typing Catching Up with Courtroom Application." *American Journal of Human Genetics* 48: 819–823.

———. 1991b. "Lander Reply." *American Journal of Human Genetics* 49: 899–903.

———. 1992. "DNA Fingerprinting: Science, Law, and the Ultimate Identifier." In Daniel J. Kevles and Leroy Hood, eds., *The Code of Codes: Scientific and Social Issues in the Human Genome Project,* 191–210. Cambridge, Mass.: Harvard University Press.

———. 1993a. "DNA Fingerprinting: The NRC Report." *Science* 260: 1221.

———. 1993b. Memorandum to Al Lazen and CLS [Commission on Life Sciences] on CLS Discussion of DNA Forensics Report, Mar. 2.

Lander, Eric S., and Bruce Budowle. 1994. "DNA Fingerprinting Dispute Laid to Rest." *Nature* 371: 735–738.

Landhuis, Esther. 2004. "Finger Points to New Evidence: Remains May Not Be Romanovs." *Stanford Report,* Mar. 3. Available at http://news.stanford.edu/news/2004/march3/romanov-33.html (accessed May 14, 2009).

Leahy, Patrick. 2004. "Justice for All Act of 2004 Section-by-Section Analysis." Available at http://leahy.senate.gov/press/200410/100904E.html (accessed Nov. 28, 2008).

Leary, Warren E. 1996. "Expert Panel Calls Evidence from DNA Fully Reliable." *New York Times*, May 3, A21.

Lee, Blewett. 1926. "Blood Tests for Paternity." *American Bar Association Journal* 12: 441.

Lee, Henry, and Jerry Labriola. 2001. *Famous Crimes Revisited from Sacco-Vanzetti to O. J. Simpson.* Southington, Conn.: Strong Books.

———. 2006. *Dr. Henry Lee's Forensic Files.* Amherst, N.Y.: Prometheus Books.

Lee, Henry C., and Frank Tirnady. 2003. *Blood Evidence: How DNA Is Revolutionizing the Way We Solve Crimes.* Cambridge, Mass.: Perseus Publishing.

Lempert, Richard. 1986. "The New Evidence Scholarship: Analyzing the Process of Proof." *Boston University Law Review* 66: 439–477.

———. 1991. "Some Caveats Concerning DNA as Criminal Identification Evidence: With Thanks to the Reverend Bayes." *Cardozo Law Review* 13: 303–341.

———. 1993a. "DNA, Science, and the Law: Two Cheers for the Ceiling Principle." *Jurimetrics: The Journal of Law, Science, and Technology* 34: 41–57.

———. 1993b. Letter to Frank Press, May 24.

———. 1993c. "The Suspect Population and DNA Identification." *Jurimetrics: The Journal of Law, Science, and Technology* 34: 1–7.

———. 1994. "Comment: Theory and Practice in DNA Fingerprinting." *Statistical Science* 9: 255–258.

———. 1997. "After the DNA Wars: Skirmishing with NRC II." *Jurimetrics: The Journal of Law, Science, and Technology* 37: 439–468.

Levy, Harlan. 1996. *And the Blood Cried Out: A Prosecutor's Spellbinding Account of the Power of DNA.* New York: Basic Books.

Lewin, Jeff L. 1998. "The Genesis and Evolution of Legal Uncertainty about 'Reasonable Medical Certainty'." *Maryland Law Review* 57: 380–504.

Lewin, Roger. 1989. "DNA Typing on the Witness Stand." *Science* 244: 1033.

Lewontin, Richard C. 1972. "The Apportionment of Human Diversity." *Evolutionary Biology* 6: 381–398.

———. 1992. "The Dream of the Human Genome." *New York Review*, May 28, 31–40.

———. 1993a. Letter to Bruce Alberts, July 7.

———. 1993b. Letter to Paul Mariano, Feb. 1.

———. 1993c. "Which Population?" *American Journal of Human Genetics* 52: 205–206 (letter).

———. 1994a. "Comment: The Use of DNA Profiles in Forensic Contexts." *Statistical Science* 9: 259–262.

———. 1994b. "DNA Forensic Typing Dispute." *Nature* 372: 398 (letter).

———. 1997. "Population Genetics Issues in the Forensic Use of DNA." In David L. Faigman, David H. Kaye, Michael J. Saks, and Joseph Sanders, eds., *Modern Scientific Evidence: The Law and Science of Expert Testimony*, 1: 686–709. St. Paul, Minn.: West Publishing.

————. 2000. *The Triple Helix: Gene, Organism, and Environment.* Cambridge, Mass.: Harvard University Press.

————. 2005. "The Fallacy of Racial Medicine." *Genewatch* 18 (July–Aug.). Available at www.gene-watch.org/genewatch/articles/18–4Lewontin.html (accessed Aug. 28, 2007).

————. 2006. "Confusions about Human Races." In *Is Race Real? A Web Forum Organized by the Social Science Research Council.* June 7. Available at http://raceandgenomics.ssrc.org/Lewontin/ (accessed Aug. 28, 2007).

Lewontin, R. C., and Daniel L. Hartl. 1991. "Population Genetics in Forensic DNA Typing." *Science* 254: 1745–1750.

Lin, Michael T., and M. Flint Beal. 2006. "Mitochondrial Dysfunction and Oxidative Stress in Neurodegenerative Diseases." *Nature* 443: 787–795.

Liptak, Adam. 2008. "Experts Hired to Shed Light Can Leave U.S. Courts in Dark." *New York Times,* Aug. 12, A1.

Litt, M., and J. A. Luty. 1989. "A Hypervariable Microsatellite Revealed by in Vitro Amplification of a Dinucleotide Repeat within the Cardiac Muscle Actin Gene." *American Journal of Human Genetics* 44: 397–401.

Liu, Peng, Stephanie H. I. Yeung, Karin A. Crenshaw, Cecelia A. Crouse, James R. Scherer, and Richard A. Mathies. 2008. "Real-Time Forensic DNA Analysis at a Crime Scene Using a Portable Microchip Analyzer." *Forensic Science International: Genetics* 2: 301–309.

Lo, Mei-chen, Horng-mo Lee, Ming-wei Lin, and Chin-Yuan Tzen. 2005. "Analysis of Heteroplasmy in Hypervariable Region II of Mitochondrial DNA in Maternally Related Individuals." *Annals of the New York Academy of Sciences* 1042: 130–135.

Lynch, Michael, Simon A. Cole, Ruth McNally, and Kathleen Jordan. 2008. *Truth Machine: The Contentious History of DNA Fingerprinting.* Chicago: University of Chicago Press.

Margolick, David. 1995. "Defense for Simpson Attacking Prosecutors' 'Rush to Judgment'." *New York Times,* Jan. 26, A1.

Marks, Jonathan. 2002. *What It Means to Be 98% Chimpanzee: Apes, People, and Their Genes.* Berkeley: University of California Press.

Marshall, Eliot. 1996. "Academy's About-Face on Forensic DNA." *Science* 272: 803–804.

————. 2005. "Flawed Statistics in Murder Trial May Cost Expert His Medical License." *Science* 309: 543.

Massie, Robert K. 1995. *The Romanovs: The Final Chapter.* New York: Random House.

Mauron, Alex. 2001. "Is the Genome the Secular Equivalent of the Soul?" *Science* 291: 831–832.

McCarthy, Michael. 1987. "Rapist in Genetic Fingerprint Case Jailed for 8 Years." *The Times* (London), Nov. 14.

McCartney, Carole. 2006. *Forensic Identification and Criminal Justice: Forensic Science, Justice, and Risk.* Cullompton, Devon, U.K.: Willan Publishing.

————. 2008. "LCN DNA: Proof beyond Reasonable Doubt?" *Nature Reviews Genetics* 9: 325.

McCormick on Evidence. 2006. 6th ed. Kenneth S. Broun, ed. St. Paul, Minn.: West Group.

McFadden, Robert D. 1989. "Reliability of DNA Testing Challenged by Judge's Ruling." *New York Times,* Aug. 15, B1.

Melton, T. 2004. "Mitochondrial DNA Heteroplasmy." *Forensic Science Review* 16: 1–20.

Melton, T., S. Clifford, M. Kayser, I. Nasidze, M. Batzer, and M. Stoneking. 2001. "Diversity and Heterogeneity in Mitochondrial DNA of North American Populations." *Journal of Forensic Sciences* 46: 46–52.

Mnookin, Jennifer. 2006. "*People v. Castro:* Challenging the Forensic Use of DNA Evidence." In Richard Lempert, ed., *Evidence Stories,* 207–237. New York: Foundation Press.

―――. 2008. "Expert Evidence, Partisanship, and Epistemic Competence." *Brooklyn Law Review* 73: 1009–1033.

Monson, Keith L., and Bruce Budowle. 1993. "A Comparison of the Fixed Bin Method with the Floating Bin and Direct Count Methods: Effect on VNTR Profile Frequency Estimation and Reference Population." *Journal of Forensic Sciences* 38: 1037–1050.

Morrison, Denton E., and Ramon E. Henkel, eds. 1970. *The Significance Test Controversy: A Reader.* London: Butterworths.

Morton, Newton E. 1992. "Genetic Structure of Forensic Populations." *Proceedings of the National Academy of Sciences (USA)* 89: 2556–2560.

―――. 1993. Letter to Frank Press, Feb. 2.

―――. 1994. "Genetic Structure of Forensic Populations." *American Journal of Human Genetics* 55: 587–588 (letter).

―――. 1995a. "Alternative Approaches to Population Structure." *Genetica* 96: 139–144.

―――. 1995b. "DNA Forensic Science, 1995." *European Journal of Forensic Science* 3: 139–144.

Morton, N. E., A. Collins, and I. Balasz. 1993. "Kinship Bioassay on Hypervariable Loci in Blacks and Caucasians." *Proceedings of the National Academy of Sciences (USA)* 90: 1892–1896.

Mountain, Joanna L., and Neil Risch. 2004. "Assessing Genetic Contributions to Phenotypic Differences among 'Racial' and 'Ethnic' Groups." *Nature Genetics* 36: S48–S53.

Mullis, Kary B. 1993. "Nobel Lecture." In Bo G. Malmström, ed., *Nobel Lectures, Chemistry, 1991–1995,* 103–113. Singapore: World Scientific Publishing.

Murphy, Erin. 2007. "The New Forensics: Criminal Justice, False Certainty, and the Second Generation of Scientific Evidence." *California Law Review* 95: 721–797.

Myerowitz, R. 1997. "Tay-Sachs Disease-causing Mutations and Neutral Polymorphisms in the HEX A Gene." *Human Mutations* 9: 195–208.

Nakamura, Y., M. Leppert, P. O'Connell, Roger Wolff, Tom Holm, Melanie Culver, Cindy Martin, et al. 1987. "Variable Number of Tandem Repeat (VNTR) Markers for Human Gene Mapping." *Science* 235: 1616–1622.

National Center for Biotechnology Information. 2004. "What Is a Genome?" In *A Basic Introduction to the Science Underlying NCBI Resources.* Available at www.ncbi.nlm.nih.gov/About/primer/genetics_genome.html (accessed June 8, 2009).

National Commission on the Future of DNA Evidence. 2000. *The Future of Forensic DNA Testing: Predictions of the Research and Development Working Group.* Washington, D.C.: National Institute of Justice.

National Research Council Commission on Life Sciences. 1993. "DNA Forensic Science: An Update—Agenda for Planning Meeting," June 14.

National Research Council Commission on Life Sciences, Commission on Behavioral and Social Sciences and Education, and Commission on Physical Sciences, Mathematics, and Applications. 1993. "DNA Forensic Science: An Update," May 18.

National Research Council Committee on DNA Forensic Science: An Update. 1996. *The Evaluation of Forensic DNA Evidence.* Washington, D.C.: National Academies Press.

National Research Council Committee on DNA Technology in Forensic Science. 1992. *DNA Technology in Forensic Science.* Washington, D.C.: National Academies Press.

National Research Council Committee on EPA's Exposure and Human Health Reassessment of TCDD and Related Compounds. 2006. *Health Risks from Dioxin and Related Compounds: Evaluation of the EPA Reassessment.* Washington, D.C.: National Academies Press.

National Research Council Committee on Evaluation of Sound Spectrograms. 1979. *On the Theory and Practice of Voice Identification.* Washington, D.C.: National Academies Press.

National Research Council Committee on Identifying the Needs of the Forensic Science Community. 2009. *Strengthening Forensic Science in the United States: A Path Forward.* Washington, D.C.: National Academies Press.

National Research Council Committee on Scientific Assessment of Bullet Lead Elemental Composition Comparison. 2004. *Forensic Analysis: Weighing Bullet Lead Evidence.* Washington, D.C.: National Academies Press.

National Research Council Committee on the Possible Effects of Electromagnetic Fields on Biologic Systems. 1997. *Possible Health Effects of Exposure to Residential Electric and Magnetic Fields.* Washington, D.C.: National Academies Press.

National Research Council Committee on the Safety of Silicone Breast Implants. 1999. *Safety of Silicone Breast Implants.* Stuart Bondurant, Virginia Ernster, and Roger Herdman, eds. Washington, D.C.: National Academies Press.

National Research Council Committee to Review the Adverse Consequences of Pertussis and Rubella Vaccines. 1991. *Adverse Effects of Pertussis and Rubella Vaccines.* Christopher P. Howson, Cynthia J. Howe, and Harvey V. Fineberg, eds. Washington, D.C.: National Academies Press.

Nesson, Charles, and John Demers. 1998. "Gatekeeping: An Enhanced Foundational Approach to Determining the Admissibility of Scientific Evidence." *Hastings Law Journal* 49: 335–341.

Neufeld, Peter J. 1993. "Practitioner Essay: Have You No Sense of Decency?" *Journal of Criminal Law and Criminology* 84: 189–202.

Newton, Giles. 2004a. "Discovering DNA Fingerprinting." Feb. 4. Available at http://genome.wellcome.ac.uk/doc_wtd020877.html (accessed Mar. 15, 2007).

———. 2004b. "DNA Fingerprinting Enters Society." Feb. 2. Available at http://genome.wellcome.ac.uk/doc_wtd020878.html (accessed Mar. 15, 2007).

Nichols, R. A., and D. J. Balding. 1991. "Effects of Population Structure on DNA Fingerprint Analysis in Forensic Science." *Heredity* 66: 297–302.

Nickerson, Raymond S. 2000. "Null Hypothesis Significance Testing: A Review of an Old and Continuing Controversy." *Psychological Methods* 5: 241–301.

Nilsson, Martina, Hanna Andréasson-Jansson, Max Ingman, and Marie Allen. 2008. "Evaluation of Mitochondrial DNA Coding Region Assays for Increased Discrimination in Forensic Analysis." *Forensic Science International: Genetics* 2: 1–8.

Noble, Deborah. 1995. "Forensic PCR: Primed, Amplified, and Ready for Court." *Analytical Chemistry* 67: 613A–615A.

Nordlinger, Jay. 2002. "Hunting Tiger: Everyone Wants a Piece of Him." *National Review Online,* Sept. 6. Available at www.nationalreview.com/nordlinger/nordlinger090602.asp (accessed Nov. 29, 2008).

Norman, Colin. 1989. "Maine Case Deals Blow to DNA Fingerprinting." *Science* 246: 1556–1558.

Oakes, Michael. 1986. *Statistical Inference: A Commentary for the Social and Behavioral Sciences.* New York: Wiley.

Odelberg, Shannon J., and Ray White. 1987. "Tandemly Repeated DNA and Its Application in Forensic Biology." In Jack Ballantyne, George Sensabaugh, and Jan Witkowski eds., *DNA Technology and Forensic Science,* 257–263. Cold Spring Harbor, N.Y.: Cold Spring Harbor Laboratory Press.

———. 1989. "Single-Locus Probes: Simple and VNTR." American Association of Blood Banks Conference, "DNA for Parentage Testing: Current State of the Art," Leesburg, Va., April 17–18, 1989.

Office of Technology Assessment. 1990. *Genetic Witness: Forensic Uses of DNA Tests.* Washington, D.C.: Government Printing Office.

Office of the Inspector General, U.S. Department of Justice. 2008. *Review of the Office of Justice Programs' Paul Coverdell Forensic Science Improvement Grants Program: Evaluation and Inspections Report* I-2008-001. Washington, D.C. Available at www.usdoj.gov/oig/reports/OJP/e0801/index.htm. Jan.

Ovchinnikov, Igor, V. Ovchinnikov, Anders Götherström, Galina P. Romanova, Vitaliy M. Kharitonov, Kerstin Lidén, and William Goodwin. 2000. "Molecular Analysis of Neanderthal DNA from the Northern Caucasus." *Nature* 404: 490–493.

Owen, Richard, and Alan Hamilton. 2006. "Captured after Four Decades as a Fugitive: Cosa Nostra's 'Boss of Bosses.'" *Times,* Apr. 12. Available at http://business.timesonline.co.uk/tol/business/law/article704594.ece (accessed May 19, 2009).

Owens, Kelly N., Michelle Harvey-Blankenship, and Mary-Claire King. 2002. "Genomic Sequencing in the Service of Human Rights." *International Journal of Epidemiology* 31: 53–58.

Pakendorf, Brigitte, and Mark Stoneking. 2005. "Mitochondrial DNA and Human Evolution." *Annual Review of Genomics and Human Genetics* 6: 165–183.

Parloff, Roger. 1989. "How Barry Scheck and Peter Neufeld Tripped up the DNA Experts." *American Lawyer,* Dec., 1–2, 53–55.

Parsons, T. J., D. S. Muniec, K. Sullivan, Nicola Woodyatt, Rosemary Alliston-Greiner, Mark R. Wilson, Dianna L. Berry, et al. 1997. "A High Observed Substitution Rate in the Human Mitochondrial DNA Control Region." *Nature Genetics* 15: 363–368.

Perlin, Mark W. 2006. "Scientific Validation of Mixture Interpretation Method." In *Proceedings of the Seventeenth International Symposium on Human Identification.* Nashville: Promega Corp. Available at www.promega.com/geneticidproc/ussymp17proc/oralpresentations/Perlin.pdf (accessed Nov. 30, 2008).

Peterson, Joseph L. 1983. "Observations on Access to Expertise." *Federal Rules Decisions* 101: 642–643.

Peterson, Joseph L., Ellen L. Fabricant, and Kenneth S. Field. 1978. *Crime Laboratory Proficiency Testing Research Program.* Washington, D. C.: U. S. Government Printing Office.

Peterson, J. L., George Lin, Monica Ho, Yingyu Chen, and R. E. Gaensslen. 2003. "The Feasibility of External Blind DNA Proficiency Testing. I. Background and Findings." *Journal of Forensic Sciences* 48: 21–31.

Phillips, C., A. Salas, J. J. Sanchez, M. Fondevila, M. Fondevila, A. Gómez-Tato, J. Álvarez-Dios, et al. 2007. "Inferring Ancestral Origin Using a Single Multiplex Assay of Ancestry-Informative Marker SNPs." *Forensic Science International: Genetics* 1: 273–280.

Polanco, J. C., and P. Koopman. 2007. "SRY and the Hesitant Beginnings of Male Development." *Developmental Biology* 302: 13–24.

Posner, Gerald. 1996. "Throwing the Books at O. J." *Esquire,* Nov., 62–69.

Radzinsky, Edvard. 1992. *The Last Tsar: The Life and Death of Nicholas II.* New York: Doubleday.

Redmayne, Mike. 1997. "Expert Evidence and Scientific Disagreement." *University of California at Davis Law Review* 30: 1027–1080.

Ridley, Matt. 2003. *The Agile Gene: How Nature Turns on Nurture.* New York: HarperCollins.

Risch, Neil, and Bernard Devlin. 1992a. "DNA Fingerprint Matches." *Science:* 256: 1744–1746.

———. 1992b. "On the Probability of Matching DNA Fingerprints." *Science* 255: 717–720.

Risinger, D. Michael, and Michael J. Saks. 2003. "A House with No Foundation." *Issues in Science and Technology* 20(Fall): 35–39.

Roberts, Leslie. 1991. "Fight Erupts over DNA Fingerprinting." *Science* 254: 1721–1723.

———. 1992a. "DNA Fingerprinting: Academy Reports." *Science* 256: 300–301.

———. 1992b. "Science in Court: A Culture Clash." *Science* 257: 732–736.

Roeder, Kathryn. 1994. "DNA Fingerprinting: A Review of the Controversy." *Statistical Science* 9: 222–247.

Romero, Simon. 2008. "Colombia Plucks Hostages from Rebels' Grasp." *New York Times,* July 3, A1.

Royall, R. M. 1997. *Statistical Evidence: A Likelihood Paradigm.* London: Chapman & Hall/CRC.

Saks, Michael J. 1987. "Accuracy v. Advocacy: Expert Testimony before the Bench." *Technology Review,* Aug.–Sept., 43–45.

———. 1989. "Prevalence and Impact of Ethical Problems in Forensic Science." *Journal of Forensic Sciences* 34: 772–793.

Saks, Michael J., and Jonathan J. Koehler. 2005. "The Coming Paradigm Shift in Forensic Identification Science." *Science* 309: 892–895.

Sanders, Joseph, and D. H. Kaye. 1997. "Expert Advice on Silicone Implants: *Hall v. Baxter Healthcare Corp.*" *Jurimetrics: The Journal of Law, Science, and Technology* 37: 113–128.

Sanger, F., S. Nicklen, and A. R. Coulson. 1977. "DNA Sequencing with Chain-Terminating Inhibitors." *Proceedings of the National Academy of Sciences (USA)* 74: 5463–5467.

Santos, F. R., A. Pandya, and C. Tyler-Smith. 1998. "Reliability of DNA-Based Sex Tests." *Nature Genetics* 18: 103.

Savage, Leonard J. 1954. *The Foundations of Statistics.* New York: John Wiley & Sons.

Scheck, Barry C. 1994. "DNA and *Daubert.*" *Cardozo Law Review* 15: 1959–1997.

———. 2006. "Lectures on Wrongful Convictions." *Drake Law Review* 54: 597–620.

Schiff, F. 1929. "Medico-legal Significance of Blood Groups." *Lancet,* Nov. 2, 921–922.

Schklar, Jason, and Shari Seidman Diamond. 1999. "Juror Reactions to DNA Evidence: Errors and Expectancies." *Law and Human Behavior* 23: 159–184.

Schmeck, Harold M., Jr. 1989. "DNA Findings Are Disputed by Scientists." *New York Times,* May 25, B1.

Seringhaus, Michael, and Mark Gerstein. 2008. "Genomics Confounds Gene Classification." *American Scientist* 96: 466–473.

Serre, D., A. Langaney, M. Chech, Maria Teschler-Nicola, Maja Paunovic, Philippe Mennecier, Michael Hofreiter, et al. 2004. "No Evidence of Neandertal mtDNA Contribution to Early Modern Humans." *PLoS Biology* 2: 313–317.

Sharp, Rob. 2007. "The DNA Cracker: Closing the Book on Jack." *Independent,* Oct. 3. Available at www.independent.co.uk/news/science/the-dna-cracker -closing-the-book-on-jack-395843.html (accessed Dec. 11, 2008).

Sheindlin, Gerald. 1996. *Blood Trail: True Crime Mysteries Solved by DNA Detectives.* New York: Ballantine Books.

Sherman, Rorie. 1992. "Genetic Testing Criticized: A Draft Report Says DNA Typing Testimony Should Not Be Admitted." *National Law Journal,* Apr. 30, 1, 45–46.

———. 1993. "New Scrutiny for DNA Testing." *National Law Journal,* Oct. 18, 3, 52.

Shriver, Mark D., and Rick A. Kittles. 2004. "Genetic Ancestry and the Search for Personalized Genetic Histories." *Nature Reviews Genetics* 5: 611–618.

Shriver, M. D., M. W. Smith, L. Jin, A. Marcini, J. M. Akey, R. Deka, and R. E. Ferrell. 1997. "Ethnic-Affiliation Estimation by Use of Population-Specific DNA Markers." *American Journal of Human Genetics* 60: 957–964.

Simpson, O. J. 2007. *If I Did It: Confessions of the Killer.* New York: Beaufort Books.

Sink, John M. 1956. "The Unused Power of a Federal Judge to Call His Own Expert Witness." *Southern California Law Review* 29: 195–214.

Slimowitz, Jennifer R., and Joel E. Cohen. 1993. "Violations of the Ceiling Principle: Exact Conditions and Statistical Evidence." *American Journal of Human Genetics* 53: 314–323.

Smith, Clive S., and Patrick D. Goodman. 1996. "Forensic Hair Comparison Analysis: Nineteenth Century Science or Twentieth Century Snake Oil?" *Columbia Human Rights Law Review* 27: 227–291.

Smith, Malcolm. 2002. "Isonomy Analysis: The Potential for Application of Quantitative Analysis of Surname Distributions to Surname Problems in Historical Research." In Malcolm Smith, ed., *Human Biology and History,* 112–133. Boca Raton, Fla.: CRC Press.

Snyder, Michael, and Mark Gerstein. 2003. "Defining Genes in the Genomics Era." *Science* 300: 258–260.

Spain, David M. 1974. *Post-mortem.* Garden City, N.Y.: Doubleday.

Stone, Richard. 2004. "DNA Forensics: Buried, Recovered, Lost Again? The Romanovs May Never Rest." *Science* 303: 753.

Stoneking, Mark. 2000. "Hypervariable Sites in the mtDNA Control Region Are Mutational Hotspots." *American Journal of Human Genetics* 67: 1029–1032.

Suro, Roberto, 1997. "DNA Now Used To Make Specific Identification; FBI Calls Lab Match 'Major Breakthrough'," *Washington Post,* Nov. 13, A4.

Sykes, Bryan. 2001. *The Seven Daughters of Eve: The Science That Reveals Our Genetic Ancestry.* New York: W. W. Norton & Co.

———. 2004. *Adam's Curse: A Future without Men.* New York: W. W. Norton & Co.

Sykes, B., and C. Irven. 2000. "Surnames and the Y Chromosome." *American Journal of Human Genetics* 66: 1417–1419.

Tang, Hua, Tom Quertermous, and Beatriz Rodriguez. 2005. "Genetic Structure, Self-Identified Race/Ethnicity, and Confounding in Case-Control Association Studies." *American Journal of Human Genetics* 76: 268–275.

Thomas, Mark G., Tudor Parfitt, Deborah A. Weiss, Karl Skorecki, James F. Wilson, Magdel le Roux, Neil Bradman, et al. 2000. "Y Chromosomes Traveling South: The Cohen Modal Haplotype and the Origins of the Lemba—The 'Black Jews of Southern Africa.'" *American Journal of Human Genetics* 66: 674–686.

Thompson, Dick. 1989. "A Trial of High-Tech Detectives." *Time,* June 5: 63.

Thompson, William C. 1993. "Evaluating the Admissibility of New Genetic Identification Tests: Lessons from the 'DNA War.'" *Journal of Criminal Law and Criminology* 84: 22–104.

———. 1996. "DNA Evidence in the O. J. Simpson Trial." *University of Colorado Law Review* 67: 827–857.

———. 1997. "Accepting Lower Standards: The National Research Council's Second Report on Forensic DNA Evidence." *Jurimetrics: The Journal of Law, Science, and Technology* 37: 405–423.

———. 2006. "Tarnish on the 'Gold Standard': Understanding Recent Problems in Forensic DNA Testing." *Champion* 30(1) (Jan.–Feb.): 14–20.

———. 2008. "Beyond Bad Apples: Analyzing the Role of Forensic Science in Wrongful Convictions." *Southwestern Law Review* 37: 971–974.

———. 2009. "Painting the Arrow around the Matching Profile: The Texas Sharpshooter Fallacy in Forensic DNA Interpretation." *Law, Probability & Risk* (in press).

Thompson, William C., and Rachel Dioso-Villa. 2008. "Turning a Blind Eye to Misleading Scientific Testimony: Failure of Procedural Safeguards in a Capital Case." *Albany Law Journal of Science and Technology* 18: 151–204.

Thompson, William C., and Simon Ford. 1989. "DNA Typing: Acceptance and Weight of the New Genetic Identification Tests." *Virginia Law Review* 75: 45–108.

Thompson, William C., Simon Ford, Travis Doom, Michael Raymer, and Dan E. Krane. 2003. "Evaluating Forensic DNA Evidence: Essential Elements of a Competent Defense Review." *Champion* 27(3) (Apr.): 16–25; 27(4) (May): 24–28.

Thompson, William C., and Dan E. Krane. 2003. "DNA in the Courtroom." In Jane Campbell Moriarity, ed., *Psychological and Scientific Evidence*, 11:1–11:138. Eagan, Minn.: West Group.

Tribe, Laurence H. 1971. "Trial by Mathematics: Precision and Ritual in the Legal Process." *Harvard Law Review* 84: 1329–1393.

Von Dungern, E., and L. Hirzfeld. 1910. "Uber Vererbung Gruppenspezifischer Strukturen des Bluts." *Zeitschrift fur Immunitatsforschung* 6: 284–292.

Wade, Nicholas. 1997. "The Revenge of the Fly People." *New York Times*, May 20, C1.

———. 1999. "Defenders of Jefferson Renew Attack on DNA Data Linking Him to Slave Child." *New York Times*, Jan. 7, A20.

———. 2007. "In the Genome Race, the Sequel Is Personal." *New York Times*, Sept. 4, F1.

Walsh, S. J., C. M. Triggs, J. M. Curran, J. R. Cullen, and J. S. Buckleton. 2003. "Evidence in Support of Self-Declaration as a Sampling Method for the Formation of Sub-population DNA Database." *Journal of Forensic Sciences* 48: 1091–1093.

Wambaugh, Joseph. 1989. *The Blooding*. London: Bantam Press.

Weaver, Courtney. 2008. "Romanov Remains Verified." *New York Times*, July 17, A12.

Weber, J. L., and P. E. May. 1989. "Abundant Class of Human DNA Polymorphisms Which Can Be Typed Using the Polymerase Chain Reaction." *American Journal of Human Genetics* 44: 388–396.

Weeks, Daniel E., Alan Young, and Ching Chun Li. 1995. "DNA Profile Match Probabilities in a Subdivided Population: When Can Subdivision Be Ignored?" *Proceedings of the National Academy of Sciences (USA)* 92: 12031–12035.

Weigand, P., M. Schurenkamp, and U. Schutte. 1992. "DNA Extraction from Mixtures of Body Fluid Using Mild Preferential Lysis." *International Journal of Legal Medicine* 104(6): 359–360.

Weiner, Alexander S. 1930. "Blood Tests for Paternity." *Journal of the American Medical Association* 95: 681.

Weir, Bruce S. 1992a. "Independence of VNTR Alleles Defined as Fixed Bins." *Genetics* 130: 873–887.

———. 1992b. "Independence of VNTR Alleles Defined as Floating Bins." *American Journal of Human Genetics* 51: 992–997.

———. 1992c. "Population Genetics in the Forensic DNA Debate." *Proceedings of the National Academy of Sciences (USA)* 89: 11654–11659.

———. 1993. "Forensic Population Genetics and the NRC." *American Journal of Human Genetics* 52: 437–439.

———. 1994. "The Effects of Inbreeding on Forensic Calculations." *Annual Review of Genetics* 28: 597–621.

———. 1999. "Court Experiences in the USA: *People v. Simpson*." In *First International Conference on Forensic Human Identification in the Millennium*, 25–36. London: Forensic Science Service.

Weir, B. S., and I. W. Evett. 1993. "Reply to Lewontin." *American Journal of Human Genetics* 52: 206 (letter).

Westin, Alan. 1989. "A Privacy Analysis of the Use of DNA Techniques as Evidence in Courtroom Proceedings." In Jack Ballantyne, George Sensabaugn, and Jan Witkowski, eds., *DNA Technology and Forensic Science*, 25–34 . Cold Spring Harbor, N.Y.: Cold Spring Harbor Laboratory Press.

White, R., M. Leppert, D. T. Bishop, D. Timothy Bishop, D. Baker, J. Berkowitz, C. Brown, et al. 1985. "Construction of Linkage Maps with DNA Markers for Human Chromosomes." *Nature* 313: 101–105.

Wigmore, John Henry. 1923. *A Treatise on the Anglo-American System of Evidence in Trials at Common Law*. 2d ed. Boston: Little, Brown & Company.

———. 1974. *Evidence in Trials at Common Law* (Chadbourn revision). Boston: Little, Brown & Company.

Willing, Richard. 2004a. "Identical Twins Complicate Use of DNA Testing." *USA Today*, Nov. 30.

———. 2004b. "Twin Suspects Spark Unique DNA Test." *USA Today*, Sept. 2, 3A.

Wilson, Colin, and Damon Wilson. 2003. *Written in Blood: A History of Forensic Detection*. London: Constable & Robinson.

Winkler, John K. 1990. "Letter on DNA Fingerprinting." *Science* 247: 1018.

Winkler, R. L., J. E. Smith, and D. G. Fryback. 2002. "The Role of Informative Priors in Zero-Numerator Problems." *American Statistician* 56: 1–4.

Wong, Z., V. Wilson, I. Patel, S. Povey, and A. J. Jeffreys. 1987. "Characterization of a Panel of Highly Variable Minisatellites Cloned from Human DNA." *Annals of Human Genetics* 51: 269–288.

Woodfill, Jerry. 2000. *The Best of NASA's Spinoffs*. Feb. 15. Available at http://www1.jsc.nasa.gov/er/seh/spinoff.html (accessed Feb. 28, 2007).

Wyman, A. R., and R. White. 1980. "A Highly Polymorphic Locus in Human DNA." *Proceedings of the National Academy of Sciences (USA)* 77: 6754–6758.

Yoffe, Emily. 1994. "Is Kary Mullis God? Nobel Prize Winner's New Life." *Esquire*, July, 68.

Zagorski, Nick. 2006. "Profile of Alec J. Jeffreys." *Proceedings of the National Academy of Sciences (USA)* 103: 8918–8920.

Zerjal, Tatiana, Yali Xue, Giorgio Bertorelle, R. Spencer Wells, Weidon Bao, Suling Zhu, Raheel Qamar, et al. 2003. "The Genetic Legacy of the Mongols." *American Journal of Human Genetics* 72: 717–721.

Zimmer, Carl. 2008. "Now: The Rest of the Genome." *New York Times,* Nov. 11, A1.

Cases and Statutes

Cases

People's Motion for Order Directing Defendant Orenthal Simpson to Supply Hair Sample; Points and Authorities and Declaration in Support Thereof, Municipal Court, June 27, 1994, available at 1994 WL 564365, 230

Transcript of Preliminary Hearing, Municipal Court, June 30, 1994a, available at 1994 WL 733981, 230–231

———, June 30, 1994b, available at 1994 WL 733982, 231

Defendant's Memorandum of Points and Authorities in Support of Defendant's Motion to Exclude DNA Evidence, Oct. 5, 1994, available at 1994 WL 568647, 137, 142–154, 163, 181–182.

People's Reply Brief to Defendant's Motion to Exclude DNA Evidence, Nov. 2, 1994, available at 1994 WL 621393, 153

Defendant's Notice of Defendant's Withdrawal of His Motion for a Pretrial in Limine *Kelly-Frye* Hearing, Jan. 4, 1995, available at 1995 WL 4298, 152–153

Transcript on Withdrawal of Defense Motion to Challenge DNA Evidence under *Kelly-Frye,* Jan. 4, 1995, available at 1995 WL 3184, 285n4

People's Proposed Questions for Personal Waiver by Defendant of Right to Scientific Evidence Admissibility Hearing, Jan. 4, 1995, available at 1995 WL 7717, 184n4

Defendant's Notice of Objections to Testimony Concerning DNA Evidence and Memorandum in Support Thereof, Mar. 20, 1995, available at 1995 WL 126286, 163, 202

Transcript of Examination of Robin Cotton, May 9, 1995, available at 1995 WL 274587, 203

Transcript of Examination of Robin Cotton, May 10, 1995, available at 1995 WL 279744, 203

Transcript of Examination of Robin Cotton, May 15, 1995, available at 1995 WL 294286, 201

Examination of Lakshmanan Sathyavagiswaran, June 13, 1995, available at 1995 WL 356944, 203

Transcript of Examination of Richard Rubin, June 16, 1995, available at 1995 WL 366155, 203

Transcript of Examination of Bruce Weir, June 22, 1995, available at 1995 WL 383835, 204, 246–247, 289n10

Transcript of Examination of Bruce Weir and William Shields, June 22, 1995, available at 1995 WL 383836, 204, 247

Transcript of Examination of Bruce Weir, June 23, 1995, available at 1995 WL 394321, 205

Transcript of Examination of Bruce Weir, June 26, 1995, available at 1995 WL 396928, 247

Transcript of Examination of Bruce Weir and Denise Lewis, June 26, 1995, available at 1995 WL 396929, 247

Transcript of Examination of Douglas Deedrick, July 6, 1995, available at 1995 WL 429478, 232

People v. Soto, 35 Cal. Rptr. 2d 846 (Ct. App. 1994), *aff'd,* 981 P.2d 958 (Cal. 1999), 153, 159, 284n28

People v. Stoughton, 460 N.W.2d 591 (Mich. Ct. App. 1990), 266n9

Statutes

Index